TOXICOLOGY IN ANTIQUITY

TOXICOLOGY IN ANTIQUITY

SECOND EDITION

Edited by

PHILIP WEXLER

Retired, National Library of Medicine's (NLM) Toxicology and Environmental Health
Information Program, Bethesda, MD, USA

ACADEMIC PRESS

An imprint of Elsevier

Academic Press is an imprint of Elsevier
125 London Wall, London EC2Y 5AS, United Kingdom
525 B Street, Suite 1650, San Diego, CA 92101, United States
50 Hampshire Street, 5th Floor, Cambridge, MA 02139, United States
The Boulevard, Langford Lane, Kidlington, Oxford OX5 1GB, United Kingdom

British Library Cataloguing-in-Publication Data
A catalogue record for this book is available from the British Library

Library of Congress Cataloging-in-Publication Data
A catalog record for this book is available from the Library of Congress

ISBN: 978-0-12-815339-0

For Information on all Academic Press publications
visit our website at https://www.elsevier.com/books-and-journals

Working together
to grow libraries in
developing countries

www.elsevier.com • www.bookaid.org

Publisher: Andre Wolff
Acquisition Editor: Rob Sykes/Kattie Washington
Editorial Project Manager: Megan Ashdown
Production Project Manager: Debasish Ghosh
Cover Designer: Matthew Limbert

Typeset by MPS Limited, Chennai, India

Dedication

For my Mom and Dad, Yetty and Will, my wife, Nancy, and our treasured dog, Gigi.

With appreciation to the Toxicology History Association and the scholars who contributed to this volume.

Many thanks, as well, to Elsevier, in particular Megan Ashdown and Kattie Washington, for expertly navigating us through the publication terrain.

Contents

List of Contributors

S.W. Ahmed Department of Pharmacognosy, Faculty of Pharmacy and Pharmaceutical Sciences, University of Karachi, Karachi, Pakistan

George Androutsos Emeritus Professor of History of Medicine, Medical School, National and Kapodistrian University of Athens, Athens, Greece

Okan Arihan Department of Physiology, Faculty of Medicine, Hacettepe University, Ankara, Turkey

Seda K. Arihan Department of Anthropology, Faculty of Letters, Van Yuzuncu Yil University, Van, Turkey

Iqbal Azhar Department of Pharmacognosy, Faculty of Pharmacy and Pharmaceutical Sciences, University of Karachi, Karachi, Pakistan

Maria D.S. Barroso Director of the Department of History of Medicine Portuguese Medical Association, Av. Almirante Gago Coutinho, Lisbon, Portugal

W. Benson Harer, Jr Independent Scholar, Seattle, Washington, United States

Valentina Borgia McDonald Institute for Archaeological Research, University of Cambridge, Cambridge, United Kingdom

Justin Bradfield University of Johannesburg, Johannesburg, South Africa

Caroline S. Chaboo University of Nebraska, Lincoln, NE, United States

Louise Cilliers Honorary Research Associate, Department of Greek, Latin and Classical Studies, University of the Free State, Bloemfontein, South Africa

Jelle Z. de Boer Department of Earth Science, Wesleyan University, Middletown, CT, United States

Carl de Borhegyi Southwestern Michigan College, Dowagiac, Michigan, United States

Suzanne de Borhegyi-Forrest B.S. Biological Sciences, Ohio State University, Columbus, Ohio, United States

John F. DeFelice Department of History, University of Maine at Presque Isle, Presque Isle, ME, United States

Walter D'Alessandro INGV, Sezione di Palermo, Palermo, Italy

Francesco M. Galassi Archaeology, College of Humanities, Arts and Social Sciences, Flinders University, Adelaide, SA, Australia

David Hillman Independent Scholar, Specializes in Ancient Pharmacy and Medicine, Madison, WI, United States

Robert K. Hitchcock Department of Anthropology, University of New Mexico, Albuquerque, NM, United States

Evelyn Höbenreich Institute of the Foundations of Law/Department of Roman Law, University of Graz, Graz, Austria

Mark A. Hoffman Independent Researcher and Author, entheomedia.org

Marianna Karamanou Associate Professor of History of Medicine, Medical School, University of Crete, Greece

Yan Liu Department of History, University at Buffalo, Buffalo, NY, United States

Zafar A. Mahmood Department of Pharmacognosy, Faculty of Pharmacy and Pharmaceutical Sciences, University of Karachi, Karachi, Pakistan

László Makra Institute of Economics and Rural Development, Faculty of Agriculture, University of Szeged, Szeged, Hungary

Yiannis Manetas Department of Biology, University of Patras, Patras, Greece

Adrienne Mayor Classics Department and Program in History and Philosophy of Science, Stanford University, Stanford, CA, United States

George Papatheodorou Laboratory of Marine Geology, University of Patras, Patras, Greece

Benny Pfanz University of Duisburg-Essen, Institute of Applied Botany and Volcano Biology, Essen, Germany

Hardy Pfanz University of Duisburg-Essen, Department of Applied Botany and Volcano Biology, Essen, Germany

P. Rammanohar Amrita Centre for Advanced Research in Ayurveda, Amrita Vishwa Vidyapeetham, Kollam, Kerala, India

Antonio Raschi CNR - IBIMET, Florence, Italy

Francois Retief Honorary Research Associate, Department of Greek, Latin and Classical Studies, University of the Free State, Bloemfontein, South Africa

Giunio Rizzelli Chair of Roman Law, University of Foggia, Foggia, Italy

Carl A.P. Ruck Department of Classical Studies, Boston University, Boston, MA, United States

Gonzalo Sanchez Medical Associates Clinic, Pierre, SD, United States

Markos Sgantzos Department of Anatomy, Faculty of Medicine, University of Thessaly, Larissa, Greece

Susan Stewart Independent Librarian, Edinburgh, United Kingdom

Alain Touwaide Brody Botanical Center, The Huntington Library, Art Collections, and Botanical Gardens, San Marino, CA, United States; Institute for the Preservation of Medical Traditions, Washington, DC, United States

Gregory Tsoucalas History of Medicine, Department of Anatomy, School of Medicine, Democritus University of Thrace, Alexandroupolis, Greece

Lyn Wadley University of the Witwatersrand, Johannesburg, South Africa

Galip Yüce Hacettepe University, Faculty of Geological Engineering, Ankara, Turkey

Foreword

Henry Ford was famously contemptuous of history. He is on record as saying, "History is more or less bunk. It's tradition. We don't want tradition. We want to live in the present and the only history that is worth a tinker's damn is the history we make today." Personally I prefer George Santayana's view: "... [W]hen experience is not retained, as among savages, infancy is perpetual. Those who cannot remember the past are condemned to repeat it." But how relevant is historical toxicology? What can we modern toxicologists (and our regulatory authorities) learn from the past? One important lesson is that toxicity can affect people anywhere, in any society, at any level, and for many years, without anyone being aware of it. Toxic effects on the brain may be responsible for misjudgments by political leaders that can have disastrous consequences. Toxicity may have caused the fall of empires. Nriagu (1983) has argued in his 1983 book that "lead poisoning contributed to the decline of the Roman empire." Louise Cilliers, in Volume 1 of this series, though, disagrees with Nriagu's conclusion. Thus, we must consider further the Roman emperors' possible exposure to other toxicants described by Cilliers and Retief in an earlier article (Cilliers and Retief, 2000). Maybe excessive wine consumption was enough to explain the emperors' self-destructive behavior. Maybe the effects of ethanol were supplemented by the effects of opium, widely used in ancient Rome as a soporific and an analgesic as well as an aid to digestion (Cilliers and Retief, 2000). There is clearly still much scope for further research in this area.

Another empire that may have suffered the consequences of toxicity at the highest level is that of China. It seems possible that the latter days and decisions of the Chinese Emperor Qin Shi Huang may have been adversely affected by exposure to mercury. According to the historian Sima Qian, this emperor was buried in a mausoleum with 100 rivers of flowing mercury in addition to his now famous "terra cotta army." Reportedly, Qin Shi Huang died as a result of ingesting mercury pills, prescribed by his alchemists and court physicians in order to make him immortal (Wright, 2001). It is not unreasonable to suppose, based on his interest in flowing mercury, which probably was to be found in his palace as well as his mausoleum, that some time before his death he was already suffering from mercury poisoning and that his mental function and judgment were impaired as a consequence. It is also likely that his son, the succeeding Emperor Qin Er Shi, had suffered exposure to mercury in his father's palace and that this led to his ill-judged decisions—for example, his command to lacquer the city walls (Hardy and Kinney, 2005). In any event, his incompetence led to revolt, he was forced to commit suicide, and the Qin Dynasty and Empire came to an end, with the Qin capital being destroyed by rebels (Hardy and Kinney, 2005).

With these thoughts in mind, all toxicologists and all those concerned with human health, the environment, and the possible influence of toxic human environments on our political leaders must welcome the insights from history that this first volume and succeeding volumes in this new series of publications will bring. Frequently, I hear toxicologists remark that they might have reached better conclusions in the past "with the benefit of hindsight." Now that this series will give us the all the benefit of hindsight, no doubt "better conclusions" will follow.

John Duffus
The Edinburgh Centre for Toxicology

References

Cilliers, L., Retief, F.P., 2000. Poisons, poisoning and the drug trade in ancient Rome. Akroterion 45, 88–100.

Hardy, G., Kinney, A.B., 2005. The establishment of the Han Empire and Imperial China. Greenwood Publishing Group, Westport, CT.

Nriagu, J.O., 1983. Saturnine gout among Roman aristocrats. Did lead poisoning contribute to the fall of the Empire? N. Engl. J. Med. 308 (11), 660–663.

Wright, D.C., 2001. The History of China. Greenwood Publishing Group, Westport, CT.

Preface to the Series and Volumes 1 and 2

In the realm of communicating any science, history, though critical to its progress, is typically a neglected backwater. This is unfortunate, as it can easily be the most fascinating, revealing, and accessible aspect of a subject which might otherwise hold appeal for only a highly specialized technical audience. Toxicology, the science concerned with the potentially hazardous effects of chemical, biological, and certain physical agents, has yet to be the subject of a full-scale historical treatment. Overlapping with many other sciences, it both draws from and contributes to them. Chemistry, biology, and pharmacology all intersect with toxicology. While there have been chapters devoted to history in toxicology textbooks, and journal articles have filled in bits and pieces of the historical record, this new monographic series aims to further remedy the gap by offering an extensive and systematic look at the subject from antiquity to the present.

Since ancient times, men and women have sought security of all kinds. This includes identifying and making use of beneficial substances while avoiding the harmful ones, or mitigating harm already caused. Thus, food and other natural products, independently or in combination, which promoted wellbeing or were found to have drug-like properties and effected cures, were readily consumed, applied, or otherwise self-administered or made available to friends and family. On the other hand, agents found to cause injury or damage—what we might call *poisons* today—were personally avoided, although sometimes employed to wreak havoc upon one's enemies.

While natural substances are still of toxicological concern, synthetic and industrial chemicals now predominate as the emphasis of research. Through the years, the instinctive human need to seek safety and avoid hazard has served as an unchanging foundation for toxicology, and will be explored from many angles in this series. Although largely examining the scientific underpinnings of the field, chapters will also delve into the fascinating history of toxicology and poisons in mythology, arts, society, and culture more broadly. It is a subject that has captured our collective consciousness.

The series is intentionally broad, thus the title *History of Toxicology and Environmental Health*. Clinical and research toxicology, environmental and occupational health, risk assessment, and epidemiology, to name but a few examples, are all fair game subjects for inclusion. Volumes 1 and 2 focus on toxicology in antiquity, taken roughly to be the period up to the fall of the Roman Empire and stopping short of the Middle Ages, with which period future volumes will continue. These opening volumes will explore toxicology from the perspective of some of the great civilizations of the past, including Egypt, Greece, Rome, Mesoamerica, and China. Particular substances, such as harmful botanicals, lead, cosmetics, kohl, and hallucinogens, serve as the focus of other chapters. The roles of certain individuals as either victims or practitioners of toxicity (e.g., Cleopatra, Mithridates, Alexander the Great, Socrates, and Shen Nung) serve as another thrust of these volumes.

History proves that no science is static. As Nikola Tesla said, "The history of science shows that theories are perishable. With every new truth that is revealed we get a better understanding of Nature and our conceptions and views are modified."

Great research derives from great researchers who do not, and cannot, operate in a vacuum, but rely on the findings of their scientific forebears. To quote Sir Isaac Newton, "If I have seen further it is by standing on the shoulders of giants."

Welcome to this toxicological journey through time. You will surely see further and deeper and more insightfully by wafting through the waters of toxicology's history.

Philip Wexler

Preface

Toxicology in Antiquity was originally published in two volumes, with 2014 and 2015 imprints, so it may seem unusual to see a new merged edition, especially of an historical tract, so soon afterwards. Several factors contributed to the decision to pursue a new edition:

- As there was no clear delineation, chronological or otherwise, between the two slender volumes, it made sense, for ease of reader use, to merge them.
- There were several topics, particularly biblical allusions to toxicology, for which contributors could not originally be found. Many of these gaps have now been filled, thus offering a more complete story of toxicology's earliest history.
- The chapters were resequenced to provide what is hopefully a better, but still inexact, flow of subject matter. The challenge was how to order chapters covering a broad spectrum of topics, which take varying perspectives, such as eras (some overlapping), geographic regions (sometimes overlapping as well), key figures, particular substances, etc.
- Even though not much time has elapsed, a new edition has given a few of the original authors a chance to revise their work to present new research.
- Although visuals had been included in a few chapters previously, the new edition offers additional figures to enhance the narrative.
- The old adage, "Don't judge a book by its cover," presents sensible advice. However, a book's cover inevitably does influence a prospective reader not already familiar with its contents in deciding whether to examine and read it. A more fitting and dramatic image was selected for the current edition of *Toxicology in Antiquity*.

So, here we have, hopefully, a more thorough and up-to-date history of toxicology's first days, perhaps more descriptive than scientific, but consistently fascinating, and setting the course for future scholars and investigations. Dating from Paleolithic times, this book's themes proceed through Egyptian, Greek, and Roman civilizations, and also encompass geographically far-flung ancient populations in Mesoamerica, South Africa, India, and China. You will read of the Paleolithic use of poisons for hunting, including an eye-opening and exquisitely detailed account of the poison-derived hunting practices of the "Bushmen" of South Africa dating back millennia. There are chapters devoted to famous figures such as Cleopatra, Socrates, and Alexander the Great and the documented or disputed toxic events in their lives. Chemical and biological warfare, so much in the news these days, dates back millennia. The hazards posed by cosmetics are both new and old news. Nor is the use of substances such as alcohol and mushrooms recreationally a phenomenon limited to the present day, or the consumption of psychoactive substances to induce spiritual experiences. Hand in hand with the recognition or development of poisons, is the perennial search for a cure-all remedy to protect against or cure their effects, a quest initiated long before the Common Era. Thankfully, we have also found an author to present an overview of toxic implications in the Hebrew and Christian bibles. Finally, a few chapters offer a look at cults and mystery religions in Greece and elsewhere. Some of the rituals described involve ingesting poisons and other biological agents, transmuting them within the body, and subsequently secreting or excreting altered substances with therapeutic properties, a metabolic alchemy of sorts. Read and be amazed.

The already published *Toxicology in the Middle Ages and Renaissance*, part of the same *History of Toxicology and Environmental Health Series*, picks up where this volume leaves off. Future volumes will concern themselves with events from 1650 or so forward.

Philip Wexler

The Prehistory of Poison Arrows

Valentina Borgia

McDonald Institute for Archaeological Research, University of Cambridge,
Cambridge, United Kingdom

OUTLINE

1.1 INTRODUCTION

The use of poisonous substances on arrows is an aspect of the study of prehistoric hunting weapons that only recently has been investigated.

The ethnographic documentation teaches us that hunters of every latitude poison their weapons with toxic substances derived from plants and animals (Bisset, 1979, 1981, 1989, 1992; Bisset and Hylands, 1977; Cassels, 1985; Jones, 2009; Heizer, 1938; Mayor, 2008; Neuwinger, 1996; Noli, 1993; Osborn, 2004; Philippe and Angenot, 2005). Indeed, the weapons could be ineffective if the tips were not poisoned (Noli, 1993). In fact, especially in large game, arrows penetrate the prey to a depth that is insufficient to kill a big animal. The suspicion that the use of poisoning arrows for hunting is an ancient practice dating back to the Paleolithic era is quite plausible.

During the Paleolithic age, the improvement of the technique of hunting at a distance, with the invention of throwing weapons such as the spear, was a revolution in hunting strategies developed by anatomically modern humans (AMH).

In terms of hunting equipment, this is characterized by the appearance of stone barbs and osseous points for hunting weapons. The desire to optimize the aerodynamic and penetrative characteristics of these artifacts suggests the use of a delivery system: the spearthrower or atlatl, and the bow.

The invention of long-range hunting weapons represents a milestone in human history since it illustrates the replacement of our phenotypic competitiveness with survival strategies based on social cooperation and the use of technology.

Chronologically, the first stone tools which are considered to be part of composite long-range weapons are the backed bladelets (called *demilunes*, segments, or lunates) found in the first South-African AMH settlements dated from 70,000 to 50,000 years BP (before present) (Lombard and Pargeter, 2008; Backwell et al., 2008). These weapons were part of the toolkit of modern humans who migrated from Africa.

Killing at a distance no longer required a direct physical confrontation but instead the use of a "strategy of deceit," which is deeply linked to our species. The deceit lies in the phases of the hunt: the silence of the ambush, the attention to every movement and wind direction, the simulation to allow the approach, and finally the launch of the weapon and capture of the prey (Brizzi, 2005).

The "coward's weapon," as the English playwright John Fletcher defined poison, is a further deceit that man uses against prey, so that it is more quickly incapacitated.

Paleolithic populations were surely able to recognize edible and toxic plants, and likely those which could be used medicinally. In sum, toxic substances were available to the prehistoric hunter/gatherers and the benefits arising from their application for hunting are notable: the safe distance of the hunter from the prey could be increased and the animal's death would be quicker.

Toxic substances prevent the animal from escaping too far, allowing the hunter to retrieve the prey more easily and, at the same time, to have meat and skin in better condition.

Finally, toxic substances are abundant in nature, and the preparation of a poison is relatively easy and not very dangerous, considering that usually (according to ethnographic data) they are handled by a particular person designated to do the job; other members of the group are safe.

1.2 WHAT DO HISTORICAL AND ETHNOGRAPHIC DOCUMENTS TELL US?

The most ancient evidence of toxic substances on arrows dates back to the Egyptian pre-dynastic period. A black compound was noticed on the tips of some arrows found in the site of Naga ed Der, dated to 2481–2050 BC (Stanley et al., 1974). The investigators of this finding proved, via rather cruel animal testing, that the mixture contained a venomous ingredient, but at that time it was difficult to confirm its identity. Samples of that material are now being analyzed at Northumbria University in Newcastle (UK), as part of a project headed by Dr. Michelle Carlin and Dr. Valentina Borgia.

Besides this unique archeological evidence, we can count on some rare literary documents addressing the use of poisons on arrows. In the Atharva Veda (900 BC), the sacred text of Hinduism, the use of aconite to poison arrows (used in war) is mentioned (Bisset, 1989).

Also, in the Greek poems, the Iliad and the Odyssey, c. 8th century BC, arrow poisons are cited. In the latter case, poisons seem to be used mainly in warfare, but the sources suggest a well-established tradition in the knowledge and use of toxic substances more generally (Mayor, 2008).

It is very interesting to note that the Greek word *toxic*, used to indicate something poisonous, has the same root of the word *toxon* (bow), and both are linked to *taxon* (yew), that is the tree used to make bows, but also one of the most toxic Mediterranean-region plants. *Aconitum napellus* is a toxic plant and *acontizo* means to hurl a javelin. These words are important since they tell us how a particular tradition of poison use (especially related to aconite/monkshood) became culturally established over an extended period. Very broadly then, it is possible to say that, during the time that ancient Mediterranean civilizations flourished, knowledge of medicinal and toxic plants was already advanced and the use of poison arrows widespread (Borgia et al., 2017).

More recently, we find the tradition of using poisonous plants in the Gaul and Celtic populations. *Limeum* (probably an extract of *Helleborus*) is the name they gave to the poison used for their arrows According to the record (Pliny, 27:76), it was used for hunting, and the part of the meat affected by poison had to be cut away.

An extensive ethnographic literature is available for the use of poisonous substances by ancient and modern hunter-gatherers (Bisset, 1979, 1981, 1989; Bisset and Hylands, 1977; Cassels, 1985; Jones, 2009; Mayor, 2008). In terms of major figures, Norman Bisset (1925–93), a former Professor of Pharmacognosy at King's College London, was the most important expert in the field of arrow poisons. Born in Glasgow (United Kingdom), Bisset had a lifelong interest in ethnopharmacology—the study of the use of natural substances as drugs in ethnic groups. He was also involved in the launch of the *Journal of Ethnopharmacology*. Much of the knowledge we have on this fascinating subject we owe to him.

From a societal point of view, throughout the world and within any group, there is typically a designated individual (e.g., shaman, group leader) who takes on the role of formulating poisons, usually in secret.

The variety of plants and animals used in the composition of poisons is huge.

Nevertheless, some plants have proven more popular than others, and the recipes have been handed down from generation to generation. *A. napellus* (Aconite, Monkshood), one of the most poisonous plants, is certainly one such. It contains alkaloids (aconitine, mesacotine, hypaconitine, and jesaconitine) having effects on the cardiovascular and respiratory systems (Haas, 1999; Bisset, 1972).

Usually the roots of the plant are dried in the sun, and pounded and mixed with water and other ingredients to form a thick consistency (Jones, 2009).

Exemplifying the famous Paracelsus dictum, *"dosis sola facit ut venenum non fit"* (only the dose permits something not to be poisonous), aconite, as many other toxic plants, also has pharmacological properties, and even today it is employed in homeopathy to treat anxiety and neuralgias, or cold and fever.

Antiaris toxicaria, the Upas tree (Ipoh in Javanese means "poison"), emits an extremely poisonous latex containing antiarine, a cardiac glycoside. This plant is used mostly in South-Eastern Asia to poison darts (Carter et al., 1997; Kopp et al., 1992; Shrestha et al., 1992).

In South America a variety of arrow poison recipes goes under the name of curare (derived from the Portuguese or Spanish and said to mean "he to whom it comes falls"). Usually curare is a mix of *Strychnos* genus plants containing curarine and turbocurarine (Bisset and Leeuwenberg, 1968; Genevíéve et al., 2004). The compound is very powerful and capable of killing an animal in a few minutes, but it only affects the blood, ensuring safety in ingesting the meat.

In Africa, many plants have been used to poison arrows. In a recent paper, Bradfield et al. (2015) collected recipes currently used by South-African Bushmen (San) hunter-gatherers. The larvae of a leaf beetle called *Diamphidia* are the main component of the mixtures, but toxic plants such as *Acokanthera, Adenium, Boophane, Euphorbia, Strophantus,* and *Swartzia madagascariensis* are also used.

Hundreds of plants have been used by hunters throughout the world to kill or weaken prey (Bisset, 1989). Furthermore, it should be noted that in poisonous preparations, while there is often a "primary source," other substances are usually added to thicken the mixture or provide ancillary properties, including magical ones.

1.3 WHAT IS THE CURRENT STATE OF RESEARCH?

The only study published to date on the use of arrow poisons among prehistoric populations concerns a wooden stick 32 cm long found in Border Cave, South Africa, and dated about 24,500 BP (d'Errico et al., 2012a,b).

The stick is similar to the poison applicators used until recently by Kalahari San populations. The results from gas chromatography analysis show traces of ricinoleic acid (castor oil, *Ricinus communis*).

The toxicity of this substance has been debated (Evans, 2012; d'Errico et al., 2012a,b); in fact although the ricin present in castor oil is indeed one of the most toxic substances in the world, this extreme potency tends only to be demonstrated with modern purification techniques; the lethal dose of castor oil is otherwise 2−15 beans, a quantity hardly compatible with an arrow. Moreover, the toxic effect appears only after several hours and causes gastroenteritis, perhaps not a first choice for an arrow poison.

There is no ethnographic evidence for the use of castor oil as an arrow poison (Bradfield et al., 2015), although castor beans have been found in more recent archeological excavations. The oil from the beans was used in facial cosmetics and in wick lamps for lighting (Al-Tamimi and Hegazi, 2008).

It is possible that in the stick analyzed by d'Errico et al., another component, more toxic, was not identified. In the same site a lump of organic material was found to contain traces of *Euphorbia tirucalli*; this can be more interesting, as *Euphorbia* is often mentioned in the ethnographic literature as an arrow poison (Bisset, 1989).

A project on the detection of plant poisons on prehistoric arrows was designed by Borgia et al. (2017); the aim of the project was twofold:

1. To form a database of the most toxic plants known in the scientific literature. Major attention has been given to plants that can be found in Europe, especially *A. napellus* (monkshood—Fig. 1.1), *Datura stramonium* (devil's snare), *Conium maculatum* (hemlock), *Veratrum album* (white veratrum), *Helleborus* (hellebore) and *Taxus* (Taxus). The purpose of this first phase was to gather current scientific data on those plants (ecology, chemical composition—using liquid chromatography/mass spectrometry—starch morphology) in order to compare the standards with the archeological (hopefully prehistoric) samples.
2. To use ethnographic samples to design more and more effective analytical methods, making use of the information gathered in the database.

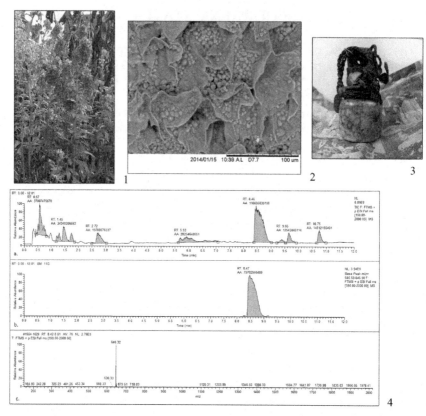

FIGURE 1.1   1—*Aconitum napellus* (monkshood); 2—Starches of the roots of the plant photographed at scanning electron microscope (photo V. Borgia); 3—Aconite pot from China at the Museum of Archaeology of Anthropology, University of Cambridge; 4—Mass spectrum for aconite.

   A noninvasive method of sampling, consisting of rubbing the archeological material with cotton (imbued with distilled water), has been developed to comply with the constraints related to preserving the archeological materials, thus precluding the possibility of abundant samples.

   The method provided positive results in the majority of the 20 analyzed samples (including poisoned arrows and darts, spatulae to apply arrow poison, and a container of poison between 50 and 100 years old—Fig. 1.2). It was fairly easy to detect the main components of the toxic plants within the samples. The origin of these samples, their geographic area, the kind of poisons usually used in the area, and in some instance the poison itself (e.g., a pot of aconite or curare) were known, thus making the overall results of the research preliminary.

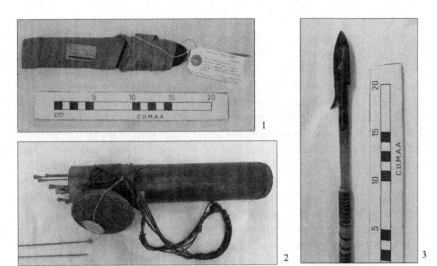

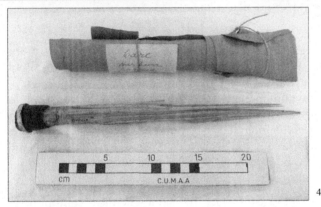

FIGURE 1.2 Various poisoned weapons at the Museum of Archaeology and Anthropology, University of Cambridge, analyzed within the project of Borgia et al. (2016). 1—Spatula covered with poison from Malaysia; 2—Quiver with darts from Borneo; 3—Single barbed iron arrow from Malaysia; 4—spearhead from Samoa with label "Care has been poisoned"; 5—Spatula for poison and iron arrowheads for crossbow China. *Photo: R. Hand, Copyright Museum of Archaeology and Anthropology, Cambridge.*

The challenge for the future is to detect unknown substances and, ideally, to reconstruct the entire toxic compound, formed by one (or more) main ingredients and other additives.

A South-African team (Bradfield et al., 2015; Wooding et al., 2017) is working toward this result. In 2015, one of the most complete and valuable collections of arrow poison recipes of San hunter-gatherers and a large database of their biochemistry were published (Bradfield et al., 2015). This impressive collection makes it clear that even in a relatively small territory an incredible variety of plants and animals with toxic properties are known and adapted to various needs: *"For example, fowl cannot be hunted using coniine, as this toxin will contaminate the meat. In contrast, a mammal poisoned with coniine can be eaten, but the milk will be contaminated"* (Bradfield et al., 2015).

In an important recent pilot study (Wooding et al., 2017), the chemical characterization of poisons, by means of ultra-high performance liquid chromatography quadrupole time-of-flight mass spectrometry (UPLC–QTOF–MS), has been performed, working on three different levels:

1. Eleven well-known toxic plants used by the San living in the Kalahari desert have been characterized from a chemical point of view, contributing to a worldwide database.
2. One of the authors recreated a recipe used by Kalahari San to poison their arrows, containing *Acokanthera oppositifolia*, *E. tirucalli*, and *Adenium multiflorum*; the compound has been analyzed in a blind test.
3. An ancient (about 100 years old) poisoned arrow from Namibia has been analyzed to detect the poisonous substances on it.

The study highlighted the difficulty of unambiguously identifying particular substances. In fact, the same chemical compounds can occur in different plant species.

Nonetheless, the results have been encouraging. In a blind test, two of three toxic plants were identified, based on unique markers. The archeological sample (very small) provided chemical signatures, although not very clear. A marker on the extract was common to *Strychnos madagascariensis*, a species that is not present in the region, but the trace can also be related to *Acokanthera oppositifolia* and *Euphorbia virosa*.

1.4 CONCLUSIONS

The investigation into the use of poisons in prehistoric periods is an innovative field of research, which adds to our understanding not only of ancient hunting techniques and rituals, but also of how the plant world was understood and exploited by ancient populations. The use of poisons to enhance a hunting weapon, in particular a composite system such as the spear/spearthrower or bow/arrow, is an important step from the point of view of cognitive evolution, as it implies a complex knowledge of the environment. Future works on ethnohistorical and archeological arrow poisons must necessarily be aimed at expanding the database of the biochemical fingerprints of the compounds, and in parallel to consider other analytical methods. This will be possible only within a network of scholars engaged in researching prehistoric hunting strategies from various points of view: archaeology, paleobotany, ethnography, chemistry, forensic toxicology, and ethnopharmacy.

References

Al-Tamimi, F.A., Hegazi, A.E.M., 2008. A case of castor bean poisoning. Sultan Qaboos Univ. Med. J. 8 (1), 83–87.

Backwell, L., d'Errico, F., Wadley, L., 2008. Middle Stone Age bone tools from the Howiesons Poort layers, Sibudu Cave, South Africa. J. Archaeol. Sci. 35, 1566–1580.

Bisset, N.G., 1972. Chemical studies on the alkaloids of Asian and African *Strychnos* species. Lloydia 35, 203–206.

Bisset, N.G., 1979. Arrow poisons in China. Part I. J. Ethnopharmacol. 1, 325–384.

Bisset, N.G., 1981. Arrow poisons in China. Part II. Aconitum −botany, chemistry and pharmacology. J. Ethnopharmacol. 4, 247–336.

Bisset, N.G., 1989. Arrow and dart poisons. J. Ethnopharmacol. 25, 1–41.

Bisset, N.G., 1992. War and hunting poisons of the New World. Part 1. Notes on the early history of curare. J. Ethnopharmacol. 36, 1–26.

Bisset, N.G., Hylands, P.J., 1977. Cardiotonic glycosides from the latex of *Naucleopsis mello-barretoi*, a dart poison plant from north-west Brazil. Econ. Bot. 31, 307–311.

Bisset, N.G., Leeuwenberg, A.J.M., 1968. The use of *Strychnos* species in Central African ordeal and arrow poisons. Lloydia 31, 208–222.

Borgia, V., Carlin, M.G., Crezzini, J., 2017. Poison, plants and Palaeolithic hunters. An analytical method to investigate the presence of plant poison on archaeological artefacts. Quat. Int. 427, 94–103.

Bradfield, J., Wadley, L., Lombard, M., 2015. Southern African arrow poison recipes, their ingredients and implications for Stone Age archaeology. South. Afr. Humanit. 27, 29–64.

Brizzi, V., 2005. Colpire a distanza: amenita' varie. Arco Antologico. ⟨http://arcontologico. blogspot.co.uk/2005/10/colpire-distanza-amenit-varie.html⟩ (accessed August 2015).

Carter, C.A., Gray, E.A., Schneider, T.L., Lovett, C.M., Scott, L., Messer, A.C., et al., 1997. Toxicarioside B and toxicarioside C. New cardenolides isolated from *Antiaris toxicaria* latex-derived dart poison. Tetrahedron 53 (50), 16959–16968.

Cassels, B.K., 1985. Analysis of a Maasai arrow poison. J. Ethnopharmacol. 14, 273–281.

d'Errico, F., Backwell, L., Villa, P., Degano, I., Lucejko, J.J., Bamford, M.K., et al., 2012a. Early evidence of San material culture represented by organic artifacts from Border Cave, South Africa. Proc. Natl. Acad. Sci. U.S.A. 109 (33), 13214–13219.

d'Errico, F., Backwell, L., Villa, P., Degano, I., Lucejko, J.J., Bamford, M.K., et al., 2012b. Reply to Evans: use of poison remains the most parsimonious explanation for Border Cave castor bean extract. Proc. Natl. Acad. Sci. U.S.A. 109 (48), E3291–E3292.

Evans, A.A., 2012. Arrow poisons in the Palaeolithic? Proc. Natl. Acad. Sci. U.S.A. 109 (48), E3290.

Geneviéve, P., Angenot, L., Tits, M., Frédérich, M., 2004. About the toxicity of some *Strychnos* species and their alkaloids. Toxicon 44, 405–416.

Haas, C., 1999. L'Aconit, poison et medicament, issu de la bave de Cerbere. Ann. Med. Interne 150, 446–447.

Heizer, R.F., 1938. Aconite arrow poison in the Old and New World. J. Wash. Acad. Sci. 28 (8), 358–359.

Jones, D.E., 2009. Poison Arrows: North American Indian Hunting and Warfare. University of Texas Press, Austin, TX.

Kopp, B., Bauer, W.P., Bernkop-Schnurch, A., 1992. Analysis of some Malaysian dart poisons. J. Ethnopharmacol. 36, 57–62.

Lombard, M., Pargeter, J., 2008. Hunting with Howiesons Poort segments: pilot experimental study and the functional interpretation of archaeological tools. J. Archaeol. Sci. 35 (9), 2523–2531.

Mayor, A., 2008. Greek Fire, Poison Arrows, and Scorpion Bombs: Biological & Chemical Warfare in the Ancient World. Penguin.

Neuwinger, H.D., 1996. African Ethnobotany: Poisons and Drugs: Chemistry, Pharmacology, Toxicology. CRC Press, Boca Raton, FL.

Noli, H.D., 1993. A Technical Investigation into the Material Evidence for Archery in the Archaeological and Ethnographical Record in Southern Africa. Unpublished Ph.D. Thesis. University of Cape Town, South Africa.

Osborn, A.J., 2004. Poison hunting strategies and the organization of technology in the circumpolar region. In: Johnson, A.L. (Ed.), Processual Archaeology: Exploring Analytical Strategies, Frames of Reference, and Culture Process. Greenwood Publishing group, Westport, pp. 134–193.

Philippe, G., Angenot, L., 2005. Recent developments in the field of arrow and dart poisons. J. Ethnopharmacol. 100 (1), 85–91.

Pliny the Elder, Naturalis Historiae, tomo II, libro XXVII: La medicina e le piante medicinali. Ed. Einaudi, Torino 1997.

Shrestha, T., Kopp, B., Bisset, N.G., 1992. The Moraceae-based dart poisons of South America. Cardiac glycosides of *Maquira* and *Naucleopsis* species. J. Ethnopharmacol. 37, 129–143.

Stanley, P., Philips, J., Clark, J.D., 1974. Interpretations of prehistoric technology from ancient Egyptian and other sources. Part I: Ancient Egyptian bows and arrows and their relevance for African Prehistory. Paléorient 2 (2), 323–388.

Wooding, M., Bradfield, J., Maharaj, V., Koot, D., Wadley, L., Prinsloo, L., et al., 2017. Potential for identifying plant-based toxins on San hunter-gatherer arrowheads. S. Afr. J. Sci. 113 (3/4), 92–102.

Further Reading

Bisset, N.G., Mazars, G., 1984. Arrow poison in South Asia Part I. Arrow poisons in ancient India. J. Ethnopharmacol. 12, 1–24.

Lombard, M., 2005. Evidence of hunting and hafting during the Middle Stone Age at Sibidu Cave, KwaZulu-Natal, South Africa: a multianalytical approach. J. Hum. Evol. 48 (3), 279–300.

CHAPTER

2

Beetle and Plant Arrow Poisons of the San People of Southern Africa

Caroline S. Chaboo[1], Robert K. Hitchcock[2], Justin Bradfield[3] and Lyn Wadley[4]

[1]University of Nebraska, Lincoln, NE, United States [2]Department of Anthropology, University of New Mexico, Albuquerque, NM, United States [3]University of Johannesburg, Johannesburg, South Africa [4]University of the Witwatersrand, Johannesburg, South Africa

OUTLINE

Toxicology in Antiquity
DOI: https://doi.org/10.1016/B978-0-12-815339-0.00002-0

11

2.1 INTRODUCTION

Arrow and dart poisons have been used for millennia by people all over the world for hunting and warfare, and as such are a source of fascination for scholars and the public. Early encounters with hunter groups using poisoned arrows, such as the San ("Bushmen") of southern Africa, lead to a morbid fear of these arrows and the mysterious poisons with which they were coated (see accounts in Theal, 1897). People, like their hominin ancestors, observed their natural environments, and experimented with plant and animal ingredients to determine what could be consumed as food, and what was poisonous. Over time hunter-gatherers identified plants and animals that could be used as fishing and hunting aids, psychotics, medicines, and for ceremonial purposes. This rich global indigenous pharmacopeia reflects the power of human observation in diverse environments and a long history of experimentation. Now, fascination for indigenous poisons is shifting to scientific exploration for new drugs (Tulp and Bohlin, 2004) and development of medicines, e.g., from curare (Wintersteiner and Dutcher, 1943; Bisset, 1992) and batrachotoxins (Spande et al., 1992). Simultaneously, we are racing against time to capture disappearing knowledge as indigenous communities are assimilated into modern society.

This chapter focuses on the indigenous plant- and animal-based arrow poisons used by diverse San nations (i.e., peoples sharing the San language family), and scattered across southern Africa—Angola, Botswana, Lesotho, Mozambique, Namibia, Zimbabwe, and South Africa (Chaboo et al., 2016) (Table 2.1).

There is great interest in modern indigenous hunting practices in Africa as they may extend far into our early evolutionary history. Hunting is widely recognized as a fundamental driver of hominin (Hominidae) evolution; it requires behavioral and cognitive innovations such as abstract thought, delayed gratification, planning, socialization, observation of animal behavior, as well as creativity and skill in the manufacture of hunting tools (Wong, 2014). The San people appear to be the earliest diverging lineage of modern humans (e.g., Schuster et al., 2010); therefore, studying San hunting techniques and the use they made of their rich fauna and flora may help us better understand the

TABLE 2.1 Arrow Poisons Used by San in Southern Africa

Group name	Poison used	Conditions	References
Hai//om	*Adenium boehmianium*	Plant found in rocky soils	Widlok (1999: 63, 72, 79, 88) and Friederich (2014: 107–111)
Ju/'hoansi	*Diamphidia nicro-ornata* and *Polyclada flexuosa* beetles	*Diamphidia* beetles are associated with *Commiphora africana.* *Polyclada* beetles are associated with *Sclerocarya caffra*	Marshall (1976: 146–152) and Lee (1979: 102, 14, 388)
Khwe	*Diamphidia nicro-ornata* and *Polyclada flexuosa beetles*	*Diamphidia* beetles associated with *Commiphora africana* and *Polyclada* beetles are associated with *Sclerocarya caffra*	Attila Paksi, Anita Heim, Julie Taylor pers. commun.
Naro	*Diamphidia nigro-ornata* or *Diamphidia simplex* and *Polyclada flexuosa* beetles	*Diamphidia* beetles associated with *Commiphora africana* and *Polyclada* beetles are associated with *Sclerocarya caffra*	Alan Barnard, Mathias Guenther, Monageng Mogalakwe personal communications
G/ui	*Diamphidia nigro-ornata* or *Diamphidia simplex* and *Polyclada flexuosa* beetles	*Diamphidia* beetles associated with *Commiphora africana* and *Polyclada* beetles are associated with *Sclerocarya caffra*	Tanaka (1980: 450), Silberbauer (1981: 206–207), and Valiente-Noailles (1993: 62–65)
G//ana	*Diamphidia nigro-ornata* or *Diamphidia simplex* and *Polyclada flexuosa* beetles	*Diamphidia* beetles associated with *Commiphora africana* and *Polyclada* beetles are associated with *Sclerocarya caffra*	Tanaka (1980: 45), Silberbauer (1981: 206–207), and Valiente-Noailles (1993: 62–65)
!Xõó	*Diamphidia simplex* beetles	*Diamphidia* beetles associated with *Commiphora africana*	H.J. Heinz personal communication
Tshwa	*Diamphidia simplex* beetles	*Diamphidia* beetles associated with *Commiphora africana*	Hitchcock field notes
/Xam	*Sceletium tortuosum, Euphorbia tirucalli*	Both are plants	Deacon (1992)
Amatola	*Sceletium tortuosum*	Plant	Mitchell and Hudson (2004: 45)

past hunting practices of early humans in Africa. Yet, studying San hunting practices is challenging because they have become marginalized communities in their countries through historical, political, and social factors, and are being pressured to give up their traditional life. Most San nations are no longer legally allowed to hunt, or are restricted to controlled annual quotas. Additionally, the conflict of interest between wildlife exploitation and conservation in southern Africa negatively impacts local subsistence hunting.

The transition from hunting with spears to hunting with bow and arrow represents an important shift in early human cognitive evolution. Spear hunting, with simple, pointed wooden staves has been dated to 300,000 years ago (Thieme, 2005; Conard et al., 2015). A similar age is available for stone points that were almost certainly used on spears in East Africa at Olorgesaile (Brooks et al., 2018). Excavations at Kathu Pan, South Africa, suggest that stone points may have been similarly employed 500,000 years ago (Wilkins et al., 2012), but the contextual evidence for the date is not as reliable as that of the more recent past in East Africa. Bow hunting may date to about 64,000 years ago based on use-wear and microresidue analyses of microlithic stone tools suitable as inserts for arrows (Lombard and Pargeter, 2008; Wadley and Mohapi, 2008; Lombard and Phillipson, 2010; Lombard, 2011; Bradfield and Lombard, 2011) as well as the discovery of bone points (that look like San bone arrowheads) from Sibudu, South Africa (Backwell et al., 2008, 2018). At Border Cave, South Africa, a lump of *Euphorbia tirucalli* latex was found mixed with beeswax that was directly dated to 35,140 ± 360 years ago (OxA-W-2455-52 [41,167−39, 194 cal BP]) (d'Errico et al., 2012a). *Euphorbia tirucalli* has very poisonous latex containing triterpenoids, so the mixture may have been used as a dual-purpose adhesive for hafting weaponry. More controversial (Evans, 2012) is the finding from Border Cave of a ∼20,000-year-old twig (possibly a poison applicator) with residues of poisonous ricin, a plant extract (d'Errico et al., 2012a), from *Ricinus communis*, a plant thought not to be indigenous to southern Africa (Wink and van Wyk, 2008). Nonetheless, the Border Cave discoveries have thrown sharp focus on the contemporary bow-and-arrow hunting of indigenous African hunter-gatherers and their repertoire of plant- and animal-based poisons. Indeed, there is interest in chemically identifying poisons on ethnographic artefacts from around the world (e.g., Borgia et al., 2016). Poison adoption could have been a crucial innovation in early human hunting technology, particularly in regions lacking the raw material to make powerful weapons capable of incapacitating prey.

Symbiotic technologies (like the bow and arrow) are defined as sets of tools that work together to perform a desired task, and where the properties of the end technology cannot be realized without the simultaneous manipulation of all the parts (Lombard and Haidle, 2012). The ability to conceptualize a set of separate, yet interdependent tools, such as the bow and arrow, implies flexibility in decision-making and a level of cognition not seen in other hominin species. The art of making poison recipes may be seen in itself as a symbiotic technology, as it could only be accomplished by individuals possessing a basic understanding of the pharmacological or adhesive properties of each plant material, the appropriate methods of preparation that would not denature the toxicity, and the sequence in which, and quantities of, ingredients must be added to a recipe (see Wadley et al., 2015). The creation of poison for arrows augments the symbiotic technology and requires a level of foresight, anticipation, planning, analogical thought, and multitasking that is unique to modern humans.

The concurrent publication of three studies (Bradfield et al., 2015; Wadley et al., 2015; Chaboo et al., 2016) synthesizing the historical literature and novel field-based data reflects renewed interest in San arrow poisons. These articles overlap in their review of the poison sources, composition, efficacy, antidotes, and chemistry practiced by San communities. Their authors conducted first-hand fieldwork to study the hunting and poison practices of two San groups, the Hai//om San in the Namibian Kalahari and the Ju/'hoan San in and around Etosha National Park and Nyae Nyae.

Here, together, we deepen the discussion of San poison practices and attempt to connect the unique modern hunter-gatherer practices with more ancient practices (e.g., Erlandson et al., 2014) that would have included the origins of arrow poisons and, by extension, medicines and insecticides (Wadley et al., 2011; d'Errico et al., 2012a,b; Evans, 2012; Langley et al., 2016; Lennox and Bamford, 2017; Lennox et al., 2017). We already know that as of 77,000 years ago, people were repelling insects from their camps by using aromatic leaves in their plant bedding (for example, *Cryptocarya woodii*), and wood with toxic properties (for example, *Spirostachys africana* and *Tarchonanthus* spp.) in their domestic fires (Wadley et al., 2011; Lennox and Bamford, 2017; Lennox et al., 2017). Incrementally, we are beginning to get a glimpse of "folk-science" as practiced in the past.

2.2 MATERIALS AND METHODS

Information on indigenous groups, hunting practices, and arrow poisons (sources, recipes) is extracted from published literature and from our own observations. Archeological, botanical, biochemical, ethnoarcheological, and ethnographic data are all relevant to this chapter. Steyn (1949) published the first major list of ~ 300 plants and 15 different animals (beetles, scorpions, spiders, and snakes) that were used as arrow poisons in central to southern Africa. Two more comprehensive works provide an entrée to the poisons of Africa: Watt and Breyer-Brandwijk (1962) focus on poisonous plants of southern and eastern Africa, while Neuwinger (1996) condenses the chemistry and pharmacology known to that date. Some of our data include direct observation of poison preparations in two Namibian San communities, Hai//om San who use the *Adenium* plant as part of their recipes, and Ju/'hoan San who use beetle poisons extensively with plants in a variety of recipes (Bradfield et al., 2015; Wadley et al., 2015; Chaboo et al., 2016). Different communities of Hai//om and Ju/'hoan San may have used other poisons in the past. Chaboo et al. (2016) took a ground-truthing approach and offer a framework for the systematic study of poison-using San groups—G|ui, G||ana, G||olo, Kua, Naro, Tsila, and X'ao-l'aen—and two non-San groups—Valley Bisa in Zambia (Marks, 1977) and the Naro in Ghanzi, Botswana (Nonaka, 1996). Bradfield et al. (2015) usefully assembled many San poison recipes and verified the use of the two poisons of Hai//om San and Ju/'hoan San in Namibia. Now, the research challenge is clear—to confirm every poison in those San communities that still practise bow-and-arrow hunting, validate scientific identifications and specimen vouchers, and conduct comparative research of the hunting kit, poison sources, recipes and chemistry, and associated cultural practices.

Taxonomy and definitions. Chaboo et al. (2016) traced the nomenclatures used variously in the historical literature and assessed the current taxonomy of San groups, plants, and animals discussed herein. We follow the online catalogs of plant names, PlantZAfrica (2018) and The Plant List Version 1.1. (2013), and Biondi and D'Alessandro (2012) for beetle taxonomic names. In this chapter, we discriminate among toxicants (any toxic substance) by the terms *poison* (a substance that causes illness or death if ingested, absorbed, or breathed into the body), *toxin* (a biologically derived substance that causes illness), and *venom* (a substance produced by animals when they bite or sting; i.e., it is injected via a specialized organ) according to the Cambridge Dictionary (2018). The San beetle poison or toxin is derived from the body fluid of the beetle ("hemolymph") and thus is not venom, like that of a wasp or snake, as it is not injected by the animal. We use "diamphidia beetle" referring to all the species of the *Blepharida*-group beetles (Chrysomelidae: Alticini), not only those of the genus *Diamphidia*, that may be used as arrow poison and are likely to share the diamphotoxin molecules. Molecular models of some of the active compounds in the San poisons are taken from PubChem, the Open Chemistry Database (2018). The San language group encompasses many local dialects, which can lead to confusing terminology. The reader can consult Sands et al. (2011) for

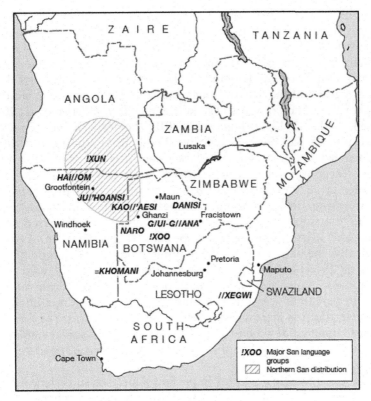

PLATE 2.1 Map of distribution of contemporary San nations in southern Africa. Prepared by M. Brouwer Burg for Chaboo et al. (2016) and used with permission.

specific hunting terminology across San dialects and Schapera (1927), Goodwin (1945), Clark (1977), Deacon (1984), and Backwell et al. (2008) for classification of San arrows.

We define the San below, and provide a map of modern distribution (Plate 2.1). In tracing patterns into antiquity, we face the challenge of referring to past groups of people living in southern Africa. We cannot assign real names though some may be ancestors of today's San. In this chapter, we cautiously refer to past groups of people as hunter-gatherers from the Stone Age or people living in the Stone Age.

2.3 RESULTS

2.3.1 History and Development

Who are the San? The San (Bushmen) of southern Africa are some of the best-known indigenous people in the world. Today, the San number around 130,000 people in seven countries—Angola, Botswana, Lesotho, Namibia, South Africa, Zambia, and Zimbabwe (Plate 2.1; Dieckmann et al., 2014; Puckett and Ikeya, 2018). San languages, which are click consonant based, are divided into three language families: Khoe-Kwadi, Ju/'hoan, and Tuu (Güldemann, 2014). Genetic and archeological evidence suggest that the San and their ancestors have resided in southern Africa for thousands of years (Tishkoff et al., 2009; d'Errico et al., 2012a; Mitchell, 2012, 2013; Schlebusch, 2010; Schlebusch et al., 2012). Historically, the San derived their livelihoods from hunting and gathering. Their neighbors in southern Africa, the Khoikhoi (Khoekhoe) were herders, keeping small stock (sheep and goats) (Elphick, 1977; Boonzaier et al., 1996; Sadr, 2013). At one time, the San occupied an area stretching from the Congo-Zambezi watershed in central Africa south to the Cape. The San were relatively widely dispersed in the region, and they numbered up to 300,000 people (Lee, 1976: 5). Historically, the San were mobile, moving about the landscape in small groups of 25–50 people depending on the availability of wild plants and animals, stone (for tools), and other resources. Their mobility related in part to the seasonality, amounts and duration of rainfall, surface water, and the locations of other groups (Lee, 1976). The San were tied into larger units which consisted of people who spoke similar languages and who were linked through kinship, marriage, friendship, and exchange ties.

In the past, foraging was an important source of subsistence, and it still is in some places, notably in the north-western and central Kalahari regions (Hitchcock et al., 2006; Puckett and Ikeya, 2018). Today, however, the majority of San have mixed economies, combining foraging with agriculture, livestock raising, craft production and sale, and working for other people (Puckett and Ikeya, 2018). These days, only a small percentage of the San depend to any significant degree on hunting and gathering. Hunting is not allowed at all in Botswana, and is restricted in Angola, Lesotho, Mozambique, Namibia, South Africa, Zambia, and Zimbabwe. As a result, finding groups that regularly use poison in hunting is not easy; some groups that used bows and poison-tipped arrows in the past now use spears, and some of their hunting is done with dogs (e.g., the Tshwa of the northeastern Kalahari and the !Xóõ of the southwestern Kalahari). More riverine-adapted San people such as the Bugakhwe and //Anikwe in the Okavango and San in the Zambezi (Caprivi) region of Namibia and southern Zambia also fish; some use plant-derived poisons to fish (Rodin, 1985; Neuwinger, 1996; Boon, 2010). All San groups continue to forage for wild plant foods and medicines, some of them exploiting over 150 species of plants. All San today also depend to at least some extent on domestic agricultural crops; about ~15% of San today have livestock (Hitchcock, field data, 2017).

Some spokespersons for San nongovernment organizations (e.g., First People of the Kalahari, Botswana Khwedom Council, Namibian San Council, South African San Council) have suggested that the term "First People" be applied to the San, building on the idea that these groups were the "first comers" or aboriginal peoples who first occupied the Kalahari Desert, and indeed most of southern Africa (Biesele and Hitchcock, 2013). It should be noted, however, that the governments of southern African countries do not consider the San to be the only indigenous people in the various states in the region (Sapignoli, 2018). Like other indigenous peoples in the 21st century, the G/ui, G//ana, Hai//om, Ju/hoansi, Naro, !Xun, Tshwa, and other San are gaining new ground in terms of land, cultural and language preservation, intellectual property rights, and political representation. In some cases, community-based natural resource management and community development programs have been initiated. It remains to be seen how effective these are in terms of increasing access to land and security of tenure. San nongovernment organizations across the region have struggled but they are making some progress in terms of gaining international recognition and awareness of issues.

Early Observations. The earliest reports from explorers in the 18th and 19th centuries briefly mentioned several San arrow poisons from plants (Wikar, 1779; Paterson, 1789; Livingstone, 1858; Livingstone and Livingstone, 1865; Passarge, 1907; Cornell, 1920); snakes (Paterson (1789); insects (Wikar, 1779; Livingstone, 1858; Livingstone and Livingstone, 1865; Passarge, 1907; Schapera, 1925); spiders (Stow, 1905; Cornell, 1920), and carrion (Cornell, 1920; Hall and Whitehead, 1927). Early reviews of dart and arrow poisons (Perrot and Vogt, 1913; Lewin, 1923; Schapera, 1925; Steyn, 1949; Bisset, 1989) are outmoded, but those of Watt and Breyer-Brandwijk (1962) and Neuwinger (1996) are more up-to-date. Below we outline what the San hunt and gather, and the methods used. Then we focus on their arrow poisons before discussing broadly the possible historical origins of poison adoption.

What do the San hunt? There are relatively few places in Africa where detailed information on San indigenous hunting and gathering is available and where there has been intensive research. One such place is the northwestern Kalahari region of Botswana and the northeastern Kalahari region in Namibia where data on wildlife utilization by San communities has been collected since the early 1950s to the present. The information below is chronologically and geographically confined to the last 60 years. Bone isotope studies on skeletons from the Western Cape suggest that people in the deep past had a diet unlike that of historical and present-day populations (Sealy, 2006). Various "San" communities distributed across the continent would have subsisted differently than the Ju/'hoan.

The San forage for about 100 different plant species (Lee, 1979), including Baobab fruit, bean, bushman truffle, edible gum, marula fruit, mongongo nuts, roots and bulbs, leafy greens, melon, mushroom, and yam. They also gather eggs (ostrich), honey, and insects. Ant eggs were a favored delicacy among certain San groups, being referred to by the early settlers as "Bushman rice" (Schapera, 1930). In the past, the San got most of their water requirements from plant roots, tubers, and desert melons found on or under the desert surface (today, they rely on bore-holes). They often stored water in the blown-out shells of ostrich eggs (which are used for many things, including storing certain poisons as shown in Chaboo, 2011: 64, Fig. 3e). It should be noted that possession of ostrich eggs by people without ostrich licenses has been punishable by arrest and fines since 1994 (see Republic of Botswana, 1994; Udall and Hitchcock, 2018). Honey is particularly prized for its sweetness and calories (Spittel, 1924; Bodenheimer, 1951; Crosse-Upcott, 1956; Smith, 1958; Brokensha and Riley, 1971; Ichikawa, 1980; Harako, 1981). Honey-collecting scenes occasionally feature in the rock art of southern Africa, indicating an important cosmological association (Mguni, 2015).

San gathering tools are the digging stick, kaross (leather blanket), small leather bag, and a mesh net (Lee, 1979). The digging stick is a multipurpose tool for digging out roots and burrowing animals and carrying items suspended or impaled on it (Lee, 1979). In the 1870s, Dunn (1931) collected a digging stick of *Olea verrucosa* from a San woman at Griquatown; Leffers (2003) reported digging sticks of *Catophractes alexandri*, *Dichrostachys cinerea*, *Grewia bicolor*, and *Mundulea sericea* (Leffers, 2003).

Fruits and vegetables are relatively easy to gather, but meat has higher caloric value. Data on hunting among the Ju/'hoansi in the Nyae Nyae region of Namibia were collected by the Marshall family and their colleagues beginning in 1950. Scavenging of dead animals is a relatively risk-free method of gathering meat that has been hypothesized for ancient hominins (Valadez, 2003) and has been observed in contemporary San—Marshall (1976) reported that San at Gautscha Pan, Nyae Nyae, Namibia scavenged a dead baby giraffe. Scavenging by "meat-robbing" is a riskier strategy employed by foragers (Binford, 1981; Silberbauer, 1981) that involves scaring the original predators off their kills—by shouting at them, running at them, or cracking a whip at them. There are cases reported in southern Africa where people find carcasses of animals in the bush; they also detect them by listening to the sounds of lions or hyenas and by seeing vultures or other birds circling over the remains of a dead animal. When asked about this kind of search technique, San say that vultures (Accipitridae), including the white-backed vulture (*Gyps africanus* Salvadori) and the lappet-faced vulture (*Torgos tracheliotus* Kaup), both of which are common residents in the northern Kalahari, are especially good as indicators of carcasses. Leffers (2003) reported that Ju/'hoansi use smoke to keep vultures away from a fresh carcass.

Active hunting is far more widely observed among contemporary San and they hunt a great variety of animals, with individual communities taking different numbers of individuals of each species. The San hunt many animals including reptiles (tortoises, python), birds (francolin, guinea fowl, hornbill, korhaan, Kori bustard, ostrich [eggs too, when allowed]), and mammals (antbear, antelope, aardwolf, buffalo, duiker, eland, gemsbok, giraffe, hare, hartebeest, hippopotamus, honey badger, kudu, porcupine, rhinoceros, roan antelope, scrub hare, springhares, steenbok, warthog, wildebeest, and zebra) (Campbell, 1964, 1968a,b; Draper, 1973; Marshall, 1976; Yellen, 1974; Marks, 1977; Lee, 1979; Parker and Amin, 1983; Wilmsen, 1989; Wawrzyniak, 2013). Before 10,000 years ago when they became extinct, giant zebra (*Equus capensis*), giant buffalo (*Pelorovis antiquus*), and the large African alcephine (*Megalotragus priscus*) were hunted in southern Africa (Helgren and Brooks, 1983). Hunting game was difficult then, and is still difficult now (Lee, 1979; Klein, 1974).

San hunting methods. The hunting method used depends on the prey species, its size and mobility, the ecological characteristics of the area, the time of the year, the number of people involved in the hunt, as well as the skill of the hunter (Manhire et al., 1985; Wawrzyniak, 2013). Men are the primary hunters in San communities, each manufacturing his own hunting tools and poison. Hunters are esteemed in their communities (Wawrzyniak, 2013), but women also help track and improve their husband's success (Biesele and Barclay, 2001), and they sometimes help to carry meat back to camp. The best hunters have great stamina and an unusual ability to orient in the wilderness and track game ("read the sands"). They have high endurance against high temperatures and lack of water. Tobias (1957: Fig. e) observed a young man with chest tattoos to indicate hunting success; hunters achieve different status with each type of hunting (Grund, 2017). Hunters have distinctive rituals; certain plants are selected as special bedding, snuff, ritual body rubs and cleansers, and foods of hunters before they hunt (Leffers, 2003).

Silberbauer (1981) reported that the G/ui in the Central Kalahari, Botswana, hunt, in order of frequency of use, by the following: shooting with bow and poisoned arrow, snaring, caching springhares by means of barbed probes thrust into warrens, running down game, spearing, clubbing, and meat robbing (scavenging). Both in the case of arrows and in the case of guns, the owner of the implement of production is the owner of the product (Wilmsen, 1989; Wiessner, 1983). Where horses have been more recently used, Tanaka (1987) notes that the owner of the horse is the owner of the meat. These various hunting methods are described below.

Spears were primarily used by Ju/hoansi as a secondary tool to finish off an animal. There were, however, Ju/'hoansi, especially younger ones, who preferred to use spears as a primary weapon. Spear hunting appears to be somewhat more productive than bow and arrow hunting (Hitchcock et al., 1996; Hitchcock and Bleed, 1997). One of the reasons given by some of the hunters we interviewed for their use of spears was that they enabled them to kill larger animals more efficiently. Lee (1979), for example, describes an elephant hunt with spears. The spear's impact, when thrown or thrust using both hands, results either in immediate death or sufficient blood loss to weaken the animal relatively quickly. Sometimes spears were used to impale or club prey such as springhare (*Pedetes capensis*) which were caught with the aid of a long, segmented stick with a hook on the end that was pushed down into burrows and used to catch and hold the animal.

Another kind of hunting strategy employed by San included the procurement of burrowing animals, notably springhare (*Pedetes capensis*), porcupine (*Hystrix africaeaustralis*), and antbear (*Orycteropus afer*). For springhare, specialized technology is employed. It is a 3–4 m long series of sticks held together with sinew that has a hook on the end which consists of steenbok horn or metal. Digging sticks are used to dig down to the animal which is obtained by hand from its lair. It is then clubbed or speared.

Encounter hunting, also referred to as persistence hunting or pursuit hunting on foot, involves running down game (Macphail, 1930) by one or two hunters. This kind of hunting was done with traditional technology (bows, arrows, spears, and clubs) and more recently with guns. Once prey was encountered, hunters could opt to sneak up on them or they could run the animals down on foot (Bartram, 1993; Silberbauer, 1981; Hitchcock et al., In review). Much of the hunting done by the Ju/'hoansi is done on foot; Wilmsen (1989: Table 6.4) observed Ju/'hoansi hunters running down small animals—birds (ostrich and guinea fowl) and mammals (duiker, steenbok, and warthog). Faunal exploitation in the Nyae Nyae and Dobe-/Du/Da areas, Namibia, includes hunting small and large mammals with the aid of bows and arrows, spears, and clubs (Marshall, 1976; Lee, 1979; Wilmsen, 1989).

Fishing with the aid of bone tools has been practiced in southern Africa for the last 18,000 years. Double-pointed bone slivers, interpreted as fishing gorges, were found associated with prolific fish remains at four coastal sites in the Western Cape Province, South Africa (Klein, 1972, 1974; Parkington, 1976, 2006; Schweitzer and Wilson, 1978). Farther inland, in the Drakensberg Mountains, three techniques of fish capture were practiced in the past. Depictions of nets occur in the rock paintings at certain shelter sites in Lesotho (Hobart, 2003), while perforated fish crania at other sites in Lesotho attest to spearing or line-and-arrow capture using a simple bone point (Plug et al., 2010). Bone hooks appear much later (Mazel, 1989). During the historic period, milky, viscous *Euphorbia* latex or other plant ingredients were sometimes used in tropical and southern Africa to poison water to capture fish (Neuwinger, 1996, 2004). The poison ingredients are mixed together and thrown into the water where fish become stupefied, die and float to the surface where they are caught by hand.

San traps include snares and other traps, such as deadfalls, logs with spear points in them placed over game trails. Wawrzyniak (2013) reported several traps used by Khoisan—holes dug near watering holes where animals came to drink, with spike inserts and grass and branch coverings; concealed snares made of *Sansieviera* twine, with food bait like small corms; smoking out animals from their resting spots. Sometimes spear points were placed in logs that were buried in game trails and hippopotamus or elephant would step on them and impale themselves. The hunting of rhinoceros, elephants, and hippopotamus was a high-risk activity that was done either on foot or from wooden canoes (*makoros*) in places such as the Okavango Delta. Poison was sometimes used on elephants, particularly plant poison, as seen, for example, among the Bisa, Manyamadze Corridor, Zambia (Stuart Marks, pers. commun. to Hitchcock, 2011). Historical Cape records describe large game trapped in pitfalls disguised with plant material (Barrow, 1801; Stow, 1905; Schapera, 1930) and some pitfalls were planted with poisoned stakes (Stow, 1905). Pitfalls lined with spears were also apparently used by the Batwa of Lake Chrissie, Mpumalanga, who drove animals to the pits (Potgieter, 1955). There is proxy archeological evidence for snare use by ancient people in southern Africa (Wawrzyniak, 2013; Manhire et al., 1985). At Sibudu, bones of small carnivores (like mongooses) were found together with blue duiker remains in occupations older than 60,000 years ago. Small carnivores are susceptible to accidental snaring, and are unlikely to have been deliberately hunted (Wadley, 2010a).

There is some evidence to support the practice of catastrophic hunting techniques, both in historic times (Bleek and Lloyd, 1911) and archeologically (Dewar et al., 2006). A Later Stone Age site from the Namaqualand coast, South Africa has produced a large sample of springbuck remains with a catastrophic mortality profile. This site is more consistent with mass harvesting or the capture of a whole herd of animals by trapping. This method of hunting is likely to have had far-reaching social implications, especially in terms of gender relations and meat sharing, which have been considered fundamental in organizing dimensions of hunter-gatherer societies (*sensu* Marshall, 1960; Lee, 1979). Archeological remains of possible hunting blinds have been found farther north, in the Gauteng region (Mason, 2012), and stone structures, resembling kites, have also been found in the southern Kalahari, although their age and exact purpose is unclear (Van der Walt and Lombard, 2018). Brooks (1984) records the building of a modern hide in Botswana so that hunters could ambush animals there. The concept of pitfall and snare was sometimes combined by G/wi who pegged a twine noose around a concealed pit to capture small creatures like steenbok or common duiker (Silberbauer, 1981). At Dobe, Ju/'hoansi hunters sometimes lay a zigzag line of branches between bushes to seal off an area. Then the line is monitored for a few days to find breaches; thereafter snares are set there (Lee, 1979). Setting successful snares involves ethology, and the same knowledge is needed to interpret hunting behavior in the past. Snares are often set and emptied by the aged, and by boys and girls in the Kalahari (Silberbauer, 1981; Kent, 1995). Snares are usually a noose made from plant fiber like *Sansieviera*. The twine noose is activated by a sapling that works like a spring (Steyn, 1971; Yellen, 1977; Lee, 1979; Silberbauer, 1981). When the bait (a ball of *Acacia* gum or a corm) is taken, the spring is released and the noose tightens around the neck of the prey (Silberbauer, 1981).

Net-hunting involved a net of cords made by pounding the bark of *Entada arenaria* and weaving the fibers into cords for the nets. Elephant sinew is thought to have been used in the past (Leffers, 2003: 104). Apart from the rock art cited here (Manhire et al., 1985), there is no archeological evidence for nets, but such organic materials would not, in any event, preserve well archeologically. Small carnivores are seldom taken in nets (Lupo and Schmitt, 2002), whereas they are often victims of snares, so the proxy evidence at Sibudu supports the use of snares in the deep past rather than nets (Wadley, 2010a).

Spears and javelins allow simple thrusting and can also serve as projectiles. The apex could be a simple carved wooden tip, or be made from a stone or bone point. Points for both spears and arrows are discussed below but it is worth noting that the use of a binding (sinew, plant) and adhesive (Boëda et al., 1996, 1999, 2008a,b) are needed to secure any point to a stick.

Hunting with bows and arrows across southern Africa. Bows and arrows are widely used and come in a variety of sizes and styles and may be used with or without poison. The hunting kit also includes a quiver, for holding the arrows, and a woven fiber bag for carrying meat, as well as materials for emergency repairs of the bow or arrows. The poisons, one of the most fascinating and mysterious aspects of San culture, are still quite understudied. Although diverse plant- and animal-based poisons have been reported, past researchers were concerned with other aspects of San culture and, apart from Marshall (1976) and Lee (1979), made only scant comments about poisoned arrows. Recently, Bradfield et al. (2015), Wadley et al. (2015), and Chaboo et al. (2016) reported on San poisons but we are far from a complete inventory of the poisons of all San groups. Certainly, time is running out to collect this kind of ground-truthed information as San have become increasingly sedentary and integrated into a modern Afro-Western culture.

At present, detailed research with two groups has revealed and corroborated that the Hai//om in and around Etosha National Park, Namibia, use a latex poison from the plant, *Adenium* (Chaboo et al., 2016), whereas the Ju/'hoansi in the Kalahari use a beetle hemolymph poison that may be mixed with various plant extracts. Below, we discuss the general practice of hunting with bow and arrows followed by an overview of the poisons.

Bow and arrow hunting usually was done in pairs, with the hunters stalking animals after carefully assessing their behavior, location in the herd, gender, age, and physical condition. The hunters would move forward, often with surprising speed, to a point where they were close enough to the animals to get off a shot with their tiny, unfletched (non-feathered) arrows (Marshall, 1976). Generally, the distance that arrows would cover was between approximately 10–30 m. Once within range, the hunters would take aim and let their arrows fly. Animals have a flight distance of 40–50 m which puts them out of range of the arrows. One of the difficulties of bow and arrow hunting in the tree-bush savanna and grasslands of the Kalahari Desert region is the lack of cover. Hunters do not have an easy time sneaking up on animals, which tend to be wary of both movement and unusual smells. Stalking requires significant expertise, and it is considered arduous by most hunters who engage in it. There is some evidence to suggest that San hunters would occasionally disguise themselves as animals or ostriches to get close enough to their prey to shoot it with their bows and arrows (Stow, 1905; Thackeray, 1979). Marshall (1976) notes that in one two-month period (September–October 1952), the people of Gautscha in Nyae Nyae either killed or were given by other hunters two warthogs, three wildebeest (one of which was a calf), two eland, an adult giraffe, a baby giraffe which was found dead and scavenged, and an ostrich.

Larger game is less abundant and more mobile than small game in restricted territories and large game hunting thus requires careful planning and preparation. Dogs (Silberbauer, 1981) and hunting from horseback (Helgren and Brooks, 1983; Tanaka, 1987) have been observed as aids in modern San communities. The Tshwa San in Zimbabwe and northeastern Botswana ambushed and used poisoned arrows and spears on buffalo, elephant, hippopotamus, and other animals near water holes from ground-based blinds or from platforms set up in trees over game trails (Alec Campbell, pers. commun. to Hitchcock, 1976, 2010, 2011;

Hitchcock, 1982). However, the Tshwa no longer use these hunting techniques, nor do they use poisons.

The effectiveness of bow and poisoned arrow hunting varies, depending on season, the toxicity of the poison used, the amount of cover available, the type and condition of the prey, and the presence of other predators and scavengers (Marshall, 1976; Lee, 1979; Chaboo et al., 2016). The gearing up phase of a hunt consists in part of making new arrows, checking one's existing arrows, renovating and straightening them, removing old, flaking poison, mixing fresh poison, and smearing it below the tip. This activity usually takes place slightly away from camp so that children cannot inadvertently step on a freshly poisoned arrow or have a fly crawl on the arrow and then into an eye, something which would cause severe pain and sometimes blindness. There are/were strict rules about the ways in which arrow poison is/was supposed to be handled, although these rules are/were not always observed. The process of gearing up, recorded in Nyae Nyae by Wadley et al. (2015), is time-consuming and requires a wide range of resources, some of which must be assembled at a considerable distance from camp.

Once the equipment is ready, the hunters go in pursuit of prey. Ideally, two or three hunters try to hit the same animal, thus doubling or tripling the amount of poison entering in the prey. More often than not, the hunters fire several arrows in the general direction of the animal selected, the object being to get as many arrows into the prey as possible. Once hit by a poisoned arrow, an animal must be tracked by the hunters, often for long distances. Sometimes hunters will note the tracks of the wounded animal, then return to camp to rest before resuming the hunt. The chances of recovery of a wounded animal depend in part upon where it was hit by the arrow, the virulence of the poison, and the animal's physical condition. If the hunters are fortunate, they will come upon the dying animal before other predators such as lions or hyenas do. They will then dispatch the animal, usually with spears but sometimes with clubs or knives. The recovery rates of animals hit by poisoned arrows varied from about 30%–70%, with the lower figures in areas where there were high densities of predators and scavengers.

According to some of the Ju/'hoansi, bow and arrow hunting has several drawbacks: First, it requires training and experience to be good at it and a man is seldom at his peak hunting skill before the age of 30. Second, it requires detailed knowledge of plants and insects as well as prey animals. Third, it is seen by some as being a less efficient means of getting meat (compared, for example, to snaring) since it often requires extensive inputs of labor in following up wounded animals, and even this labor expenditure does not guarantee prey recovery. It generally takes between 3 and 24 hours for an animal to die from the effects of the poison, by which time scavengers may be on the trail of the animal. The chances of recovery of prey hit by spears is much greater than is the case for prey hit by poisoned arrows, in part because death by spear is more immediate. There are, however, some drawbacks to spear hunting, not least of which is the fact that hunters had to get relatively close to the animals to dispatch them. This is a problem when it comes to hunting large dangerous animals, such as African buffalo [*Syncerus caffer* (Sparrman)] or those antelopes that use their horns to good effect, as is the case with gemsbok [*Oryx gazelle* (L.)].

Every account and observation supports the view that when the poisoned animal is felled, the meat around the arrow puncture is cut out and discarded and the rest of the animal is eaten.

2.3.2 The San Bow and Arrow Hunting Kit

The bow and arrow hunting kit of San hunters varies across the region but generally comprises four major parts—the bow, the arrow, the poison-making materials, and the quiver to carry the poisoned arrows, with the hunting bag to contain all the equipment (Maingard, 1937; Goodwin, 1945; Clark, 1977; Steyn, 1984; Backwell et al., 2008). Each item is really a compound structure, assembled from multiple sources; wooden pieces are notched to insert parts together; a stronger better weapon is produced by stabilizing these parts together with adhesives and sinew bindings. Altogether the San club, spear, quiver, bow, and poisoned arrows are very light-weight, and appear almost like toys. We discuss each tool below.

Quivers. Several kinds of quivers were used by San groups. Leffers (2003: 22) indicated that San removed the core wood from *Acacia* root (Fabaceae) to leave the empty, intact bark to make the quiver. Sometimes wet rawhide was molded around a piece of wood and sewn together; the dried rawhide thus formed a skin quiver. These two quiver types have been found in the Drakensberg region (Vinnicombe, 1971; Oosthuizen, 1977). Quivers also have a lid of bark or hide that closes the quiver to secure the poisoned arrows; a rope strap helps to hang the quiver cross-body style. Other less-common quivers consist of tanned hide of steenbok or other small mammals (Steyn, 1984). Today, men in the Tsumkwe area use pieces of plastic plumbing pipe for their quivers (Wadley et al., 2015).

Bow. Bowstaves are cut from *Grewia flava* wood and fire is used to bend or straighten the wood. The bow string can be animal sinew such as gemsbok or kudu (Steyn, 1984; Anderson, 2001; Wawrzyniak, 2013; Wadley et al., 2015) or twig and root fibers that were smashed and fibers used as binding to hold parts together (e.g., *Rhus lancea* (L.f.) F.A. Barkley or African sumac). Rootlets of *Combretum zeyheri* Sond. are smashed and the fibers used as emergency bow strings (Leffers, 2003: 70); see Plate 2.2, Fig. 8 of a San man preparing fibers for twine. Fat or leaf extracts are rubbed on the strings (Steyn, 1984) and simple plant glue and compound adhesives are also applied as layers to arrow parts to secure the bindings (Lee, 1979; Wadley et al., 2015).

Arrows. Early European explorers to southern Africa observed different types of arrows being used to hunt different species of game (Stow, 1905; Goodwin, 1945). The diversity of San arrow types has been interpreted as signaling different hunting strategies (Deacon, 1976) or emblematic signaling of group identity (Wiessner, 1983). Regional adaptations (*sensu* Hughes, 1998) may be discounted given the occurrence of several arrow types in a single quiver. A hunting kit, including a quiver of arrows found in a cave in the Drakensberg, contained 19 arrows of three varieties, all poisoned (Vinnicombe, 1971). Arrows used to hunt springbuck (*Antidorcas* Sundevall), a type of antelope, were different from those used to hunt larger game, like oryx (*Oryx* de Blainville), another antelope. At least 11 arrow types have been identified (Bosc-Zanardo et al., 2008) and some are described in detail by Bradfield (2015).

Some arrows consist of a single piece of sharpened wood (see illustrations in Dunn, 1931 and Deacon, 1984). The San also have innovated a unique three-part compound arrow (van Riet Lowe, 1954) with the main shaft, a link shaft, and the arrow-head. Wadley et al. (2015) observed four arrow components in Nyae Nyae because a short grass collar is glued to the metal arrowhead for attachment to the linkshaft, which in turn connects to the detachable shaft.

Main shaft. This may be stems of *Sesamum triphyllum* Welw. ex Asch., *Grewia* L. or reeds (Maingard, 1937; Steyn, 1984). In Nyae Nyae, blue grass culms (*Andropogon gayanus*) are used for shafts (Wadley et al., 2015), twine is laid over the shaft on compound adhesive made from latex of *Ozoroa schinzii* (Engl.) R.Fern. & A.Fern. mixed with grass charcoal. Later, *Terminalia sericea* Burch. ex DC. gum is used as a clear, upper layer (like varnish) to seal the twine.

Link shaft. This is a double-pointed lozenge made from wood (*Grewia flava* in Nyae Nyae) or a piece of bone (gemsbok or ostrich; Steyn, 1984). The linkshaft and collar are permanently fixed to the metal arrowhead in Nyae Nyae (Wadley et al., 2015) and the three-part arrowhead is designed to separate from its shaft and remain embedded in the prey. Thus, the poison vector is not dislodged during flight.

Arrowhead. Today arrowheads in Nyae Nyae are made from fencing wire and nails beaten flat on a metal anvil and shaped with a metal file (Wadley et al., 2015; Chaboo et al., 2016). In the past, various types of arrowheads were used. Simple carved, pointed tips of bone or wood were used, as well as triangular designs (Dunn, 1931; Goodwin, 1945; Deacon, 1984; Robbins et al., 2012). Smith and Poggenpoel (1988) describe the great labor of producing arrowheads from bone; this is not a trivial task. The arrowhead is designed with consideration of size, shape, and force needed to penetrate animal hide and may be barbed or barbless. We now know that barbs were even used on weaponry in the Middle Stone Age, at least by 64,000 years ago (De la Peña et al., 2018). The arrowhead or hafted points are made from bone (e.g., of giraffe), shell, stone (Goodwin, 1945; Marshall Thomas, 2006; Lee, 1979, Silberbauer, 1981: 206; Wiessner, 1983; Wawrzyniak, 2013), horn (Robbins et al., 2012), fish mandible (Clark et al., 1974), or more recently, from glass (Deacon, 1984) or metal. Dunn (1931) was shown by a San woman how two stone fragments could be hafted to form a pointed arrow head. A single transversely mounted fragment of stone formed an arrowhead at the recent Later Stone Age site, Adam's Krans Cave (Binneman, 1994).

Fletching is uncommon on San arrows, although examples do exist, primarily from the /Xam, Ju'hoansi, and Hai//om tribes of the Northern Cape and Namibia, where a single tangential feather was sometimes used on arrows made from slightly stouter, heavier reed shafts (Schapera, 1927; Logie, 1935). Fletching is necessary only if the arrow shaft is poorly balanced, caused either through light-weight tips or heavier shafts (Hughes, 1998). The presence of fletching and heavier arrows implies the use of heavier, more powerful bows. Bows and arrows were not exclusive to Bushmen hunter-gatherers, but were also used by various pastoral Khoi-speaking groups (Stow, 1905). Among the Bushmen, however, Stow witnessed short bows in use among the tribes of the Stormberg, Orange, and Caledon Rivers, whilst those tribes living in the Langeberg, as well as those to the north and west of the Vaal River, are said to have possessed bows of a longer, sturdier and, therefore, more powerful variety. Bearing testament to this observation are several rock paintings in the Western Cape and in Namibia depicting what look like fully recurved bows (Walton, 1954; Rudner and Rudner, 1978; Manhire et al., 1985; Noli, 1993). Similar examples may also be present at rock art sites in Lesotho (Mitchell, pers. commun., to Hitchcock, 2013). The presence of this sturdier, more powerful equipment has been taken as signaling the influence of Bantu-speaking farmers (Vinnicombe, 1971). Although Maingard (1932) considers the bow and arrow to be the invention of the Bushmen, he does note that Venda bows from southern Zimbabwe turn up at both ends, thus displaying what could be considered a slight recurvature.

Adhesives. The above discussion focuses on the tools of hunting of contemporary San and briefly traces the archeological evidence of similar technologies in southern Africa. Adhesives and poisons used by contemporary San and ancient humans in southern Africa also provide insights into the technical organization, planning, multitasking, and technological skills.

Leffers (2003) indicated the use of plant-based glues by Ju/'hoan hunters in Namibia to hold snares in place (*Grewia*) or extracts of *Brachiaria nigropedata*, *Hernannia tomentosa*, and *Ozoroa paniculosa* in the glue mixture to bind sinews of arrows (Leffers, 2003: 44, 124, 152). Wadley et al. (2015) documented three plant-based adhesive recipes and preparations of Ju/'hoan San hunters in Namibia that employed bulb scales of *Ammocharis* Herb. (Amaryllidaceae), gum of *Terminalia* L. (Combretaceae), and latex from the root of *Ozoroa* Delile (Anarcardiaceae) mixed with burnt grass. *Ammocharis* bulb scales and notched *Ozoroa* roots were heated over cool coals. The resulting glues and adhesives are used to hold kudu sinew bindings in place and secure the arrowhead to the arrow shaft.

Glues have a long history. The first compound structures—attaching bone or stone points to a stick—are currently dated to ~200,000 years ago and were used by both Neanderthals and modern humans (Boëda et al., 1996; Mazza et al., 2006). Simple glues (i.e., with one ingredient) have been dated to ~200,000 years ago based on birch-bark tar at an Italian site (Mazza et al., 2006). Other ancient glues include bitumen at a Syrian site (Boëda et al., 1996), pitch in two German sites (Pawlik and Thissen, 2011; Koller et al., 2001), and red-ochre compound adhesives at Kenyan (Ambrose, 1998) and South African sites (Lombard, 2007b). These later compound adhesives illustrate deep knowledge and skills to identify suitable plant sources and to experiment with recipes to produce viable adhesives. Several research groups are identifying and evaluating the ingredients and implications of collections, and mechanical and chemical effects of different recipes for adhesives and glues that use plants (e.g., *Acacia* gum, and *Betula* and *Podocarpus* resin), ochre minerals, beeswax, and fat, and involve heating by fire (d'Errico, 2008; Wadley, 2010b; Wadley et al., 2009; Villa et al., 2015; Charrié-Duhaut et al., 2013; Kozowyk et al., 2016, 2017).

2.3.3 San Arrow Poisons

We discuss San poison sources below, first, those reported in the literature but poorly documented, and second, a more detailed account of those we have personally confirmed. We list simple poisons (made of one ingredient) and compound poisons (two or more ingredients). The chemistry of the various San arrow poisons has been researched extensively since the 1890s (Watt and Breyer-Brandwijk, 1962; Neuwinger, 1996), yet these still have not been fully characterized and there are no known antidotes (Bisset, 1989). Hopefully, modern analytical chemistry techniques will eventually offer a more satisfying explanation.

2.3.4 Plant-Based Poisons

The two well-documented and major plant poisons used by San are extracts of *Acokanthera* (Plate 2.2, Fig. 1) and *Adenium* (Plate 2.2, Figs. 2–3). Lesser-known plant poisons are derived from the genera *Acacia, Abrus, Acanthosicyos, Bobunnia, Boophane* (Plate 2.2, Fig. 4), *Boscia, Elephantorrhiza, Euphorbia* (Plate 2.2, Fig. 5), *Hyaenanche, Pachypodium* (Plate 2.2, Fig. 6), *Sansierveria* (Plate 2.3, Figs. 1–2), *Sceletium, Spirostachys, Strychnos, Strophanthus* (Plate 2.3, Fig. 3), and *Swartzia*. Finally, we review *Ricinus* (source of ricin) as it has been implicated as a possible ancient arrow poison in South Africa (d'Errico et al., 2012a) even though some botanists consider the plant to be alien to southern Africa.

Acokanthera **G. Don (Apocynaceae)** (Plate 2.2, Fig. 1). This small genus of ~five species is a widespread source of arrow poisons in Africa, particularly East Africa (Steyn (1949). For example, the Ogiek of Kenya (Blackburn, In prep.) cut small branches of the trees, *Acokanthera friesiorium* and *Acokanthera schimperi*, which are boiled, and may be mixed with branches of *Vepris simplicifolia* (Engl.) Mziray (Rutaceae; *Teclea simplicifolia* used in earlier texts). This poison is applied to different kinds of arrows, each one specific to certain animals, including elephants. The extraordinary potency of *Acokanthera* across the plant's distribution range explains its use by numerous southern African indigenous groups to poison arrows, spears, and traps—the Bemba in Zambia, Zimbabwe (Dornan, 1916), San and Zulu in South Africa (Schapera, 1925), Khoi (Lewin, 1923; Shaw et al., 1963) San in the Cape (Thunberg, 1772; Shaw et al., 1963) and Kalahari (Dornan, 1925; see Chaboo et al., 2016), Tshwa in Zimbabwe and northeastern Botswana (Hitchcock, 1982), and Valley Bisa in Zambia (Marks, 1977). Its trade and sales, even across nation borders, complicates accurate species identifications. Neuwinger (1996: 68) lists numerous documented cases of murders and accidental deaths by *Acokanthera*-based poisoned arrows. It is also used in mild concentrations medicinally (Hutchings et al., 1996). *Acokanthera oppositifolia* is the "Bushman's Poison Bush" used in southern Africa.

Remarkably, though not unexpectedly for such potent molecules, an unusual ecosystem has evolved around *Acokanthera* species. The African crested rat, *Lophiomys imhausi* (Muridae), gnaws on the roots and bark of the plant and rubs this masticate on its flank to produce a "poison

PLATE 2.2   Plant poisons of San people in Southern Africa.
1. *Acocanthera oppositfolia* (Lam.) Codd, South Africa (Family Apocynaceae;
   Photo: L. Wadley).
2. *Adenium boehmianum* Roem. & Schutlt, Namibia (Family Apocynaceae;
   Photo: R.K. Hitchcock).
3. *Adenium obesum* (Forsk.) Roem. et Schult., Tanzania (Family Apocynaceae;
   Photo: R.K. Hitchcock).
4. *Boophane* Herb., South Africa (Family Amaryllidaceae; Photo: L. Wadley).
5. *Euphorbia tirucalli* L., South Africa (Family Euphorbiaceae; Photo: L. Wadley).
6. *Pachypodium saundersii* N.E.Br., South Africa (Family Apocynaceae; Photo: L. Wadley).

PLATE 2.3 Plant poisons of San people in Southern Africa.
1. *Sansievieria* Thunb. with strap-like leathery leaves and fruits (Family Asparagaceae; Photo: L. Wadley).
2. San man making twine from fibers of *Sansievieria*, Nyae Nyae Conservancy, Namibia (Photo: G. Trower).
3. *Strophanthus* L., South Africa (Family Apocynaceae; Photo: C. Sievers).

fur." Study of the hairs revealed a specialized perforated structure to absorb the plant juice. When threatened, the animal erects these hairs as a defense mechanism (Kingdon et al., 2011). Such multispecies interactions around natural toxins are not uncommon in nature—note the arthropod/frog/arrow poison system of South America and the insect/poison-bird system in New Guinea (Daly, 1998; Alto, 2011).

Poison preparation (recipes). All parts of *Acokanthera* are regarded as poisonous but leaves, stems, and roots are commonly used. Schapera (1925: 201) indicated that the wood was pounded then boiled for many hours to a "glutinous fluid" in the northern Transvaal (now Limpopo) area and the fruit was used as poison. He also commented that "no serpent poison or other substance is added." Plants are boiled for many hours for a simple poison or other plants may be added to moderate the poison effect. Soil is added to increase the amount of poisonous material.

Toxin involved. Research on the extract began in the 1880s and a French chemist, Leon-Albert Arnaud (1888, 1898a,b) isolated and identified a cardiac glycoside as the active toxic agent which he named ouabain, following the term used by hunters. Louis Lewin (Lewin, 1893b; 1907, 1923), a trained toxicologist, continued research on this molecule, and so it is fortunate that information grew steadily in the next century. Today, many more glycosides of *Acokanthera* species have been identified (Raymond, 1936, 1939; Kingston and Reichstein, 1974; Neuwinger, 1996), all exhibiting extreme toxicity, especially to the heart.

Pharmacological implications. Like the *Adenium* glycosides discussed above, ouabain is a cardiac glycoside and it has become a model for biomedical research towards treatment of heart attacks and ventricular insufficiency (Fuerstenwerth, 2010) and towards a male contraceptive (Sultana Syeda et al., 2018). This molecule has also been isolated in extract of *Strophanthus gratus* (Apocynaceae) which we discuss below as a putative southern African arrow poison.

***Adenium* Roem. & Schutlt (Apocynaceae)** (Plate 2.2, Figs. 2–3). This small genus, commonly called "desert rose" or Bushman's rose, comprises five known species of succulent plants that are distributed across sub-Saharan Africa and the Arabian Peninsula (Plazier, 1980). In southern Africa, various San and non-San groups use *Adenium boehmianum* Schinz as an arrow poison—Dama and Hai//om in and around Etosha National Park, Bergdama around Windhoek, people in Kaokoland (Boehm, 1890; Lewin, 1894; Fourie, 1926; Shaw et al., 1963; Neuwinger, 1996) and Ovambo, northern Namibia (Schinz, 1891; Rodin, 1985). Chaboo et al. (2016) confirmed this usage through fieldwork and interviews with Hai//om hunters in Namibia where they call the plant *!kores* (San word). The Hai//om use two types of arrows, with poison and without poison, for prey of different sizes, range, and time of day (e.g., resting in the hottest part of the day). It is unclear if two kinds of bows are used, and how they distinguish the two arrows within their quivers.

Poison preparation. Sap from the plant root tubers and trunk is used to poison weaponry for hunting animals and fish (Steyn, 1949, 1957; Watt and Breyer-Brandwijk, 1962; Karimi, 1973; Gerhadt and Steiner, 1986; Omino and Kokwaro, 1993; Neuwinger, 1996; Bartram, 1997; Schmelzer and Gurib-Fakim, 2008; Peters et al., 2009; Friederich, 2014; Chaboo et al., 2016). Sap is extracted and boiled to make a thick black latex that is applied to larger-sized arrows of Hai//om hunters in and around Etosha National Park, Namibia. Spare arrow shafts and poison arrowheads are kept separately in a carrying bag known as |hôagaos. Leffers (2003) indicated that Ju/'hoan hunters in Namibia compound the plant poison with other additives: extracts of *Acacia* and *Sansievieria aethiopica* and chewed-bark spittle of *Boscia albitrunca*, pounded seeds of *Elephantorrhiza elephantina*, and fruit extracts of *Bobunnia madagascariensis*

The toxicity of the *A. boehmianum* arrow poison apparently varies seasonally (Peters et al., 2009). Hai//om hunters using *A. boehmianum* estimated that it took 2−12 hours for larger animals to die, depending on the body part hit and the condition of the animal. The knowledge of the plant and how to prepare the poison was passed down from one generation to the next but only to a limited extent among contemporary Hai//om (Chaboo et al., 2016).

Chemistry. Yamauchi and Abe (1990) isolated 20 different cardenolide glycosides and two steroidal compounds from *A. boehmianum*; these include digitalin (=gitoxigenin 3-*O*-glucosyldigitaloside) (also known in foxglove or *Digitalis* L.), honghelin (=digitoxigenin b-d-thevetoside), and somalin (=digitoxigenin b-d-cymaroside). Drugs based on digoxin and digitoxin, first isolated in *Digitalis* plants, are commercially available for heart, liver, and kidney functions; more are under clinical trials by the US Food and Drug Administration (ClinicalTrials.gov, 2018). Several appear as murder weapons in literature (e.g., Agatha Christie's "Appointment with Death") and in various television shows. Other biomolecules of *A. boehmianum* are being explored for antiviral, antitumor, and cyctotoxic effects (see Versiani et al., 2014).

Toxicology. Cardenolide glycosides are steroid molecules that affect cardiac cells by binding to and inhibiting the enzyme, adenosine triphosphate (Na^+K^+ = ATPase, commonly called the "sodium pump"), on cell wall membranes. The ATPase enzyme controls intracellular K^+ and Na^+ to maintain the normal resting potential of cells. Inhibiting the enzyme, blocking the pore channel, activating pore channels, or even opening new pore channels changes the influx and efflux of ions, and thus alters the relative concentration gradients. Different cardenolide glycosides can have different ways of interfering with the cell membrane, but with similar outcomes of cell apoptosis, cardiac arrhythmia, hyperventilation, hypothermia, heart block, and death (Akinmoladun et al., 2014; Versiani et al., 2014).

2.3.5 Minor Plant-Based Poisons

Acacia **Martius (Fabaceae).** This is a widespread and speciose genus (∼1000 spp.) in Africa and Australia. African acacias are physically protected against herbivores by their thorns, but the leaves have an additional line of defense—cyanogenic poison or tannin in lethal quantities (Furstenburg and Vanhoven, 1994). Herbivory on one plant (e.g., by antelope, kudu, and giraffes) appears to stimulate a plant alarm signal of ethylene gas that alerts neighboring plants to up their manufacture of this poison in leaves (Attenborough, 1995). Many species also contain toxic amines and psychoactive alkaloids, which may explain their use in some traditional African medicines (Stafford et al., 2008).

Abrus **Adans. (Fabaceae).** This genus contains about 20 species, but one species, *Abrus precatorius*, has been indicated as an arrow poison ingredient in southern Africa (Watt and Breyer-Brandwijk, 1962), but was not discussed by Neuwinger (1996) and has not been confirmed again. The seeds have the highly toxic molecules, abrine A and C, which are similar chemically to ricin, but far more potent. They inhibit protein synthesis, and can kill humans (Wei et al., 1974; Van Wyk et al., 2002; Dickers et al., 2003). It also increases blood capillary permeability, which leads to leakage (called "vascular leak syndrome") (Dickers et al., 2003). Infusions from this plant are also used medicinally and as a mild sedative (Adesina, 1982; Hutchings et al., 1996; Stafford et al., 2008).

Acanthosicyos **Welw. ex Benth. & Hook.f. (Curcubitaceae).** This genus comprises just two species. *Acanthosicyos horridus* Welw. ex Hook. f. is an important food of nara or tsamma melons of the Toopnar people on the Namibian coast. *Acanthosicyos naudinianus* is the Gemsbok cucumber that is eaten by indigenous people in western Botswana, eastern Namibia and northern South Africa. *Acanthosicyos horridus* is used as an arrow poison by !Kho bushmen in south-east Angola who boil the tap root for several hours (Guerreiro, 1968; Neuwinger, 1996). Little else is known about this poison though Curcubitaceae contain distasteful curcubitacin glycosides.

Boophone **Herb. (Amaryllidaceae)** (Plate 2.2, Fig. 4). Two species of *Boophane* Herb. are recognized. *Boophone disticha* (Linne f.) Herbert is commonly called "cattle killer," underscoring its use as an arrow poison across its distribution in eastern and southern Africa, Angola, Namibia, and Zimbabwe (Creighton Wellman, 1907; Dornan, 1916; Schapera, 1925). The bulb extract is dried in the sun, forming a latex that is used singly or mixed with *Euphorbia* extract as an arrow poison. Farini (1886) reported it as an ingredient mixed with venom from snakes and spiders. Since then, more than 100 alkaloids, including several unique to this family, Amaryllidaceae, have been identified and some are neurotoxic. Some alkaloids show particular specificity to serotonin transporters (Sandager et al., 2005; Stafford et al., 2008). Weak infusions from the bulb are used by traditional healers to treat mental illness (Neuwinger, 1996; Sobiecki, 2002). The plant properties were well-known in prehistory; bulb scales were used to line pits for storing oil-rich *Pappea capensis* seeds at sites like Boomplaas (Deacon, 1984).

Boscia **Lam. (Capparaceae; as "Capparidaceae" in** Neuwinger, 1996). *Boscia* comprises ∼35 species; *Boscia albitrunca* (Burch.) Gilg & Ben. occurs in South Africa and is a common tree of the Kalahari. Two species are reported as base ingredients of arrow poisons in Niger, where other plant and snake venom ingredients are added and boiled down to a paste that is applied to arrows (Kaziende, 1943). Various alkaloids, new flavonol glycosides (e.g., bosenegaloside A), and stachydrine are found in *Boscia* species (Morgan et al., 2017).

Elephantorrhiza Benth. (Burch.) Skeels (Fabaceae). The genus has only nine known species. *Elephantorrhiza elephantina* (Burch.) Skeels is a small shrub confined to southern Africa. Maroyi (2017) reviewed the widely used traditional medicines made from the root of *E. elephantina* to treat many diseases; this plant has numerous interesting phytochemicals (figured in his text). However, excessive dosages produce negative effects, including lethality in rabbits (Hutchings et al., 1996; Jansen, 2005; Maphosa et al., 2010). This plant was not cataloged as an arrow poison before (Watt and Breyer-Brandwijk, 1962).

Euphorbia L. (Euphorbiaceae) (Plate 2.2, Fig. 5). *Euphorbia* is a large and diverse genus (~2000 spp.), whose species all exhibit a white milky sap. This genus of plants was noted early as a source of arrow poisons (Wikar, 1779; Livingstone, 1858; Stow, 1905; Cornell, 1920; Lewin, 1923; Dornan 1925; Lebzelter, 1996; Hall and Whitehead, 1927; Raymond, 1947; Steyn, 1949; Shaw et al., 1963; Neuwinger, 1996) and poison water to capture game (Schapera, 1925). The poison could be used singly, thickened by drying in the sun or heated; its consistency could be improved by adding mimosa extract (Schapera, 1925). *Euphorbia* extract was also noted as an additive added to the San beetle poison discussed below (Lewin, 1912a,b). There is some confusion in the literature between the *Euphorbia*-derived arrow poison and the *Acokanthera*-derived poisons and both have been referred to commonly as ouabain. Neuwinger (1996) indicated different species being used as arrow poisons in Burkina Faso, Ivory Coast, and Somalia (Pax, 1895), and as a fishing poison in Kenya and Zimbabwe (Drummond et al., 1975). *Euphorbia tirucalli* L. is used as a fish poison (Boon, 2010) and sometimes boiled latex is used to catch birds for food (Gildenhuys, 2006: 14). Several San groups are reported to use *Euphorbia* species as an arrow poison but these have not been confirmed by recent research. Remarkably, an ancient 40,000-year-old lump of resin of *Euphorbia tirucalli* mixed with beeswax may be the oldest evidence of poison use in hunting implements in southern Africa (d'Errico et al., 2012a,b; Mitchell, 2012). *Euphorbia* are extremely interesting plants; they even have a specific organ, the laticifer, to store their toxic latex (Mahlberg et al., 1987). Their chemistry is greatly underexplored.

Hyaenanche Lamb. (Picrodendraceae). The single species of this genus, *Hyaenanche globosa* Lamb. & Vahl, is a small tree that is very narrowly restricted (and thus vulnerable) to one mountain, Van Rhynsdrop in southern Namaqualand, Van Rhynsdrop in southern Namaqualand, South Africa (Momtaz et al., 2010). The term *"Hyaenanche"* is Greek for hyena poison; the fruits were used to poison carcass baits and so control hyenas and other vermin (Van Wyk et al., 2002). It contains some toxic sesquiterpene lactones (e.g., isodihydrohyaenanchine mellitoxin, tutin, and urushiol III) which cause convulsions and coma in humans (Momtaz et al., 2010).

Pachypodium **Lindl. (Family Apocynaceae).** This genus contains 21 recognized species from southern Africa and Madagascar (Burge et al., 2013). *Pachypodium lealii* Welw. has been used as an arrow poison in Namibia, and it contains a glucoside, pachypodiin, with a digitalis action (Watt and Breyer-Brandwijk, 1962). Like the other highly poisonous genera, *Adenium*, *Acokanthera*, and *Strophanthus*, *Pachypodium* is in the Apocynaceae. *Pachypodium* species are overcollected by plant enthusiasts and their endemic habitats are being destroyed; thus, the genus is protected under the international Convention on International Trade in Endangered Species of Wild Fauna and Flora. Collecting plants to make poison is illegal.

Ricinus **L. (Euphorbiaceae).** This is a monotypic genus and *Ricinus communis* L. is considered one of the most poisonous plants in the world due to the toxic molecule, ricin, in the castor bean seeds. It has not been reported as being used as an arrow poison today, but we include it here because of the historical interest.

A stick found in Border Cave, South Africa, has been shown to have residual traces of ricin. This has been interpreted as a poison applicator (d'Errico et al., 2012a,b; Mitchell, 2012); some consider this controversial (d'Errico et al., 2012b; Evans, 2012). Neuwinger (1996) did not treat *Ricinus* as an arrow poison source nor mention any related drugs and poisons, however both *Ricinus* and *Euphorbia* are poisonous (Scarpa and Guerci, 1982; Kang et al., 1985) and the wax may have been a binder for a hunting poison, but Cheikhyoussef et al. (2011) indicate the use of *Ricinus* extract to treat epilepsy in Namibia.

Ricin is extracted from the seeds, and is similar chemically to the abrin toxin from the *Abrus* plant discussed above as another arrow poison source (Watt and Breyer-Brandwijk, 1962). It is unclear if *Ricinus* was growing in South Africa about 24,000 years ago, when the Border Cave materials are dated (Bradfield et al., 2015) but the plant's range has been expanding since then and it has spread also from its natural distribution in north Africa and India to a wide distribution today due to its ornamental and commercial cultivation. Ricin could have been used in the past, and not currently.

Ricin is produced as waste from the industrial manufacture of castor oil. It can be inhaled or ingested. Once it gets inside the cell, ricin molecules bind with cell carbohydrates and interfere with protein synthesis (Magnusson et al., 1993). This can lead to a variety of symptoms, including respiratory failure, depending on dosage. No antidotes for ricin poisoning are known.

There is global concern about its potential to be used in terrorism and warfare (because of the ease of production). It was used famously to assassinate a Bulgarian journalist in London via injection from the tip of an umbrella (Schep et al., 2009).

Sansievieria **Thunb. (Asparagaceae)** (Plate 2.3, Figs. 1–2). This genus comprises ~70 species in Africa and Asia and their distinctive appearance with strap-like leathery leaves explains their popularity as household plants. Reports of Zhu and Juǀhoan San mixing extract with *diamphidia* beetle poison (Neuwinger, 1996; Nadler, 2006; Robbins et al., 2012) are unconfirmed. *Sansievieria* exhibit many alkaloids, flavonoids, saponins, glycosides, terpenoids, tannins, and proteins that may have medicinal value (Anbu et al., 2009) or may be toxic or hemolytic (Neuwinger, 1996; Bradfield et al., 2015).

Sceletium **(Aizoaceae).** This is a small genus of ~8 species. The diaries of Van Riebeeck from 1662 (see Pooley, 2009) provide the first indication that the dried plants were chewed by indigenous people to produce a euphoric effect. Today, we know it is chewed and smoked for its narcotic and sedative effects and other medicinal applications by one San group as well as the non-San groups, Nama of Namaqualand and Khoi pastoralists in the Little Karoo, South Africa (Gericke and Viljoen, 2008). *Sceletium* contains alkaloids such as mesembrine, mesembrenone, mesembranol, and tortuosamine that may be psychoactive, likely through its action as serotonin-uptake inhibitors (Smith et al., 1996; Gericke and Van Wyk, 2001; Stafford et al., 2008; Harvey et al., 2011).

The Amatola (Amatole, Abatwa) San of southeastern Africa, particularly the Maluti-Drakensberg Mountains region of Lesotho and South Africa, may have used *Sceletium tortuosum* (L.) N.E. Br. plants, which can be highly toxic, depending on preparation and use (Mitchell and Hudson, 2004:45). Amatola San descendants told author Hitchcock in multiple interviews that their ancestors used plant poison on their arrows in warfare against commando groups in the 19th century, as well as for hunting (for discussions of the Amatola, see Orpen, 1874; How, 1970; Prins, 2009; Challis, 2012). Microliths have been found in rock shelters in the Maluti-Drakensberg Mountains that may have been used with poison (David Ambrose, pers. commun. to Hitchcock, 2012). Analyses of microliths from Lesotho shelters are underway to determine whether they were used as poison. Some analyses have revealed the presence of *S. tortuosum*.

The /Xam San of the Northern Cape region, South Africa, also used *S. tortuosum* poison on their arrows, according to /Xam descendants and to archival records in the Bleek and Lloyd collections at the University of Cape Town (Deacon, 1992; Skotnes, 2002; Thomas Dowson, pers. commun. to author Hitchcock, 2004). Additionally,

Hitchcock has assembled over a dozen interviews about the use of this plant as a poison in Lesotho. Both the Amatola and the /Xam resided in areas that are far from the savannas of the Kalahari and adjacent areas; it would be useful, therefore, to obtain a greater understanding of their hunting and poison technologies and the factors that conditioned them.

Spirostachys Sond. (Euphorbiaceae). The two species (Govaerts et al., 2000) are commonly called tambuti or tamboti. The white milky sap is reportedly boiled to make a compound arrow poison with *Euphorbia latex* by Hai//om San (Fourie, 1964; Neuwinger, 1996), with *Adenium boehmiamum* (Wilhem, 1963; Neuwinger, 1996), with *Aloe* and *Spirostachys africana* in northern Namibia (Vedder, 1923; Neuwinger, 1996), and as a simple fishing poison along the Angola—Namibia border and in Zimbabwe (Rodin, 1985; Neuwinger, 1996). Anthocyanins, diterpenes, saponins, steroids, tannins, and volatile oils have all been detected but we know little about the active molecule(s) of the arrow poison.

Strychnos L. (Loganiaceae). *Strychnos* poisons are used widely circum-tropically. It is the source of curare in American curare blowdarts and of strychnine in Asian arrow poisons. In Africa, about 50 species of *Strychnos* are known (Plant List Version 1.1., 2013), and they are diverse and widespread. They are used by many African indigenous groups as arrow poisons and ordeal poisons (to determine guilt; the trial is often fatal). The African poison is made by boiling a reduction that can be stored. It also is used in some traditional medicines. The active chemicals increase nervous system activity, muscle paralysis, and convulsions, then death. More than 250 alkaloids have been identified from the African *Strychnos* (Massiot and Delaude, 1989) and their modes of action are under study.

Strophanthus D.C. (Apocynaceae) (Plate 2.3, Fig. 3). This genus of about 40 species includes vines, shrubs, and trees. Livingstone and Livingstone (1865) first mentioned its use in the Zambesi river area as an arrow poison that felled most game except elephant and hippopotamus. Fraser (1872) described this Kombé poison as a cardiac poison after experiments with pigeon, frog, rabbit, dogs, and cats. This poison is widely used in West Africa (Chevalier, 1950) by San and non-San peoples from Tanzania to southern Africa, with the poison having many local names (Neuwinger, 1996). These plants have many interesting chemicals including cardiac glycosides like strophantin and there was a period of chemical exploration to determine medical applications however it is not in clinical use today. About 1 mg of ouabain can be lethal when injected: it causes heart glycoside poisoning (Wink and van Wyk, 2008).

Swartzia Schreb. (Fabaceae). About 200 species are described in this African and neotropical genus. *Swartzia madagascariensis* is a fishing poison in Africa (Dalziel, 1937; Raponda-Walker and Sillans, 1961; Watt and Breyer-Brandwijk, 1962; Schultes, 1979). The plant ranges into northern Namibia and Neuwinger (1996) observed poison preparation where this plant extract was added to the *Diamphidia* beetle base poison and, in rare cases, the plant was used singly as a weaker poison on small animals. He commented that it may also be used by San in Zambia and Zimbabwe. Various saponines, oleanolic acid, and glycosides have been identified in the fruit, beans, and seeds, while the stems have variations on a pterocarpane structure that have been demonstrated as hemolytic and toxic to fish. Interestingly, this plant is also widely used as medicine, underscoring its interesting chemistry (Watt and Breyer-Brandwijk, 1962). At Nyae Nyae, *Swartzia* pod shavings are mixed with *Diamphidia* beetle poison as one potential arrow poison (Wadley et al., 2015).

2.3.6 San Animal-Based Arrow Poisons

Today, the best documented arrow poison of southern Africa is from beetles. Other poisons from scorpions, spiders, and snakes are widely cited but unconfirmed, without even a photograph in popular and scientific literature. Schapera (1925) indicated that these poisons were "ground up into powder."

2.3.7 Arthropod-Based Poisons

Insects: Beetles (Insecta: Coleoptera) (Plate 2.4, Figs. 1–6). By far the most iconic hunting poison used in southern Africa is that which the San extract from beetles. A San insect poison was mentioned first by Wikar (1779). Early explorers, Lichtenstein (1930; traveling in the early 1800s), Mentzel (1944; traveling in 1735–1741), and Livingstone (1858) discussed arrow poisons. Later, beetles were indicated as the poison (Schinz, 1891; Fairmaire, 1893; Schultze, 1907; Trommsdorff, 1911; Schapera, 1925; Koch, 1958; Marshall, 1958). Today, several beetle species in four genera from two different families, leaf beetles or Chrysomelidae and ground beetles or Carabidae, are reported as San arrow poisons:

Chrysomelidae (leaf beetles): *Blepharida* Chevrolat (73 species), *Diamphidia* Gerstaecker (nine species), and *Polyclada* Chevrolat (12 species).
Carabidae (ground beetles): *Lebistina* Motschulsky (14 species).

The use of species of *Diamphidia* and *Polyclada* Chevrolat as arrow poisons by Ju'hoansi in Nyae Nyae Conservancy, Namibia

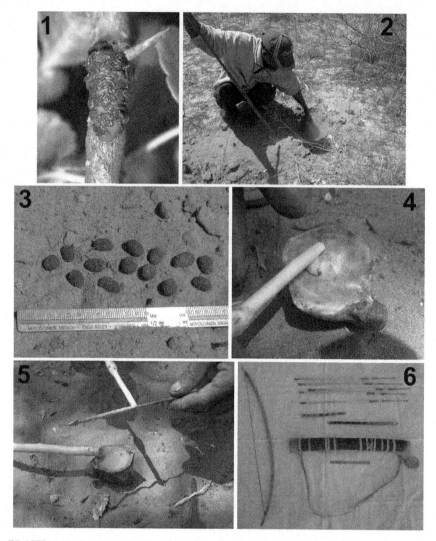

PLATE 2.4 Beetle poison source, poison preparation, and arrow preparation in Nyae Nyae Conservancy, Namibia (Photos: C.S. Chaboo).

1. *Diamphidia* beetle larva, a mature stage coated in its feces on its host plant, *Commiphora*, in South Africa. This larva migrates down the plant and burrows into the soil and forms a cocoon. It stays in a quiescent phase until triggered to rapidly pupate and metamorphose to the adult stage. Adults emerge from the ground and live freely on the same species of host plant.
2. San man using a digging stick to collect the beetle cocoons from the loose sand around the base of the host plant.
3. Numerous cocoons are sifted out the sand. The cocoon is made of sand grains glued together and hardened by secretions of the leaf-beetle larva.
4. Cocoons containing beetle larvae. More than 100 cocoons may be collected around the base of the host plant, stored in an ostrich egg shell or a plastic container, brought back to camp and stored for a while. When the rainy season starts, the larvae rapidly pupate and metamorphose into the adults that emerge to feed on the host plants.
5. San hunter preparing beetle poison by squeezing larval contents with a wood "pestle" into a bone "mortar" that is stabilized in the sand. About 7−8 larvae (not pupae) are squeezed to release body tissue that are macerated with saliva into a thick paste that is applied with the stick applicator to the shaft behind the arrow head.
6. San hunting kit (ruler for scale): quiver, bow, three-part arrows, and fire-starting sticks.

(Chaboo et al., 2016), and Tsodilo Hills, Botswana (Robbins et al., 2011, 2012), has been confirmed through fieldwork. *Blepharida vittata* (Baly) was also indicated as arrow poison [Lewin, 1912b; Schapera, 1925; then as *Blepharida evanida* (Baly)]; *Blepharida* is closely related to *Diamphidia* and *Polyclada* (these genera belong to the *Blepharida*-group lineage within their tribe Alticini; Heikertinger and Csiki, 1940), and occurs in southern Africa, however there have been no reports or observations of members of this genus being used as arrow poison.

Many authors, including recent chemists, have claimed that a "grub" or beetle pupa is used to extract poison. It is important to be certain of the stage of the insect life cycle as the chemistry may be different from one stage to another. "Grub" is an imprecise term for juvenile stages of insects as many people confuse the different life stages and apply the term to even noninsect invertebrates. The reports of pupae are likely an erroneous assumption—typically, cocoon-making insect larvae construct, secrete, or weave a cocoon that protects the pupal stage, so one would expect normally to find pupae in the underground cocoons collected by San. However, in *Diamphidia* and *Polyclada* (and perhaps all the *Blepharida*-group), the larval morphology is apparently maintained from one season to the next (and perhaps for years) until seasonal conditions (e.g., spring rainfall) trigger the larvae to quickly pupate and metamorphose. The collected cocoons (Plate 2.4, Fig. 3) primarily have larvae, and rarely other stages. The harsh arid conditions may drive this atypical life cycle for chrysomelid beetles. Furthermore, Chaboo et al. (2016) observed that Ju/'hoansi hunters reject pupal and adult stages for poison preparation and use only the larval stage for poison.

The poison is prepared by squeezing the larval hemolymph onto a bone mortar and macerating it with human saliva (Plate 2.4, Figs. 4 and 5; Bisset, 1989; Wadley et al., 2015; Chaboo et al., 2016) or with plant extracts—*Urginea sanguinea* Schinz (Liliaceae: bulb section, Caprivi Strip Bushmen) (Bisset, 1989: 10), *Sansevieria* Thunb. (Asparagaceae) (Robbins et al., 2012), *Asparagus* root sap (Wadley et al., 2015), or *Swartzia madagascariensis* (de la Harpe et al., 1983). Others have reported that the beetles are sun-dried, powdered, and mixed with plant juices before application. All reports agree that the thick beetle paste is applied with a stick behind the arrowhead (see Koch, 1958; Bradfield et al., 2015; Wadley et al., 2015; Chaboo et al., 2016).

Toxicity. The poison is extremely effective, but it should be noted, however, that the efficacy of the poison varied considerably. It was not easy to find the arrow-poison beetles—one must know where the host plants are located under *Commiphora* (Burseraceace) bushes or *Sclerocarya* (Anacardiaceae) trees and then dig 0.5–3 m under the sand around the plants to sift out cocoons. Simply learning the plants identity and finding them in the savanna habitat are challenging. The toxicity tended to decline in the late dry season, and in drought years and extremely wet years the beetles were often not available at all (Lee, 1979; Hitchcock and Bleed, 1997). The poison was shown still to be effective in museum specimens that were over 80 years old (Shaw et al., 1963).

Chrysomelid beetles produce many noxious defensive chemicals in all life stages; some are sequestered from their host plants or modified a little from host-plant compounds, while others are the outcome of de novo synthesis. They advertise their distastefulness in vivid color patterns or concentrate nasty compounds in defensive fecal coatings. The larvae of the *Blepharida*-group are noted for this kind of fecal defense (Chaboo et al., 2007, 2016; Chaboo, 2011). Additionally, these three genera used by San are herbivores on two plant families, Anacardiaceae (mango family) and Burseraceace (frankincense and myrrh family), which are themselves distinct chemically.

Toxin involved. The poison from this San beetle larva is slow-acting, with large animals taking as long as three days to succumb (Marshall, 1960). Various chemists have analyzed the beetle poison (Boehm, 1890, 1897; Heubner, 1907; Trommsdorff, 1911; Hall and Whitehead, 1927; Breyer-Brandwijk, 1937; Bijlsma and De Waard, 1957; Koch, 1958; Kündig, 1978; Mebs et al., 1982; Woollard, 1986; Woollard et al., 1984; Kao et al., 1989; Kann, 1989; Jacobsen et al., 1990). The toxic molecule was described and characterized eventually as a low-molecular-weight peptide, and named diamphotoxin by De la Harpe et al. (1983) in the *Journal of Biological Chemistry*, where many toxic molecules of the San have been described (see our citations for this journal). Like the snake and scorpion peptide poisons, diamphotoxin interferes with cell membrane permeability. Today, Gary Trower (pers. commun., 2018) is conducting dissertation research on the chemistry of one of the beetle genera, *Polyclada*, so we await new updates on the nature of this molecule(s).

2.3.7.1 Minor Animal-Based Poisons

Beetles, *Lebistina* Motchulsky (Coleoptera: Carabidae: Lebiinae). In addition to the diamphidia beetles above, carabid beetles of the African genus *Lebistina* have also been reported as an arrow poison (Mebs et al., 1982; Neuwinger, 2004). Apparently, a juvenile stage is extracted from pupation cocoons and macerated and applied to arrows. *Lebistina* and other members of the tribe Lebiinae have an atypical life cycle among beetles; the eggs are laid on the ground but the larvae migrate underground, enter the cocoons of diamphidia beetles, and become ectoparasitoids on the latter, developing on and eventually killing their hosts. The adult *Lebistina* are arboreal (not on the ground), living with and likely mimicking the aposematic coloration of their leaf-beetle hosts.

Observation and lack of photographic confirmation casts doubt on the use of *Lebistina* beetle larva as an arrow poison. It is also rarely found, in fewer than 5% of leaf-beetle cocoons (Chaboo et al., 2016). Generally, carabid beetles manufacture many potent defensive chemicals in a pair of abdominal glands and which they ooze and spray (e.g., bombardier beetles) and the chemistries of ~400 species have been characterized (see Will et al., 2001). Many molecules are acids (e.g., formic acid) and none appear to be toxic.

Several authors have attempted to illustrate tritrophic relationships between the host plants, the diamphidia leaf-beetle herbivores, and their *Lebistina* carabid ectoparasitoids (see Nadler, 2006: 63−64) but the genera and species of these carabid ectoparasitoids await confirmation through genomics and rearing since we have no library of associated and identified specimen collections. There may be several genera of lebiine carabids involved (Assmann et al., 2017).

Ants (Hymenoptera: Formicidae). Ants are not implicated in San arrow poisons. However, we comment briefly on ants because of other evolutionary models where it was ultimately revealed that ants are the source of other potent poisons that were long known in vertebrates that consume the ants. As the ecology surrounding San animal-based poisons is unraveled, we may yet determine other trophic links in the food chains that influence the producers of the toxic molecules.

Butterflies and moths (Insecta: Lepidoptera). Grubs and caterpillars appear rarely in the literature as a poison source (e.g., Steyn, 1984), but are unconfirmed. As noted above, "grub" is an imprecise vernacular term that can apply to many small crawling insects. Many Lepidoptera have offensive chemistry that is frequently advertised in their vivid aposematic color patterns so it is not impossible that certain caterpillars may be poison sources. Many species also are quite host-specific and may sequester compounds from their host plants—each poison plant species may have a distinct herbivorous fauna that must deal with their host's toxic compounds (to denature, neutralize, or co-opt these molecules).

Scorpions (Arthropoda: Scorpiones). Scorpion venoms are indicated as arrow poison of hunter-gatherer groups in Lesotho (Schapera, 1925; Liebenberg, 1990) and in the Kalahari (Nadler, 2006 Bradfield et al., 2015), but this has not been verified. All scorpions produce venoms from their telson sting to subdue invertebrate prey; like other animal-based venoms, these are cocktails of molecules and some are neurotoxic [i.e., impacting ion channels and so interfering with nervous signaling (Schaffrath et al., 2018)]. Scorpion stings are intensely painful and localized, but can lead to tachycardia; only a few species of scorpions have toxic stings. In southern Africa, the two scorpion families, Buthidae and Scorpionidae, exhibit a spectrum of toxicities and painful reactions, from trembling and pain to respiratory failure, tachycardia, and death (Müller, 1993). Scorpion venoms as a San arrow-poison raises the question of if and how these largely invertebrate-targeting poisons might impact game animals and if such molecules retain their functionality when dried on an arrow head. These small proteins may be easily denatured in air.

Spiders (Arthropoda: Araneae). San poisons based on spider venoms have been indicated in multiple sources (Livingstone, 1858; Farini, 1886; Stow, 1905; Schapera, 1925; Steyn, 1984; Nadler, 2006), yet are poorly documented or photographed. Only Steyn (1984) indicates an *Argiope* species as a poison source. Schapera (1925) noted that both the spider and eggs were ground and mixed with *Euphorbia* extract (for consistency) by Cape Bushmen. Spider venoms originated to subdue prey and are not considered neurotoxic and are rarely fatal. However, the reaction profile of a spider bite is different from that of scorpions and snakes, and varies within spiders. Typically, spider bites are painful, with a few mild symptoms in a condition termed arachnidism. The Theridiidae spiders (including black widow spiders in North America and button spiders in Africa) do have a neurotoxic venom, and their unique injury pattern is termed lactrodectism. Recluse spiders have a different cocktail mix, with different damage in a condition called loxoscelism (Richard and Isbister, 2008). Spider venoms contain a diverse range of compounds with peptides being dominant. Spider venoms are primarily cytolytic and affect insect nervous systems (King and Hardy, 2013) though atypical spider venoms (e.g., widow spiders) have neurotoxins dominating their venoms. It is difficult to imagine how spider venoms would be harvested to use as an arrow poison; today, venom glands are dissected to extract the venoms for research.

Snakes (Reptilia: Serpentes). Snake venoms are remarkably varied, range in cell toxicity, and have become models of medical research. In Africa, snake-based arrow poisons of San poisons were cataloged (Schapera, 1925; Steyn, 1949; Watt and Breyer-Brandwijk, 1962; Neuwinger, 1996) and are widely cited (e.g., Liebenberg, 1990). Snake venoms are also reported as an additive to primarily San plant poisons (Dornan, 1916 cited in Schapera, 1925; Thunberg, 1986). Schapera (1925) considered the snake poisons to be "the most widely distributed." The snake was decapitated, poison glands extracted, dried, and ground (and the snake eaten). The powder was mixed with *Acokanthera* or *Euphorbia* extract, then boiled and reduced to a thick lumpy paste that was stored. Stow (1905) reported a variation that extract of *Amarylis* was added. The snakes indicated as sources are black mamba (Elapidae: *Dendroaspis polylepis* Günther), Cape cobra (Elapidae: *Naja nivea* L.), and African puff adder (Viperidae: *Bitis arietans* (Merrem) (Schapera, 1925; Steyn, 1984; Shebuski et al., 1989; Nadler, 2006).

Toxicology. Snake venoms are complex cocktails of proteins and peptides that have evolved probably from salivary enzymes and molecules to immobilize and kill vertebrate prey. They range in cytotoxity, cardiotoxicity, and neurotoxicity. Depending on the amount and site of injection, the severity of impact on a victim can also vary (De Weille et al., 1991; Currier et al., 2010). The specific snake species used for southern Africa poisons are particularly potent, containing multiple peptides that inhibit platelet formation and therefore blood clotting; some of these include bitistatin, trigramin, and echistatin (Shebuski et al., 1989).

Unlike scorpion and spider venoms that primarily target invertebrate prey, snake venom molecules directly target vertebrates and thus could be more feasible arrow poisons for San game animals. Reliable verification and photographic evidence, even in tourism materials, of snake-derived arrow-poisons are unavailable.

2.4 DISCUSSION

The diverse San poisons and venoms are the result of different evolutionary processes in distantly related taxa. These toxins are structurally very different yet appear to ultimately affect ion (e.g., Ca^{2+}, Cl^-, K^+, and Na^+) transport across cell membranes, regulating cell processes. Some appear to be highly specific to cardiac muscles, neurons, or cells of tissues around injury site.

The plant-derived poisons are commonly cardenolide glycosides. These are closely related structurally to bufadienolides (e.g., bufotalin and bufotoxin), originally described from the frog family, Bufonidae (Stoll et al., 1933; Chen and Kovaříková, 1967), and now known in some plant families (Crassulaceae and Hyacinthaceae; Kolodziejczyk-Czepas and Stochmal, 2017). These molecules only differ in the attached lactone ring-structure—cardenolide glycosides have a five-member ring and bufadienolides have a six-member ring; both act similarly by interfering with cell membrane ion pumps and induce a similar toxidrome. Cardenolide glycosides and bufadienolide molecules appear to be psychoactive in small doses but lethal in larger doses. Convergent evolution of toxic biomolecules is not unique; Catterall and Beress (1978) demonstrated that scorpion toxins and sea anemone toxins have similar polypeptide structures that inhibit inactivation of sodium ionophores.

Convergent evolution across such distantly related taxa in the evolutionary tree of life suggests that toxic molecules are far more common than we know. Fry et al. (2009) discuss the molecular modifications of proteins that can produce toxic proteins and the convergent origins of toxins in Metazoa. Clearly, the San are using a subset of toxins and poisons from their environment and, although these are from diverse sources, it appears that they might be similar structurally and/or similar in action and toxidrome.

Upward sequestration of these molecules in the food chain appears possible—grasshoppers feeding on milkweeds can store their host's cardiac glycosides (Euw et al., 1967). In Africa, the koppie foam grasshopper [Pyrogomorphidae: *Dictyophorus spumans* (Thunberg)] uses cardiac glycoside toxins in its foam defenses (Picker et al., 2004). We should look out for such trophic patterns (e.g., the diamphidia beetles and their carabid parasiotids) as they may help us discriminate the original producers of certain toxins.

The San poison and venom molecules may be structurally similar or dissimilar, simple or complex, and derived from different evolutionary processes. They may work similarly or differently and yet may produce similar cellular responses (e.g., irregular ion transport and signaling) and similar toxidromes of disorientation, irregular heartbeat, stumbling movement, and slow paralysis. These effects include the delay between the animal being hit and dropping down before the human hunter can finish it off; this delay is at the root of the San tracking culture.

Arthropods manufacture numerous distasteful and irritant compounds that serve diverse functions, as nuptial gifts, in sexual signaling, as defenses, to subdue prey, and to deter predators (Gwynne, 2008). Trophic relationships around the American dart poisons from frogs have been revealed to be part of a large ecological relationship where frog and bird predators sequester toxins from their mite, ant, and beetle prey in their skin and feathers, respectively (Dumbacher et al., 1992, 2004; Alto, 2011). The frogs and birds appear to have a way to neutralize these arthropod biomolecules so they are not affected. The dart frog's renowned chemical defense is co-opted from alkaloids of the prey (captive poison-dart frogs eat raised crickets and do not develop this defense). Similarly, insect herbivores (like the koppie foam grasshopper above) are capable of co-opting plant-manufactured biomolecules. There are numerous examples of mimicry complexes involving multiple taxa evolving around certain irritating molecules, e.g., cantharidin.

The physiological and ecological roles of diamphotoxin in the life of the San poison beetles is entirely unknown. Yet, it is compelling to consider tritrophic connections between the chemically interesting host plants of Anacardiaceae and Burseraceae, the leaf-beetle herbivores that may sequester plant chemicals (typical in leaf beetles), and chemically interesting carabid beetle ectoparasitoids (Nadler, 2006: 63–64; Chaboo, 2011: 64). A great deal of fieldwork and chemistry is needed to assess linkages but this hypothetical model is not evolutionarily impossible. Our point is that life on Earth has evolved a vast library of molecules that await discovery. It is remarkable, yet perhaps not unsurprising, that these potent toxins have been selected by the San and that earlier hominins probably also used poisons.

How likely was adoption of poison? The San subsistence strategies reflect a very sophisticated understanding of their complex desertic and arid environment, its seasonality, and the patchiness of resources (Biesele, 1978; Barnard 1979; Cashdan et al., 1983; Wiessner, 1983; Steyn, 1984). Their religion is animistic and they are respectful of nature (Wawrzyniak, 2013; Hitchcock, in prep.).

It is clear from the great diversity of plant and animal ingredients used to poison hunting weapons that the San had an intimate procedural knowledge of the chemical properties of the species in their environment (sensu Wynn and Coolidge, 2007). Except in a few cases, San arrow poisons consist of recipes incorporating multiple ingredients (Bradfield et al., 2015). When these ingredients are mixed together they create a new substance that has both toxic and adhesive properties. This blending of substances to create something new, and that is irreversibly altered, is one signifier of complex cognition (Wadley, 2013). Transformative technologies are common-place among hunter-gatherer groups and can be traced back at least 100,000 years ago (Henshilwood et al., 2011).

As far as we know, the use of poisons for hunting purposes is unique to *Homo sapiens*. The ability to mix toxins effectively implies long attention spans, abstract thought, and the capacity to conceptualize synergistic technologies (Bradfield et al., 2015). The identification of poison recipes on archeological artefacts can therefore tell us much about the cognitive abilities of the people at that time.

What was the likelihood of plant-based arrow poison? There is a vast array of toxic plants in southern Africa (Van Wyk et al., 2002), of which only a fraction is known to have been used by the hunter-gatherers. There may have been many poisons used in the past but which have already disappeared as indigenous people became more westernized. The tentative identification of ricin at Border Cave 24,000 years ago, and *Euphorbia tirucalli* at ~40,000 years ago, highlights this potential, as does the earlier (77,000 years ago) presence of insecticidal wood and leafy material at Sibudu.

What was the likelihood of a beetle-based arrow poison? A most intriguing question about the San beetle-based poison is how they first identified these as poisonous. The San search both the surface and subsurface of their arid habitats for water as well as numerous foods (e.g., roots, tubers, and the Kalahari truffles; Leffers, 2003: 185). They also hunt certain animals from underground. Like many aspects of their knowledge, the San also have a deep knowledge of the smaller animals of their environment. They eat honey and insects (Green, 1998; Nonaka, 2005, 2009; Chen et al., 2009; Lesnik, 2014; McGrew, 2014; Raubenheimer et al., 2014). In several southern African rock paintings, honey-hunting is depicted (Pager, 1971, 1973); some appear to show the use of smoke to evict the bees (Namibia: Toghwana Dam, Matopo Hills). In ancient times, beeswax has been argued as a binding ingredient used with plant-based poisons, based on residues on a wood stick that is interpreted as a poison applicator in a hunting kit in southern Africa (d'Errico et al., 2012a,b; Mitchell, 2012). Insecticidal bedding of insect-repelling plant leaves, twigs, and branches are known in modern indigenous groups and probably date into the past (Villa et al., 2009; Wadley et al., 2011).

The diamphidia beetle adults and larvae are abundant above ground on their host plants, the low-growing shrubs and treelets of *Commiphora* as well as *Sclerocarya*, yet these are not utilized as poison. Chemical analyses have detected diamphotoxin in these aboveground stages (Woollard et al., 1984; Woollard, 1986; De la Harpe and Dowdle, 1980; De la Harpe et al., 1983) but the San do not collect these more visible stages, possibly due to a less potent toxin. The suspended larval stage is creamy-colored, not possessing any vivid colors to typically advertise distastefulness and danger, and is enclosed in an underground blackish-brown cocoon. It is extraordinary that the San figured out to dig 0.5–3 meters underground to get beetle cocoons containing such a potent poison.

Another fascinating possibility is that diamphidia beetles may have been used widely by hunter-gatherers living in the Stone Age. The beetles and the host plants occur as far afield as KwaZulu-Natal, meaning that this poison source could have been available to ancient communities, such as those occupying Border Cave and Sibudu. Correctly analyzing the diamphotoxin molecular structure and its byproducts might allow us to identify traces of this poison on archeological artefacts.

How old is the San poison practice? The San poisons we observe today need not necessarily be ancient, but they hold clues to what hominins were doing before them. In fact, the environments of southern Africa have changed and older poison sources may have changed in distribution or gone extinct.

San bow-and-arrow hunting uses a complex projectile weapon system that comprises separate and multicomponent structures, the bow, the arrow, the quiver, and the poison-making equipment. An obvious question is when did it originate? Early humans observed plants and animals closely, determining what was edible, nutritious, distasteful, medicinal, and narcotic. Meat-eating and hunting appeared early in hominin history (Wrangham, 1977), when ancient hominins added higher caloric meat-eating and ambush hunting to their foraging, gathering, and scavenging of food. We can trace in the fossil record that hunting methods and associated technology and behaviors (e.g., group hunting and signaling) became more complex. The shift in food acquisition and hunting was probably in tandem with or stimulated social, cultural, and religious shifts; for example, as hunters became more esteemed and certain game became more valued than others.

Hafted stone points, presumably as the tips for spears and arrowheads, provide direct evidence of ancient hunting. While these examples may seem precociously early, diagnostic impact fractures and/or hafting residues have been found on convergent stone flakes and segments from the pre-Still Bay at more than 100,000 years ago through to the post-Howiesons Poort industries of the Middle Stone Age at Blombos Cave, Sibudu, Umhlatuzana, and Klasies River Cave 2, all in South Africa (Lombard, 2005, 2006, 2007a,b, 2008; Wurz and Lombard, 2007).

Preparations of simple and compound adhesives have been documented in the archeological record (for example, d'Errico et al., 2012a; Charrié-Duhaut et al., 2013) and were used in the manufacture of javelins and spears—awareness of adhesives and bindings predate the origin of the bow-and-arrow hunting technology that may have appeared by 64,000 years ago in South Africa (Backwell et al., 2008, 2018). The time of earliest origins of stone-tipped "projectile" weapons (to throw and kill from a distance) has been pushed back to ~60,000 years ago (Shea, 1993, 2006; Langley et al., 2016). The collection and processing of diverse wood and plant materials to create spears, bows, arrows, and other tools indicate a detailed knowledge of plant diversity and, significantly, the behavior of these plant materials (e.g., durability, bending with aid of fire, projectability, etc.). The controlled use of fire to treat rocks and sticks has been dated ~170,000 years ago (Wadley, 2013; Aranguren et al., 2018). Organic parts of hunting tools (wood, rope, bone, sinew, plant fibers, feathers) are unlikely to survive long periods of time and have not been preserved except for the points, some with residues. It is difficult to determine the coevolution of each item in the ancient hunting kit (a set of symbiotic technology) or, importantly, how changes in weapon design stimulated changes in hominin physiology and morphology (see Wong, 2014).

The archeological record in Africa suggests that bow-and-arrow hunting originated at least 60,000 years ago (Shea, 1993, 2006; Backwell et al., 2008, 2018; Lombard and Phillipson, 2010). There may have been multiple origins. This technology is sophisticated; bows are designed to take advantage of the latent energy of a bent stick and arrows are fletched to fly better. The different technological advances in this weapon system are being studied: comparisons of gross design of contemporary and ancient stone points (Shott, 1997; Kuhn and Stiner, 2001); style (Wiessner, 1983), ballistic design, aerodynamics, and performance (Odell and Cowan, 1986; Brooks et al., 2006; Shea, 1988, 2006, 2009; Sisk and Shea, 2009; Waguespack et al., 2009); stone points (Lombard and Phillipson, 2010); wear patterns on stone points (Shea, 1988; Lombard,

2005; Lombard and Phillipson, 2010); residues (Lombard and Phillipson, 2010); and adhesives (Wadley et al., 2009). San use of metal points dates to about 100 years ago (Dunn, 1931; Wiessner, 1983).

Indirect evidence of bow-and-arrow hunting is recorded in scenes of hunting of ancient rock paintings (Walton, 1954), but few paintings are more than a few hundred years old in southern Africa, so the rock art is not useful for informing us of the origins of bow and arrow hunting.

Ancient evidence for poison use in hunting. Direct association of poisons with hunting weapons is scant in the archeological record. Poison degrades over time and may not be recognized easily on arrow- or spearheads found archeologically. Poisonous plant remains that are recovered are difficult to interpret as a poison ingredient, since many of the plants used to make hunting poisons are also used for medicinal purposes and to make glues (see Neuwinger, 1996; Wadley et al., 2015). Macrobotanical remains of plants with known medicinal and toxic properties have been found in several Later Stone Age contexts. For example, scales of *Boophane disticha*, a known arrow poison used by numerous San groups, were used to wrap a stone tool 6000 years ago at Melkhoutboom (Deacon and Deacon, 1999), and the mummified remains of a woman 2000 years ago (Binneman, 1999).

It is during the Howiesons Poort period at Sibudu, South Africa, that we find the first circumstantial evidence for the use of poisons. Here, a single bone point was recovered whose morphology matches that of 20th century San arrowheads, and which has evidence of having experienced longitudinal impact, consistent with use as a projectile hunting weapon (Backwell et al., 2008, 2018; Bradfield and Lombard, 2011). Morphological congruency with San poisoned arrows is cited as evidence that 43,000-year-old bone points from Border Cave, South Africa, and 35,000-year-old bone points from White Paintings Shelter, Botswana, may have been intended to be used with poison (Robbins et al., 2012). This interpretation is supported by the presence of incised decoration on some of these points, which resemble marks of ownership that the Ju/hoansi applied to their arrows. Among the Ju/hoansi these decorations played an intricate role in meat distribution rites, and are tied to poison bow-and-arrow hunting (sensu Lee, 1979).

Traces of a ricin-based poison have been identified on a wooden applicator found in the 24,000 year levels from Border Cave, South Africa (d'Errico et al., 2012a,b). So far, this is the oldest securely dated plant poison associated with "arrow" points, but this is controversial (Evans, 2012). The castor bean plant, from which ricin derives, is unknown as an arrow poison in southern Africa, and it is possible that the chemical signature of ricin may have been confused with abrin, a closely related highly toxic compound that occurs naturally in the vicinity of Border Cave (Bradfield et al., 2015). The historic distribution of

Ricinus communis, and whether it grew naturally in this area, is relevant to how we interpret this finding. Either way, this evidence at Border Cave demonstrates the diversity of plant-based toxins that were used in the deep past, and highlights the need to explore chemically any residues present on archeological artefacts.

Adoption of poisons on spears and arrowheads likely had multiple impacts in the other hunting tools, as one would anticipate with such symbiotic technology. Poison probably reduced the need for heavy technology and so influenced subsequent weapon design. This hypothesis is also relevant to blow dart poisons—adoption of a potent poison could drive design towards lighter weight, smaller designs. The slow-paralyzing effect of diamphotoxin and the delay in capturing the hit animal could also drive the iconic tracking culture of the San.

2.5 CONCLUSIONS

A great deal of ground-truthing of current poison practices needs to be done, but this will be challenging as the San are losing their traditions in the face of enormous historical and contemporary challenges and restrictions. It is unclear whether this knowledge of poisons is transmitted down generations. Additional poisons may have been used in the past. Identifying those and refining the estimated dates of certain developments is an ongoing process, especially as new tools are applied.

The San knowledge of their environment and their ability to live in an environment of limited water holds many lessons for contemporary life as we grapple with issues of diminishing drinking water supply (through increased pollution) and increasing aridification in Africa. Some of the poison sources are restricted due to overcollection and/or habitat alteration; these poison sources are part of a larger issue of conservation of biodiversity and indigenous knowledge.

San philosophy could help greatly as we move into the future uncertain world. Examining San practices can also provide a window to the past and to the evolution of humans in southern Africa. The San have already given the world an appetite suppressant from the plant, *Hoodia* Sweet ex Decne. (Apocynaceae; MacLean and Luo, 2004). The active molecules of arrow poisons have furthermore yielded ouabain, strophanthidin, cymarin, and digoxin, which have become model molecules for medical and pharmacological research. The remarkable precision of some San toxicants suggests that a little structural modification could turn them into protective medicines. Some, like digoxin, marketed under the tradename Lanoxin for heart conditions, are on the World Health Organzation's WHO Model List of Essential Medicines (2017).

Many toxic biomolecules have become important models for evolutionary, biochemical, and medical research (e.g., conotoxins of the marine *Conus* seashells; see Gray et al., 1988). It is hard to predict what other biomolecules we may develop beneficially, but it is crucial that the San pharmacopeia is practiced, conserved, and studied. We hope this chapter spurs the search for, analyses of, and reconsideration of evidence of hunting behavior, so we can drill down into the origins, sequence, and divergence of early human hunting technology.

Acknowledgments

The research of Chaboo and Hitchcock is funded by a 2018 National Geographic Society Explorers Award (#NGS–181R–18; PI C.S. Chaboo, and co-PIs M. Biesele, T. Bird, R. Hitchcock, and C. Nyamukondiwa). We are grateful to the San people for their inspiration and sharing their culture. We thank editor Philip Wexler for the invitation to contribute to this volume, reviewing the manuscript, and to the Elsevier editorial team for their support. Author Hitchcock thanks David Ambrose, Attila Paksi, Anita Heim, Julie Taylor, Alan Barnard, Mathias Guenther, H. J. Heinz, and Monageng Mogalakwe for sharing their field observations with various San communities. We acknowledge the use of photographs by Christine Sievers and Gary Trower and thank Athula Attygalle and Lorenzo Prenzini for discussion about poisons as well as Wills Flowers for a careful reading of the manuscript. We thank our respective families for their unwavering support.

References

Adesina, S.K., 1982. Studies on some plants used as anticonvulsants in Amerindian and African traditional medicine. Fitoterapia 5–6, 147–162.

Akinmoladun, A.C., Olaleye, M.T., Farombi, E.O., 2014. 13—Cardiotoxicity and cardioprotective effects of African medicinal plants. In: Kuete, V. (Ed.), Toxicological Survey of African Medicinal Plants. Elsevier Inc, Amsterdam, pp. 395–421.

Alto, E., 2011. Effects of dietary specialization on chemical defense of poison dart frogs. Eukaryon 7, 84–86.

Ambrose, S.H., 1998. Chronology of the later stone age and food production in East Africa. J. Archaeol. Sci. 25 (4), 377–392.

Anbu, J.S.J., Jayaraj, P., Varatharajan, R., Thomas, J., Jisha, J., Muthappan, M., 2009. Analgesic and antipyretic effects of *Sanseviera trifasciata* leaves. Afr. J. Traditional Complement. Alternat. Med. 6, 529–533.

Anderson, T., 2001. A Beginner's Guide to the Plants of Kimberley and Surrounds. With Special Reference to Magersfontein Battlefield. McGregor Museum, Kimberly.

Aranguren, B., Revedin, A., Amico, N., Cavulli, F., Giachi, G., Grimaldi, S., et al., 2018. Wooden tools and fire technology in the early Neanderthal site of Poggetti Vecchi (Italy). Proc. Natl. Acad. Sci. 115 (9), 2054–2059.

Arnaud, A., 1888. Sur la matière cristallisée active des flèches empoisonnées des Comalis. C. R. 106, 1011–1014.

Arnaud, A., 1898a. Recherches sur l'ouabaine. C. R. 127, 346–369.

Arnaud, A., 1898b. Sur une heptacetine cristalisée, derivée de l'ouabaine. C. R. 127, 1654–1656.

Assmann, T., Boutaud, E., Drees, C., Marcus, T., Nolte, D., Starke, W., et al., 2017. Two new *Lebistina* Motschulsky, 1864 species from Kenya and Tanzania (Coleoptera, Carabidae, Lebiini). Afr. Invertebrates 58 (1), 9–21.

Attenborough, D., 1995. The Private Life of Plants: A Natural History of Plant Behavior, first ed. BBC Books, London, 320 pp.

Backwell, L., d'Errico, F., Wadley, L., 2008. Middle Stone Age bone tools from the Howiesons Poort layers, Sibudu Cave, South Africa. J. Archaeol. Sci. 35 (6), 1566–1580.

Backwell, L., Bradfield, J., Carlson, K.J., Jashashvili, T., Wadley, L., d'Errico, F., 2018. The antiquity of bow-and-arrow technology: evidence from Middle Stone Age layers at Sibudu Cave. Antiquity 92 (362), 289–303.

Barnard, A., 1979. Kalahari Bushman settlement patterns. In: Burnham, P.C., Ellen, R.F. (Eds.), Social and Ecological Systems. Academic Press, London, pp. 131–144.

Barrow, J., 1801. An Account of Travels into the Interior of South Africa in the Years 1797 and 1798. Cadell and Davies, London.

Bartram, L., 1993. An Ethnoarchaeological Analysis of Kua San (Botswana) Bone Food Refuse (Ph.D. dissertation). University of Wisconsin, Madison, Wisconsin.

Bartram, L.E., 1997. A comparison of Kua (Botswana) and Hadza (Tanzania) bow and arrow hunting. In: Knecht, Heidi (Ed.), Projectile Technology. Plenum Press, New York and London, pp. 321–343.

Biesele, M., 1978. Sapience and scarce resources: communication systems of the !Kung and other foragers. Soc. Sci. Inf. 17, 921–947.

Biesele, M., Barclay, S., 2001. Ju/'hoan women's tracking knowledge and its contribution to their husband's hunting success. African Study Monographs (Suppl. 26), 67–84. Published by Kyoto University, Japan. Available from: < http://jambo.africa.kyoto-u. ac.jp/asm/index.html > .

Biesele, M., Hitchcock, R.K., 2013. The Ju/'hoansi San of Nyae Nyae and Namibian Independence: Development, Democracy, and Indigenous Voices in Southern Africa, Paperback edition Berghahn Books, New York, NY.

Bijlsma, U.G., De Waard, F., 1957. 'N pylvergif van die Boesmans in Suid-Afrika. S. Afr. Med. J. 31, 115–120.

Binford, L.R., 1981. Behavioral archaeology and the "Pompeii Premise". Am. Antiq. 37, 195–208.

Binneman, J., 1999. Mummified human remains from Kouga Mountains, Eastern Cape. The Digging Stick 16, 1–2.

Binneman, J.N.F., 1994. A unique stone tipped arrowhead from Adam's Kranz Cave, Eastern Cape. South. Afr. Field Archaeol. 3, 58–60.

Biondi, M., D'Alessandro, P., 2012. Afrotropical flea beetle genera: A key to their identification, updated catalogue and biogeographical analysis (Coleoptera, Chrysomelidae, Galerucinae, Alticini). ZooKeys (Special issue) 253, 1–158.

Bisset, N.G., 1989. Arrow and dart poisons. J. Ethnopharmacol. 25, 1–41.

Bisset, N.G., 1992. Curare. In: Pelletier, W.S. (Ed.), Alkaloids: Chemical and Biological Perspectives, vol. 8. Springer, Berlin, pp. 3–150.

Blackburn, R.H., (in preparation) The Okek: Kenya Forst Foragers. Manuscript in Author's possession.

Bleek, W.H.I., Lloyd, L.C., 1911. Specimens of Bushman Folklore. George Allen & Company, London, 630 pp.

Bodenheimer, F.S., 1951. Insects as Human Food, A Chapter of the Ecology of Man. Dr. W. Junk Publishers, The Hague.

Boëda, É., Connan, J., Dessort, D., Muhesan, S., Mercier, N., Valladas, H., et al., 1996. Bitumen as a hafting material on Middle Palaeolithic artefacts. Nature 380 (6572), 336–338.

Boëda, E., Geneste, J.M., Griggo, C., Mercier, N., Muhesen, S., Reyss, J.L., et al., 1999. A Levallois point embedded in the vertebra of a wild ass (*Equus africanus*): hafting, projectiles and Mousterian hunting weapons. Antiquity 73, 394–402.

Boëda, É., Bonilauri, S., Connan, J., Jarvie, D., Mercier, N., Tobey, M., et al., 2008a. New evidence for significant use of bitumen in Middle Palaeolithic technical systems at Umm El Tlel (Syria) around 70,000 bp. Paléorient 34 (2), 67–83.

Boëda, E., Bonilauri, S., Connan, J., Jarvie, D., Mercier, N., Tobey, M., et al., 2008b. Middle Palaeolithic bitumen use at Umm el Tlel around 70000 BP. Antiquity 82 (318), 853–861.

Boehm, R., 1890. Ueber das Echujin. Ein Beitrag zur Kenntniss der afrikanishen Pfeilgifte. Arch. Exp. Pathol. Pharmakol. 26, 165–176.

Boehm, R., 1897. Ueber das gift der larven von Diamphidia locusta. Arch. Exp. Pathol. Pharmakol. 38, 424–427.

Boon, R., 2010. Pooley's Trees of Eastern South Africa. Flora and Fauna, Durban, 624 pp.

Boonzaier, C.M., Smith, A., Berens, P., 1996. The Cape Herders: A History of the Khoikhoi of Southern Africa. David Philip and Athens/Ohio University Press, Cape Town and Johannesburg/Ohio.

Borgia, V., Carlin, M.G., Crezzini, J., 2016. Poison, plants and Palaeolithic hunters. An analytical method to investigate the presence of plant poison on archaeological artefacts. Quat. Int. 427, 94–103.

Bosc-Zanardo, B., Bon, F., Fauvelle-Aymar, F.X., 2008. Bushmen arrows and their recent history: crossed outlooks of historical, ethnological and archaeological sources. Palethnologie 1, 341–357.

Bradfield, J., 2015. Use-trace analysis of bone tools: a brief overview of four methodological approaches. S. Afr. Archeological Bull. 70, 3–14.

Bradfield, J., Lombard, M., 2011. A macrofracture study of bone points used in experimental hunting with reference to the South African Middle Stone Age. S. Afr. Archaeological Bull. 66, 67–76.

Bradfield, J., Wadley, L., Lombard, M., 2015. Southern African arrow poison recipes: their ingredients and implications for Stone Age archeology. South. Afr. Humanities 27, 29–64.

Breyer-Brandwijk, M.G., 1937. A note on the Bushman arrow poison, Diamphidia simplex Péringuey. Bantu Stud. 11, 279–284.

Brokensha, D., Riley, B.W., 1971. Bee-keeping among the Mbeere and some notes on Tharaka. Mila 2 (1), 13–24.

Brooks, A., 1984. In: Hall, M., Avery, D.M., Avery, G., Wilson, M., Humphreys, A.J.B. (Eds.), San Land Use Patterns, Past and Present Implications for Southern African Prehistory. Southern African Archaeology, Frontiers, pp. 40–52. , BAR 207, Oxford.

Brooks, A.S., Newell, L., Yellen, J.E., Hartmann, G.N., 2006. Projectile technologies of the African MSA: implications for modern human origins. In: Hovers, E., Esrella, E., Kuhn, S. (Eds.), Transitions before the Transition: Interdisciplinary Contributions to Archaeology. Springer, New York, pp. 233–255.

Brooks, A.S., Yellen, J.E., Potts, R., Behrensmeyer, A.K., Deino, A.L., Leslie, D.E., et al., 2018. Long-distance stone transport and pigment use in the earliest Middle Stone Age. Science 630, 90–94.

Burge, D.O., Mugford, K., Hastings, A.P., Agrawal, A.A., 2013. Phylogeny of the plant genus Pachypodium (Apocynaceae). Peer J. 1, e70. Available from: https://doi.org/10.7717/peerj.70.

Cambridge Dictionary, 2018. Cambridge Dictionary, Cambridge University Press, Cambridge, UK.

Campbell, A.C., 1964. A few notes on the Gcwi Bushmen of the Central Kalahari Desert, Bechuanaland. NADA 9 (1), 39–47.

Campbell, A.C., 1968a. Gcwi Bushmen: some notes on hunting with poisoned arrows. Botsw. Notes. Rec. 1, 95–96.

Campbell, A.C., 1968b. Central Kalahari game Reserve: II. Afr. Wildl. 22, 321.

Cashdan, E., Barnard, A., Bicchieri, M.C., Bishop, C.A., Blundell, V., Ehrenreich, J., et al., 1983. Territoriality among human foragers: ecological models and an application to four Bushman groups. Curr. Anthropol. 24 (1), 47–66.

Catterall, W.A., Beress, L., 1978. Sea anemone toxin and scorpion toxin share a common receptor site associated with the action potential sodium ionophore. J. Biol. Chem. 253 (20), 7393–7396.

Chaboo, C.S., 2011. Defensive behaviors in leaf beetles: from the unusual to the weird. Chem. Biol. Tropics 59–69.

Chaboo, C.S., Grobbelaar, E., Larsen, A., 2007. Fecal ecology in leaf beetles: novel records in the African arrow-poison beetles, Diamphidia Gerstaecker and Polyclada Chevrolat (Chrysomelidae: Galerucinae). Coleopterists Bull. 61 (2), 297–309.

Chaboo, C.S., Biesele, M., Hitchcock, R.K., Weeks, A., 2016. Beetles and plant arrow poisons of the Ju|hoan and Hai||om San peoples of Namibia (Insecta, Coleoptera, Chrysomelidae; Plantae, Anacardiaceae, Apocynaceae, Burseraceae). Zookeys 558, 9–54.

Challis, S., 2012. Creolisation on the nineteenth century frontiers of southern Africa: a case study of the AmaTola 'Bushmen' in the Maloti-Drakensberg. J. South Afr. Stud. 38 (2), 265–280.

Chen, X., Feng, Y., Chen, Z., 2009. Common edible insects and their utilization in China. Entomol. Res. 39, 299–303.

Charrié-Duhaut, A., Porraz, G., Cartwright, C.R., Igreja, M., Connan, J., Poggenpoel, C., et al., 2013. First molecular identification of a hafting adhesive in the late Howiesons Poort at Diepkloof rock shelter (Western Cape, South Africa). J. Archaeol. Sci. 40 (9), 3506–3518.

Cheikhyoussef, A., Shapi, M., Matengu, K., Ashekele, H.M., 2011. Ethnobotanical study of indigenous knowledge on medicinal plant use by traditional healers in Oshikoto region, Namibia. J. Ethnobiol. Ethnomed. 7, 11 pp.

Chen, K.K., Kovaříková, A., 1967. Pharmacology and toxicology of toad venom. J. Pharm. Sci. 56 (12), 1535–1541.

Chevalier, A., 1950. Les Strophanthus comme plantes toxiques spécialement dans les savanes de l'Afrique Occidentale. Rev. int. bot. appliquée et d'agriculture trop. 337–338, 578–588.

Clark, J.D., 1977. Interpretations of prehistoric technology from ancient Egyptian and other sources. Part II: Prehistoric arrow forms in Africa as shown by surviving examples of the traditional arrows of the San Bushmen. Paleorient 3, 127–150.

Clark, J.D., Phillips, J.L., Staley, P.S., 1974. Interpretations of prehistoric technology from ancient Egyptian and other sources. Part I: Ancient Egyptians bows and arrows and their relevance to African prehistory. Paleorient 2 (2), 323–388.

ClinicalTrials.gov, 2018. U.S. National Library of Medicine. ⟨www.clinicaltrials.gov⟩ (accessed April 29, 2018).

Conard, N.J., Serangeli, J., Böhner, U., Starkovich, B.M., Miller, C.E., Urban, B., et al., 2015. Excavations at Schöningen and paradigm shifts in human evolution. J. Hum. Evol. 89, 1–17.

Cornell, F.C., 1920. The Glamour of Prospecting: Wanderings of a South African Prospector in Search of Copper, Gold, Emeralds, and Diamonds. T. Unwin Fisher, London, 334 pp.

Creighton Wellman, F., 1907. Uber pfeilgifte in Westafrika und besonders eine Kaferlarve als pfeilgift in Angola. Dtsch. Entomol. Z. 17, 17–19.

Crosse-Upcott, A.R.W., 1956. Social aspects of Ngindo bee-keeping. JRAI 86, 81–109.

Currier, R., Harrison, R., Rowley, P., Laing, G., Wagstaff, S., 2010. Intra-specific variation in venom of the African puff adder (Bitis arietans): differential expression and activity of snake venom metalloproteinases (SVMPs). Toxicon 55, 864–873.

d'Errico, F., 2008. Le rouge et le noir: implications of early pigment use in Africa, the Near East and Europe for the origin of cultural modernity. S. Afr. Archaeological Soc. Goodwin Ser. 10, 168–174.

d'Errico, F., Backwell, L., Villa, P., Degano, I., Lucejko, J.J., Bamford, M.K., et al., 2012a. Early evidence of San material culture represented by organic artifacts from Border Cave, South Africa. Proc. Natl. Acad. Sci. 109 (33), 13214–13219.

d'Errico, F., Backwell, L., Villa, P., Degano, I., Lucejko, J.J., Bamford, M.K., et al., 2012b. Reply to Evans: use of poison remains the most parsimonious explanation for Border Cave castor bean extract. Proc. Natl. Acad. Sci. 109 (48), E3291–E3292.

Daly, J., 1998. Thirty years of discovering arthropod alkaloids in amphibian skin. J. Nat. Proc. 61, 162–172.

Dalziel, J.M., 1937. The Useful Plants of West Tropical Africa. Published under the Authority of the Secretary of State for the Colonies by the Crown Agents for the Colonies, London, 612 pp.

De la Harpe, J., Dowdle, E., 1980. Isolation and characterization of diamphotoxin. S. Afr. J. Sci. 76, 428.

De la Harpe, J., Reich, E., Reich, K.A., Dowdle, E.B., 1983. Diamphotoxin—the arrow poison of the !Kung Bushmen. J. Biol. Chem. 258 (19), 11924–11931.

De la Peña, P., Taipale, N., Wadley, L., Rots, V., 2018. A techno-functional perspective on quartz micro-notches in Sibudu's Howiesons Poort indicates the use of barbs in hunting technology. J. Archaeol. Sci. 93, 166–195.

De Weille, J.R., Schweitz, H., Maes, P., Tartar, A., Lazdunski, M., 1991. Calciseptine, a peptide isolated from black mamba venom, is a specific blocker of the L-type calcium channel. Proc. Natl. Acad. Sci. 88, 2437–2440.

Deacon, H., 1976. Where Hunters Gathered: A Study of Stone Age People in the Eastern Cape. South African Archaeological Society Monographs 1. South African Archaeological Society, Claremont, xiii + 232 pp.

Deacon, H.J., Deacon, J., 1999. Human Beginnings in South Africa: Uncovering the Secrets of the Stone Age. David Philip, Cape Town.

Deacon, J., 1984. The Later Stone Age of southernmost Africa. Archaeopress, Oxford.

Deacon, J., 1992. Arrows as agents of belief among the /Xam bushmen. In: Margaret Shaw Lecture, 3. South African Museum, Cape Town, pp. 21.

Dewar, G., Halkett, D., Hart, T., Orton, J., Sealy, J., 2006. Implications of a mass kill site of springbok (*Antidorcas marsupialis*) in South Africa: hunting practices, gender relations, and sharing in the Later Stone Age. J. Archaeol. Sci. 33, 1266–1275.

Dickers, K., Bradberry, S., Rice, P., Griffiths, G., Vale, J., 2003. Abrin poisoning. Toxicol. Rev. 22 (3), 137–142.

Dieckmann, U., Thiem, M., Dirkx, E., Hays, J. (Eds.), 2014. "Scraping the Pot": San in Namibia Two Decades After Independence. Legal Assistance Centre and Desert Research Foundation of Namibia, Windhoek.

Dornan, S.S., 1916. Some notes on Rhodesian native poisons. Rep. S. Afr. Assoc. Adv. Sci. 356–361.

Dornan, S.S., 1925. Pygmies and Bushmen of the Kalahari: An Account of the Hunting Tribes Inhabiting the Great Arid Plateau of the Kalahari Desert, Their Precarious Manner of Living, Their Habits, Customs and Beliefs, With Some Reference to Bushman Art, Both Early and of Recent Date, and to the Neighbouring African Tribes. Seeley, Service and Co, London, p. 318.

Draper, P., 1973. Crowding among hunter gatherers: The !Kung bushmen. Science 182 (9109), 301–303.

Drummond, R.B., Gelfand, M., Mavi, S., 1975. Medicinal and other uses of succulents by the Rhodesian African. Excelsa 5, 51–56.

Dumbacher, J.P., Beehler, B.M., Spande, T.F., Garraffo, H.M., Daly, J.W., 1992. Homobatrachotoxin in the genus *Pitohui*: chemical defense in birds? Science 258, 799–801.

Dumbacher, J.P., Wako, A., Derrickson, S.R., Samuelson, A., Spande, T.F., Daly, J.W., 2004. Melyrid beetles (Choresine): A putative source for the batrachotoxin alkaloids found in poison-dart frogs and toxic passerine birds. Proc. Natl. Acad. Sci. U.S.A. 101, 15857–15860.

Dunn, E.J., 1931. The Bushmen. C. Griffin & Company, London, 130 p.

Elphick, R., 1977. Kraal and Castle: Khoikhoi and the Founding of White South Africa. Yale University Press, New Haven, 116 pp.

Erlandson, J., Watts, J., Jew, N., 2014. Darts, arrows, and archaeologists: distinguishing dart and arrow points in the archaeological record. Am. Antiquity 79 (1), 162–169.

Euw, J.V., Fishelson, L., Parsons, J.A., Reichstein, T., Rothschild, M., 1967. Cardenolides (heart poisons) in a grasshopper feeding on milkweeds. Nature 214, 35–39.

Evans, A.A., 2012. Arrow poisons in the Palaeolithic? Proc. Natl. Acad. Sci. 109 (48), E3290.

Fairmaire, L., 1893. Coleopteres de l'Ouganghi, recueillis par Crampel. Ann Soc entomologique Fr. 62, 133–156.

Farini, G.A. 1973 (1886). Through the Kalahari Desert. Cape Town: C. Struik.

Fourie, L., 1926. Preliminary notes on certain customs of the Hei-||om Bushmen. J. S. West Afr. Sci. Soc. 1, 49–63.

Fourie, L., 1964. Preliminary notes on certain custons of the Hei-kom Bushmen. J. S. West Afr. Wiss. Gesell. 19–34.

Fraser, T.R., 1872. On the Kombè arrow-poison (Strophanthus Hispidus, D. C.) of Africa. J. Anat. Physiol. 7 (Pt 1), 139–155.

Friederich, R., 2014. In: Lempp, H. (Ed.), Etosha: Hai//om Heartland: Ancient Hunter-Gatherers and their Environment. Namibia Publishing House, Windhoek.

Fry, B.G., Roelants, K., Champagne, D.E., Scheib, H., Tyndall, J.D.A., King, G.F., et al., 2009. The toxicogenomic multiverse: convergent recruitment of proteins into animal venoms. Annu. Rev. Genomics Hum. Genet. 10, 483–511.

Fuerstenwerth, H., 2010. Ouabain—the insulin of the heart. Int. J. Clin. Pract. 64, 1591–1594.

Furstenburg, D., Vanhoven, W., 1994. Condensed tannis as anti-defoliate agent against browsing by giraffe (Giraffa camelopardalis) in Kruger National Park. Comp. Biochem. Physiol. 107A, 425–431.

Gerhadt, K., Steiner, M., 1986. An Inventory of a Coastal Forest in Kenya at Gedi National Monument Including a Checklist and a Nature Trial: Report from a Minor Field Study. Swedish University of Agricultural Sciences, Uppsala, 52 pp.

Gericke, N., Viljoen, A.M., 2008. Sceletium—a review update. J. Ethnopharmacol. 119 (3), 653–663.

Gericke, N.P., Van Wyk, B.-E., 2001. United States Patent 6,288,104: Pharmaceutical compositions containing mesembrine and related compounds. Inventors: Gericke, N.P., Van Wyk, B.E. Assignee: African Natural Health CC.

Gildenhuys, S., 2006. The three most abundant tree Euphorbia species of the Transvaal (South Africa). Euphorbia World 2, 9–14.

Goodwin, A.J.H., 1945. Some historical Bushman arrows. S. Afr. J. Sci. 41, 429–443.

Govaerts, R., Frodin, D.G., Radcliffe-Smith, A., 2000. World Checklist and Bibliography of Euphorbiaceae (and Pandanaceae). The Board of Trustees of the Royal Botanic Gardens, Kew. Vol. 1–4, University of Chicago Press, Chicago, pp. 1–1622

Gray, W.R., Olivera, B.M., Cruz, L.J., 1988. Peptide toxins from venomous Conus snails. Annu. Rev. Biochem. 57, 665–700.

Green, S.V., 1998. The bushman as an entomologist. Antenna 22 (1), 4–8.

Grund, B.S., 2017. Behavioral ecology, technology, and the organization of labor: How a shift from spear thrower to self bow exacerbates social disparities. Am. Anthropol. 119 (1), 104–119.

Guerreiro, M.V. 1968. Bochimanes !Kho de Angola. Lisboa.

Güldemann, T., 2014. Khoisan' linguistic classification today. In: Guldemann, T., Fehn, A. M. (Eds.), Beyond 'Khoisan': Historical Relations in the Kalahari Basin. John Benjamins Publishing, Amsterdam, pp. 1–44.

Gwynne, D.T., 2008. Sexual conflict over nuptial gifts in insects. Annu. Rev. Entomol. 53 (1), 83–101.

Hall, I.C., Whitehead, R.W., 1927. A pharmaco-bacterologic study of African poisoned arrows. J. Infect. Dis. 41, 51–69.

Harako, R., 1981. Ecological and sociological importance of honey to the Mbuti net hunters Eastern Zaire, Kyoto. Afr. Study Monogr. (Kyoto Univ.) 1, 55–68.

Harvey, A.L., Young, L.C., Viljoen, A.M., Gericke, N.P., 2011. Pharmacological actions of the South African medicinal and functional food plant Sceletium tortuosum and its principal alkaloids. J. Ethnopharmacol. 137, 1124–1129.

Heikertinger, F., Csiki, E., 1940. Partes 166 et 169. Chrysomelidae: Halticinae, Volumen 25. In: Schenkling, S. (Ed.), Coleopterorum Catalogus. Dr. W. Junk, Gravenhage, pp. 1–635.

Helgren, D.M., Brooks, A.S., 1983. Geogarcheology at /Gi: a Middle Stone Age and Later Stone Age site in the north-west Kalahari. J. Archeol. Sci. 10, 181–197.

Henshilwood, C.S., d'Errico, F., van Niekerk, K.L., Coquinot, Y., Jacobs, Z., Lauritzen, S.-E., et al., 2011. A 100,000-year-old ochre-processing workshop at Blombos Cave, South Africa. Science 334, 219–222.

Heubner, W. von, 1907. Über das pfeilgift der Kalahari. Arch. Exp. Pathol. Pharmakol. 57, 358–366.

Hitchcock, R.K., in preparation. Challenges Facing the San of Southern Africa: A Personal Exploration. The Evolution of San Religion: Issues and Debates.

Hitchcock, R.K., 1982. The Ethnoarchaeology of Sedentism: Mobility Strategies and Site Structure among Foraging and Food Producing Populations in the Eastern Kalahari Desert, Botswana (Ph.D. dissertation). University of New Mexico, Albuquerque. New Mexico.

Hitchcock, R.K., Yellen, J.E., Gelburd, D.J., Osborn, A.J., Crowell, A.L., 1996. Subsistence hunting and resource management among the Ju/'hoansi of Northwestern Botswana. African Study Monographs 17 (4), 153–208.

Hitchcock, R.K., Bleed, P., 1997. Each according to need and fashion: spear and arrow use among San Hunters of the Kalahari. In: Knecht, H. (Ed.), Projectile Technology. Plenum Press, New York, NY, pp. 345–368.

Hitchcock, R.K., Crowell, A.L., Brooks, A.S., Yellen, J.E., Ebert, J.I., Osborn, A.J., In review. The ethnoarchaeology of ambush hunting: a case study of GI Pan, Western Ngamiland, Botswana. Afr. Archaeol. Rev.

Hitchcock, R.K., Ikeya, K., Biesele, M., Lee, R.B. (Eds.), 2006. Updating the San: Image and Reality of an African People in the 21st century. National Museum of Ethnology, Osaka, Japan.

Hobart, J.H., 2003. An old fashioned approach to a modern hobby: fishing in the Lesotho Highlands. In: Mitchell, P.J., Haour, A., Hobart, J.H. (Eds.), Researching Africa's Past: New Contributions from British Archaeologists. Oxford University School of Archaeology, Oxford, pp. 44–53.

How, M.W., 1970. The Mountain Bushmen of Basutoland. J.L. Van Schaik, Ltd., Pretoria.

Hughes, S.S., 1998. Getting to the point: evolutionary change in prehistoric weaponry. J. Archaeological Method Theory 5, 345–408.

Hutchings, A., Scott, A.H., Lewis, G., Cunningham, A., 1996. Zulu Medicinal Plants. An Inventory. University of Natal Press, Pietermarizburg, South Africa.

Ichikawa, M., 1980. Beekeeping of the Suiei Dorobo in East Africa (in Japanese). Kikan-Jinruigaku II (2), 117–152.

Jacobsen, T.F., Sand, O., Bjøro, T., Karlsen, H.E., Iversen, J.G., 1990. Effect of *Diamphidia* toxin, a Bushman arrow poison, on ionic permeability in nucleated cells. Toxicon 28 (4), 435–444.

Jansen, P.C.M., 2005. Elephantorrhiza elephantina. In: Jansen, P.C.M., Cardon, D. (Eds.), Plant Resources of Tropical Africa 3: Dyes and Tannins. PROTA Foundation. Backhuys Publishers, Wageningen, The Netherlands, pp. 75–76.

Kang, S.S., Cordell, G.A., Soejarto, D.D., Fong, H.H.S., 1985. Alkaloids and flavonoids from *Ricinus communis*. J. Nat. Prod. 4 (1), 155–156.

Kann, N., 1989. Further purification of the Basarwa arrow poison. Unpublished manuscript on file with the Kalahari Peoples Fund, Austin, Texas, 21 pp.

Kao, C.Y., Salwen, M.J., Hu, S.L., Pitter, H.M., Woollard, J.M.R., 1989. *Diamphidia* toxin, the bushman's arrow poison: Possible mechanism of prey-killing. Toxicon 27, 1351–1366.

Karimi, M., 1973. The Arrow Poisons. East African Literature Bureau, Nairobi, 97 pp.

Kaziende, L., 1943. Poison de fleches. In: Notes Africaines Nr. 18, L'Institut Francais d'Afrique Noire. http://signare.free.fr/tables/notes1939-1948.pdf.

Kent, S., 1995. Does sedentism promote gender inequality? A case study from the Kalahari. J. R. Anthropol. Inst. 1, 513–536.

King, G.F., Hardy, M.C., 2013. Spider-venom peptides: structure, pharmacology, and potential for control of insect pests. Annu. Rev. Entomol. 58, 475–496.

Kingdon, J., Agwanda, B., Kinnaird, M., O'Brien, T., Holland, C., Gheysens, T., et al., 2011. A poisonous surprise under the coat of the African crested rat. Proc. R. Soc. B . Available from: https://doi.org/10.1098/rspb.2011.1169.

Kingston, D.G., Reichstein, T., 1974. Cytotoxic cardenolides from *Acokanthera longiflora* Stapf. and related species. J. Pharm. Sci. 63 (3), 462–464.

Klein, R.G., 1972. The Late Quaternary mammalian fauna of Nelson Bay Cave (Cape Province, South Africa): its implications for megafaunal extinctions and environmental and cultural change. Quat. Res. 2, 135–142.

Klein, R.G., 1974. Environment and subsistence of prehistoric man in the southern Cape Province, South Africa. World Archaeol. 5, 249–284.

Koch, C., 1958. Preliminary notes on the coleopterological aspect of the arrow poison of the Bushmen. S. Afr. Biol. Soc. Pamphlet 20, 49–54.

Koller, J., Baumer, U., Mania, D., 2001. High-tech in the middle Palaeolithic: Neandertal-manufactured pitch identified. Eur. J. Archaeol 4 (3), 385–397.

Kolodziejczyk-Czepas, J., Stochmal, A., 2017. Bufadienolides of *Kalanchoe* species: an overview of chemical structure, biological activity and prospects for pharmacological use. Phytochem. Rev. 16, 1155–1171.

Kozowyk, P.R.B., Langejans, G.H.J., Poulis, J.A., 2016. Lap shear and impact testing of ochre and beeswax in experimental Middle Stone Age compound adhesives. PLoS ONE. 11 (3), e0150436.

Kozowyk, P.R.B., Soressi, M., Pomstra, D., Langejans, G.H.J., 2017. Experimental methods for the Palaeolithic dry distillation of birch bark: implications for the origin and development of Neandertal adhesive technology. Sci. Rep. 7, 8033.

Kuhn, S.L., Stiner, M.C., 2001. The antiquity of hunter-gatherers. In: Panter-Brick, C., Layton, R.H., Rowley-Conwy, P. (Eds.), Hunter-gatherers: An Interdisciplinary Perspective. Cambridge University Press, Cambridge, pp. 99–142.

Kündig, H., 1978. The pharmacology of *Diamphidia* arrow poison. PhD Thesis. Department of Pharmacology, University of Witwatersrand, Johannesburg, 180 pp.

Langley, M.C., Prendergast, M.E., Shipton, C., Quintana Morales, E.M., Crowther, A., Boivin, N., 2016. Poison arrows and bone utensils in late Pleistocene eastern Africa: Evidence from Kuumbi Cave, Zanzibar, Tanzania. Archaeological Res. Afr. 51 (2), 155–177.

Lebzelter, V., 1996. Eingeborenen Kulturen von Südwestafrika: Die Buschmänner. Super Print, Swakopmund.

Lee, R.B., 1976. Introduction. In: Lee, R.B., DeVore, I. (Eds.), Kalahari Hunter-Gatherers: Studies of the !Kung San and Their Neighbors. Harvard University Press, Cambridge, pp. 3–24.

Lee, R.B., 1979. The !Kung San: Men, Women, and Work in a Foraging Society. Cambridge University Press, Cambridge, 556 pp.

Leffers, A., 2003. Gemsbok Bean and Kalahari Truffle: Traditional Plant Use by Ju ǀ 'hoansi in North-Eastern Namibia. Gamsberg MacMillan Publishers, Windhoek, 202 pp.

Lennox, S.J., Bamford, M.K., 2017. Identifying Asteraceae, particularly *Tarchonanthus parvicapitulatus*, in archaeological charcoal from the Middle Stone Age. Quat. Int. 6759. Available from: https://doi.org/10.1016/j.quaint.2017.03.074.

Lennox, S.J., Bamford, M.K., Wadley, L., 2017. Middle Stone Age wood use 58 000 years ago in KwaZulu-Natal: charcoal analysis from two Sibudu occupation layers. South. Afr. Humanities 30, 247–286.

Lesnik, J.J., 2014. Termites in the hominin diet: a meta-analysis of termite genera, species and castes as a dietary supplement for South African robust australopithecines. J. Hum. Evol. 71, 94–104.

Lewin, L., 1894. Die pfeilgifte: historische und experimentelle unterschungen. Virschows Arch. Pathologische, Anatomie Physiol. Klin. Med. 138, 283–346.

Lewin, L., 1912a. Uber die Pfeilgifte der Bushmanner. Z. Ethnologie Bd 44 (5), 831–837.

Lewin, L., 1912b. *Blepharida evanida*, cin neuer Pfeilgiftkafer. Arch exp. Pathol. Pharmakol. 69, 60–66.

Lewin, L., 1923. Die Pfeilgifte. Verlag von Johann Ambrosius Barth, Leipzig, 517 pp.

Lichtenstein, H., 1930. Travels in Southern Africa in the Years 1803, 1804, 1805 and 1806. Van Riebeeck Society, Cape Town, Vol. 11 (Vol. I: 1812; Vol. II: 1815).

Liebenberg, L., 1990. The Art of Tracking. David Philip Publishers, Claremont, SA.

Livingstone, D., 1858. Missionary Travels and Researches in South Africa Including a Sketch of Sixteen Years' Residence in the Interior of Africa. J. Murray, London, 732 pp.

Livingstone, D., Livingstone, C., 1865. A Narrative of an Expedition to the Zambesi. J. Murray, London, 608 pp.

Logie, A., 1935. Preliminary notes on some Bushman arrows from South-West Africa. S. Afr. J. Sci. 32, 553–559.

Lombard, M., 2005. Evidence of hunting and hafting during the Middle Stone Age at Sibidu Cave, KwaZulu-Natal, South Africa: a multi-analytical approach. J. Hum. Evol. 48 (3), 27–300.

Lombard, M., 2006. First impressions of the functions and hafting technology of Still Bay pointed artefacts from Sibudu Cave. South. Afr. Humanities 18 (1), 27–41.

Lombard, M., 2008. Finding resolution for the Howiesons Poort through the microscope: micro-residue analysis of segments from Sibudu Cave, South Africa. J. Archaeol. Sci. 35, 26–41.

Lombard, M., 2007a. The gripping nature of ochre: The association of ochre with Howiesons Poort adhesives and Later Stone Age mastics from South Africa. J. Hum. Evol. 53, 406–419.

Lombard, M., 2007b. Evidence for change in Middle Stone Age hunting behaviour at Blombos Cave: results of a macrofracture analysis. S. Afr. Archaeol. Bull. 62, 62–67.

Lombard, M., 2011. Quartz-tipped arrows older than 60 ka: further use–trace evidence from Sibudu, KwaZulu-Natal, South Africa. J. Archaeol. Sci. 38, 1918–1930.

Lombard, M., Haidle, M.N., 2012. Thinking a bow-and-arrow set: cognitive implications of Middle Stone Age bow and stone-tipped arrow technology. Cambridge Archaeological J. 22 (02), 237–264.

Lombard, M., Pargeter, J., 2008. Hunting with Howiesons Poort segments: pilot experimental study and the functional interpretation of archaeological tools. J. Archaeol. Sci. 35 (9), 2523–2531.

Lombard, M., Phillipson, L., 2010. Indications of bow and stone-tipped arrow use 64,000 years ago in KwaZulu-Natal, South Africa. Antiquity 84, 635–648.

Lupo, K.D., Schmitt, D.N., 2002. Upper Paleolithic net-hunting, small prey exploitation, and women's work effort: a view from the ethnographic and ethnoarchaeological record of the Congo Basin. J. Archaeological Method Theory 9 (2), 147–179.

MacLean, D.B., Luo, L.G., 2004. Increased ATP content/production in the hypothalamus may be a signal for energy-sensing of satiety: studies of the anorectic mechanism of a plant steroidal glycoside. Brain Res. 1020 (1–2), 1–11.

Macphail, J.G.S., 1930. The Bandala method of hunting elephant on foot. Sudan. Notes Rec. 13, 279–283.

Magnusson, S., Kjeken, R., Berg, T., 1993. Characterization of two distinct pathways of endocytosis of ricin by rat liver endothelial cells. Exp. Cell Res. 205 (1), 118–125.

Mahlberg, P.G., Davis, D.G., Galitz, D.S., Manners, G.D., 1987. Laticifers and the classification of Euphorbia: the chemotaxonomy of *Euphorbia esula* L. Bot. J. Linn. Soc. 94 (1–2), 165–180.

Maingard, L., 1932. History and distribution of the bow and arrow in South Africa. S. Afr. J. Sci. 24, 711–723.

Maingard, L.F., 1937. The weapons of the /ʔAuni and the ꞊Khomani Bushmen. In: Rheinallt Jones, J.D., Doke, C.M. (Eds.), Bushmen of the Southern Kalahari. University of the Witwatersrand, Johannesburg, pp. 277–283.

Manhire, T., Parkington, J., Yates, R., 1985. Nets and fully recurved bows: rock paintings and hunting methods in the Western Cape, South Africa. World Archaeol. 17 (2), 161–174.

Maphosa, V., Masika, P.J., Moyo, B., 2010. Toxicity evaluation of the aqueous extract of the rhizome of *Elephantorrhiza elephantina* (Burch.) Skeels. (Fabaceae), in rats. Food Chem. Toxicol. 48 (1), 196–201.

Marks, S., 1977. Hunting behavior and strategies of the Valley Bisa in Zambia. Hum. Ecol. 5 (1), 1–36.

Maroyi, A., 2017. *Elephantorrhiza elephantina*: traditional uses, phytochemistry, and pharmacology of an important medicinal plant species in Southern Africa. Hindawi Evid Based Complement Alternat Med 2017, 18 pp.

Marshall, J., 1958. Huntsmen of Nyae Nyae: The !Kung still practice Man's oldest craft. Part II. Nat. Hist. 67, 291–309. 376–395.

Marshall, L., 1960. !Kung Bushman Bands. Africa: J. Int. Inst. 30, 325–355.

Marshall, L., 1976. The !Kung of Nyae Nyae. Harvard University Press, Cambridge, Mass.

Marshall Thomas, E., 2006. The Old Way: A Story of the First People. Farrar, Straus and Giroux.

Mason, R., 2012. A built stone alignment associated with an LSA artefact assemblage on MIA Farm, Midrand, South Africa. S. Afr. Archaeological Bull. 67, 214–230.

Massiot, G., Delaude, C., 1989. African Strychnos Alkaloids. Alkaloids: Chem. Pharmacol. 34, 211–329.

Mazel, A., 1989. People making history: the last ten thousand years of hunter-gatherer communities in the Thukela Basin. Nat. Museum J. Humanities 1, 1–168.

Mazza, P.P.A., Martini, F., Sala, N., Magi, M., Colombini, M.P., Giachi, G., et al., 2006. A new Palaeolithic discovery: tar-hafted stone tools in a European Mid-Pleistocene bone-bearing bed. J. Archaeol. Sci. 33 (9), 1310–1318.

McGrew, W.C., 2014. The 'other faunivory' revisited: insectivory in human and non-human primates and the evolution of human diet. J. Hum. Evol. 71, 4–11.

Mebs, D., Brüning, F., Pfaff, N., Neuwinger, H.D., 1982. Preliminary studies on the chemical properties of the toxic principle from *Diamphidia nigroornata* larvae, a source of Bushman arrow poison. J. Ethnopharmacol. 6 (1), 1–11.

Mentzel, O., 1944. A complete and Authentic Geographical and Topographical Description of the Famous and Remarkable African Cape of Good Hope. van Riebeeck Society No. 25, Cape Town.

Mguni, S., 2015. Termites of the Gods: San Cosmology in Southern African Rock Art. Wits University Press, Johannesburg.

Mitchell, P., 2013. Southern African Hunter-Gatherers of the last 25,000 years. In: Mitchell, P., Lane, P. (Eds.), The Oxford Handbook of African Archaeology. Oxford University Press, Oxford, pp. 473—488.

Mitchell, P., Hudson, A., 2004. Psychoactive plants and southern African hunter-gatherers: a review of the evidence. South. Afr.Humanities 16, 39—57.

Mitchell, P.J., 2012. San origins and transition to the Later Stone Age: New research from Border Cave, South Africa. S. Afr. J. Sci. 108 (11—12), 1—2.

Momtaz, S., Lall, N., Hussein, A., Nasser Ostad, S., Abdollahi, M., 2010. Investigation of the possible biological activities of a poisonous South African plant; Hyaenanche globosa (Euphorbiaceae). Pharmacogn. Mag. 6 (21), 34—41.

Morgan, A.M.A., Kim, J.-H., Park, S.-U., Kim, Y.-H. 2017. Chemical constituents of the leaves of *Boscia senegalensis*. In: Abstract, Eighth International Conference and Exhibition on Metabolomics & Systems Biology, May 08—10, 2017, Singapore.

Müller, G., 1993. Scorpionism in South Africa, a report of 41 serious scorpion envenomations. S. Afr. Med. J. 83, 405—411.

Nadler, O., 2006. Die Pfeilgifte der !Kung-Buschleute. Oserna-africana-Verlag, Heidelberg, 101 pp.

Neuwinger, H.D., 1996. African Ethnobotany: Poisons and Drugs: Chemistry, Pharmacology, Toxicology. Chapman and Hall, London.

Neuwinger, H.D., 2004. Plants used for poison fishing in tropical Africa. Toxicon 44, 417—430.

Noli, H.D. 1993. A Technical Investigation into the Material Evidence for Archery in the Archaeological and Ethnographical Record in Southern Africa. Unpublished Ph.D. Thesis. University of Cape Town, 252 pp.

Nonaka, K., 1996. Ethnoentomology of the Central Kalahari San. Afr. Stud. Monogr. 22, Lagos, Nigeria, 29—46.

Nonaka, K., 2005. Ethnoentomology—Insect Eating and Human—Insect Relationship Tokyo. University of Tokyo Press.

Nonaka, K., 2009. Feasting on insects. Entomol. res. 39, 304—312.

Odell, G.H., Cowan, F., 1986. Experiments with spears and arrows on animal targets. J. Field Archaeol. 13, 194—212.

Omino, E.A., Kokwaro, J.O., 1993. Ethnobotany of Apocynaceae species in Kenya. J. Ethnopharmacol. 40 (3), 167—180.

Oosthuizen, M., 1977. The description of an unusual hunting kit considered to be of southern Bushman origin. Ann Nat Museum 23, 75—85.

Orpen, J.M., 1874. A glimpse into the mythology of the Maluti Bushmen. Cape Mon. Mag. 9, 1—13.

Pager, H., 1971. Ndedema. Akademische Druck-u. Verlagsanstadt, Graz, Austria.

Pager, H., 1973. Rock paintings in southern Africa showing bees and honey hunting. Bee World 54, 61—68.

Parker, I., Amin, M., 1983. Ivory Crisis. Chatto and Windus Ltd., London, 184 pp.

Parkington, J., 2006. Shorelines, Strandlopers and Shell Middens. Creda, Cape Town.

Passarge, S., 1907. Die Buschmanner der Kalahari (The Bushman of the Kalahari). Translated by Wilmsen, E.N. (Ed.) The Kalahari Ethnographies (1869—1898) of Siefried Passarge. Research in Khoisan Studies 13: 127—218. Rüdiger, Köppe, Verlag, Köln, 332 pp.

Paterson, W., 1789. A Narrative of Four Journeys into the Country of the Hottentots and Caffaria, in the Years 1777, 1778, 1779. J. Johnson, London, 205 pp.

Pawlik, A.F., Thissen, J.P., 2011. Hafted armatures and multi-component tool design at the Micoquian site of Inden-Altdorf, Germany. J. Archaeol. Sci. 38 (7), 1699—1708.

Pax, F., 1895. Euphorbiaceae africanae. Bot. Jahrbucher f. Systematik (A. Engler) 19, 124.

Perrot, E., Vogt, E., 1913. Poisons de Flèches et Poisons d'épreuves. Vigot Frères, Paris.

Peters, J., Dieckmann, U., Vogelsang, R., 2009. Tracking down the past: ethnohistory meets archaeozoology. In: Grupe, G., McGlynn, G., Peters, J. (Eds.), Losing the Spoor: Hai//om Animal Exploitation in the Etosha Region. Verlag Marie Leidorf GmbH, Rahden/Wesf, pp. 81–102.

Picker, M., Griffiths, C., Weaving, A., 2004. Field Guide to Insects of South Africa. Random House Struik, Cape Town, 440 pp.

PlantZAfrica. 2018. South African National Biodiversity Institute, Pretoria. ⟨http://pza.sanbi.org/⟩ (accessed April 29, 2018).

Plazier, A.C., 1980. A revision of *Adenium* Roem. and Schult. and of *Diplorhynchus* Welw. ex Fic. and Hiern (Apocynaceae). Mededelingen Landbouwhogeschool 80 (12), 1–40.

Plug, I., Mitchell, P., Bailey, G., 2010. Late Holocene fishing strategies in southern Africa as seen from Likoaeng, highland Lesotho. J. Archaeol. Sci. 37, 3111–3123.

Pooley, S., 2009. Jan van Riebeeck as pioneering explorer and conservator of natural resources at the Cape of Good Hope (1652–62). Env. His. 15, 3–33.

Potgieter, E.F., 1955. The Disappearing Bushmen of Lake Chrissie: A Preliminary Survey. J. L. van Schaik, Pretoria.

Prins, F.E., 2009. Secret San of the Drakensberg and their rock art legacy. Crit. Arts 23 (2), 190–208.

PubChem, the Open Chemistry Database. 2018. U.S. National Library of Medicine. ⟨https://pubchem.ncbi.nlm.nih.gov⟩ (accessed April 30, 2018).

Puckett, F., Ikeya, K., 2018. Research and Activism Among the Kalahari San today: Ideals, Challenges, and Debates. Senri Ethnological Studies 99. National Museum of Ethnology, Osaka.

Raponda-Walker, A., Sillans, R., 1961. Les plantes utiles du Gabon. Q. J. Crude Drug Res. 1 (1), 27.

Raubenheimer, D., Rothman, J.M., Pontzer, H., Simpson, S.J., 2014. Macronutrient contributions of insects to the diets of hunter-gatherers: a geometric analysis. J. Hum. Evol. 71, 70–76.

Raymond, W.D., 1936. The composition and examination of Tanganyika arrow poisons. Analyst 61, 100–103.

Raymond, W.D., 1939. The m-dinitrobenzene reaction of ouabain and its application to the examination of East African arrow poison. Analyst 64, 113–115.

Raymond, W.D., 1947. Tanganyika arrow poisons. Tanganyika Notes Rec. 23, 49–65.

Republic of Botswana, 1994. Ostrich Management Plan Policy. Government Paper No. 1 of 1994. Government Printer, Gaborone, Botswana.

Richard, R.S., Isbister, G.K., 2008. Medical aspects of spider bites. Annu. Rev. Entomol. 53, 409–429.

Robbins, L., Campbell, A., Brook, G., Murphy, M., Hitchcock, R., 2012. The antiquity of the bow and arrow in the Kalahari Desert: Bone points from White Paintings Rock Shelter, Botswana. J. Afr. Archaeol. 10, 7–20.

Rodin, 1985. Missouri Botanical Garden Press, USA, publishes this journal. Available from: < http://www.mobot.org/MOBOT/Research/publications.shtml >.

Rudner, J., Rudner, I., 1978. Bushman Art. In: Tobias, P.V. (Ed.), The Bushmen: San Hunters and Herders of Southern Africa. Human & Rousseau, Cape Town and Pretoria.

Sadr, K., 2013. A short history of early herding in southern Africa. In: Bolling, M., Schnegg, M., Wotzka, H.-P. (Eds.), Pastoralism in Africa: Past, Present, and Future. Berghahn Books, Oxford and New York, NY, pp. 171–197.

Sandager, M., Nielsen, N.D., Stafford, G.I., van Staden, J., Jäger, A.K., 2005. Alkaloids from *Boophane disticha* with affinity to the serotonin transporter in rat brain. J. Ethnopharmacol. 98 (3), 367–370.

Sands, B., Chebanne, S., Shah, S., 2011. Hunting terminology in = /Hoan. In: Paper Presented at the International Symposium on Khoisan languages and linguistics, July11–13, 2011, Riezlern/Kleinwalsertal.

Sapignoli, M., 2018. Hunting Justice: Displacement, Law and Activism in the Kalahari. Cambridge University Press, Cambridge.

Scarpa, A., Guerci, A., 1982. Various uses of the castor oil plant (*Ricinus communis* L.) a review. J. Ethnopharmacol. 5 (2), 117−137.

Schaffrath, S., Prendini, L., Predel, R., 2018. Intraspecific venom variation in southern African scorpion species of the genera *Parabuthus, Uroplectes* and *Opistophthalmus* (Scorpiones: Buthidae, Scorpionidae). Toxicon 144, 83−90.

Schapera, I., 1925. Bushman arrow poisons. Bantu Stud. 2, 199−214.

Schapera, I., 1927. Bows and arrows of Bushmen. Man 71/72, 113−117.

Schapera, I., 1930. The Khoisan Peoples of South Africa. George Routledge and Sons Ltd., London.

Schep, L.J., Temple, W.A., Butt, G.A., Beasley, M.D., 2009. Ricin as a weapon of mass terror—separating fact from fiction. Environ. Int. 35 (8), 1267−1271.

Schinz, H., 1891. Deutsch-Südwestafrika, Forschungsreisen Durch Die Deutschen Schutzgebiete Groß- Nama- und Hereroland, Nach Dem Kunene, Dem Ngamisee und Kalahari, 1884−1887. Research Expedition of Herero and Nama Country, the Kunene Region, Lake Ngami and the Kalahari-1884−1887, German South West Africa.

Schlebusch, C., Skoglund, P., Sjodin, P., Gattepaille, L., Hernandez, D., Jay, F., et al., 2012. Genomic nariation in seven Khoe-San groups reveals adaptation and complex African history. Science 338, 374−379.

Schlebusch, C.M., 2010. Genetic Variation in Khoisan Speaking Populations from Southern Africa. Ph.D. Dissertation. Faculty of Health Sciences, University of the Witwatersrand, Johannesburg, South Africa.

Schmelzer, G.H., Gurib-Fakim, A. (Eds.), 2008. Plant Resources of Tropical Africa 11(1): Medicinal Plants I. PROTA Foundation/Backhuys Publishers/CTA, Wageningen, 790 pp.

Schultes, R.E., 1979. De plantis toxicariis e mundo novo tropicale commentiones XX. Medicinal and toxic uses of *Swartzia* in the northwest Amazon. J. Ethnopharmacol. 1, 79−87.

Schultze, L., 1907. Aus Namaland und Kalahari, Bericht an die königlich Preussische Akademie der Wissenschaften zu Berlin über eine Forschungsreise im westlichen und zentralen Südafrika in den Jahren 1903−1905. Jena: Gustav Fischer Vlg, pp.−102-104.

Schuster, S.C., Miller, W., Ratan, A., Tomsho, L.P., Giardine, B., Kasson, L.R., et al., 2010. Complete Khoisan and Bantu genomes from Southern Africa. Nature 463 (7283), 943−947.

Schweitzer, F., Wilson, M., 1978. A preliminary report on excavations at Byneskranskop, Bredarsdorp district, Cape Province. S. Afr. Archaeol. Bull. 33, 134−140.

Sealy, J., 2006. Diet, Mobility, and Settlement Pattern among Holocene Hunter-Gatherers in Southernmost Africa. Curr. Anthropol. 47 (4), 569−595.

Shaw, E.M., Woolley, P.L., Rae, F.A., 1963. Bushman arrow poisons. Gmbebasia, Windhoek 7, 1−41.

Shea, J.J., 1988. Spear points from the Middle Paleolithic of the levant. J. Field Archaeol. 15, 441−450.

Shea, J.J., 2006. The origins of lithic projectile technology: evidence from Africa, the Levant, and Europe. J. Archaeol. Sci. 33 (6), 823−846.

Shea, J.J., 2009. The impact of projectile weaponry on late Pleistocene human evolution. In: Hublin, J.-J., Richards, M. (Eds.), The Evolution of Hominid Diets: Integrating Approaches to the Study of Paleolithic Subsistence. Springer, New York, NY, pp. 187−197.

Shebuski, R.J., Ramjit, D.R., Bencen, G.H., Polokoff, M.A., 1989. Characterization and platelet inhibitory activity of bitistatin, a potent arginine−glycine−aspartic acid-containing peptide from the venom of the viper *Bitis arietans*. J. Biol. Chem. 264 (36), 21550−21556.

Shott, M. J., 1997. Stones and shafts redux: the metric discrimination of chipped-stone dart and arrow points. Am. Antiquity 62, 86−101.

Silberbauer, G.B., 1981. Hunter and Habitat in the Central Kalahari Desert. Cambridge University Press, New York,NY.

Sisk, M.L., Shea, J.J., 2009. Experimental use and quantitative performance analysis of triangular flakes (Levallois points) used as arrowheads. J. Archaeol. Sci. 36, 2039–2047.

Skotnes, P., 2002. The art of repossession: the place of the !Xu and Khwe in South Africa. In: Krempel, U., Eisenhofer, S. (Eds.), Bushman Art: Contemporary Art from Southern Africa. Arnoldsche Art Publishers, Stuttgart, pp. 28–45.

Smith, A.B., Poggenpoel, C., 1988. The technology of bone tool fabrication in the South-Western Cape, South Africa. World Archaeol. 20, 103–115.

Smith, F.G., 1958. Beekeeping observations in Tanganyika 1949–1957. Bee World 39 (2), 29–36.

Smith, M.T., Crouch, N.R., Gericke, N., Hirst, M., 1996. Psychoactive constituents of the genus Sceletium N.E.Br. and other Mesembryanthemaceae: a review. J. Ethnopharmacol. 50, 119–130.

Sobiecki, J., 2002. A preliminary inventory of plants used for psychoactive purposes in southern African healing traditions. Trans. R. Soc. S. Afr. 57, 1–24.

Spande, T.F., Garraffo, H.M., Edwards, M.W., Yeh, H.J.C., Pannell, L., Daly, J.W., 1992. Epibatidine: a novel (chloropyridyl) azabicycloheptane with potent analgesic activity from an Ecuadoran poison frog. J. Am. Chem. Soc. 114 (9), 3475–3478.

Spittel, R.L., 1924. Wild Ceylon, Describing in Particular the Lives of the Present day Veddas, Ceylon (H.R.A.F.). The Colombo Apothecaries, Ceylon.

Stafford, G., Pedersen, M.E., van Stadena, J., Jäger, A.K., 2008. Review on plants with CNS-effects used in traditional South African medicine against mental diseases. J. Ethnopharmacol. 119 (3), 513–537.

Steyn, H.P., 1949. Vergiftiging Van Mens en Dier Met Gifplante, Voedsel en Drinkwater. Van Schaik, Pretoria, 264 pp.

Steyn, H.P., 1957. The bushman arrow poison. S. Afr. Med. J. 31, 119–120.

Steyn, H.P., 1971. Aspects of the economic life of some nomadic Nharo groups. Ann. S. Afr. Museum 56 (6), 275–322.

Steyn, H.P., 1984. Southern Kalahari San subsistence ecology: a reconstruction. S. Afr. Archaeol. Bull. 39, 117–124.

Stoll, A., Suter, E., Kreis, W., Bussemaker, B.B., Hofmann, A., 1933. Die herzaktiven Substanzen der Meerzwiebel. Scillaren A. Helv. Chim. Acta 16, 70.

Stow, G.W., 1905. The Native Races of South Africa: A history of the Intrusion of the Hottentots and Bantu into the Hunting Grounds of the Bushmen, the Aborigines of the Country. Swan Sonnenschein & Co., Ltd., London.

Sultana Syeda, S., Sanchez, G., Hong, K.H., Hawkinson, J.E., Georg, G.I., Blanco, G., 2018. Design, synthesis, and in vitro and in vivo evaluation of Ouabain analogues as potent and selective Na,K-ATPase α4 isoform inhibitors for male contraception. J. Med. Chem. 61 (5), 1800–1820.

Tanaka, J., 1980. The San, Hunter-Gatherers of the Kalahari: A Study in Ecological Anthropology. Tokyo University Press, Tokyo, 200 pp.

Tanaka, J., 1987. The recent changes in the life and society of the Central Kalahari San. Afr. Study Monogr. 7, 37–51.

Thackeray, J.F., 1979. An analysis of faunal remains from archaeological sites in Southern South West. Africa. S. Afr. Archaeol. Bull. 34 (129), 18–33.

Theal, G.M., 1897. History of South Africa Under the Administration of the Dutch East India Company, 1652 to 1795. Swan Sonnenschein & Co., London.

The Plant List Version 1.1., 2013. Royal Botanic Gardens, Kew and Missouri Botanical Garden. ⟨http://www.theplantlist.org⟩ (accessed April 27, 2018).

Thieme, H., 2005. Die ältesten Speere der Welt—Fundplätze der frühen Altsteinzeit im Tagebau Schöningen. Archäologisches Nachrichtenblatt 10, 409–417.

Thunberg C., (1772-1775). Travels at the Cape of Good Hope, 1986, Van Riebeeck Society, Cape Town. < https://books.google.com/books/about/Travels_at_the_Cape_of_Good_Hope_1772_17.html?id = ocliNOUc20oC > .

Thunberg, C., 1986. [1772-1775]. Travels at the Cape of Good Hope. Van Riebeeck Society, Cape Town.

Tishkoff, S.A., Reed, F.A., Friedlaender, F.R., Ehret, C., Ranciaro, A., Froment, A., et al., 2009. The genetic structure and history of Africans and African Americans. Science 324, 1035–1044.

Tobias, P.V., 1957. Bushmen of the Kalahari. Man 57, 33–40.

Trommsdorff, H., 1911. Experimentale Untersuchung ueber eine von Buschleten zum vergiften der Pfeilspitz benutzte Käferlarve. Archiv für Schiffs- und Tropen. Hygiene 15 (19), 617–630.

Tulp, M., Bohlin, L., 2004. Unconventional natural sources for future drug discovery. Drug Discovery Today 9, 450–458.

Udall, L., Hitchcock, R.K., 2018. Status of San families living in the Central Kalahari Game Reserve. Trip Report to Botswana-Central Kalahari Game Reserve. Sacharuna Foundation.

Valadez, A.R., 2003. La domesticación animal. UNAM, Instituto de Investigaciones Antropológicas, Plaza y Valdés Editores, México City.

Valiente-Noailles, C., 1993. The Kua: Life and Soul of the Central Kalahari Bushmen. A.A. Balkema, Amsterdam, 232 pp.

Van der Walt, J., Lombard, M., 2018. Kite-like structures in the Nama Karoo of South Africa. Antiquity 92 (363), E3. Available from: https://doi.org/10.15184/aqy.2018.96.

van Riet Lowe, C., 1954. An interesting Bushman arrowhead. S. Afr. Archaeological Bull. 35, 88.

Van Wyk, B.-E., Van Geerden, F., Oudtshoorn, B., 2002. Poisonous Plants of South Africa. Briza Publications, Pretoria.

Vedder, H., 1923. The Bergdama. L. Friederichsen and Co., Hamburg, 199 pp.

Versiani, M.A., Ahmed, S.K., Ikram, A., Ali, S.T., Yasmeen, K., Faizi, S., 2014. Chemical constituents and biological activities of Adenium obesum (Forsk.) Roem. et Schult. Chem. Biodivers. 1 (2), 171–180.

Villa, P., Soressi, M., Henshilwood, C.S., Mourre, V., 2009. The Still Bay points of Blombos Cave (South Africa). J. Archaeol. Sci. 36, 441–460.

Villa, P., Pollarolo, L., Degano, I., Birolo, L., Pasero, M., Biagioni, C., et al., 2015. A milk and ochre paint mixture used 49,000 years ago at Sibudu, South Africa. PLoS ONE. 10 (6), e0131273.

Vinnicombe, P., 1971. A Bushman hunting kit from the Natal Drakensberg. Ann. Nat. Museum 20, 611–625.

Wadley, L., 2010a. Were snares and traps used in the Middle Stone Age and does it matter? A review and a case study from Sibudu, South Africa. J. Hum. Evol. 58 (2), 179–192.

Wadley, L., 2010b. Compound-adhesive manufacture as a behavioral proxy for complex cognition in the Middle Stone Age. Curr. Anthropol. 51 (Supplement 1), S111–S119.

Wadley, L., 2013. Recognizing complex cognition through innovative technology in stone age and Palaeolithic sites. Cambridge Archaeological J. 23 (2), 163–183.

Wadley, L., Mohapi, M., 2008. A segment is not a monolith: evidence from the Howiesons Poort of Sibudu, South Africa. J. Archaeol. Sci. 35, 2594–2605.

Wadley, L., Hodgskiss, T., Grant, M., 2009. Implications for complex cognition from the hafting of tools with compound adhesives in the Middle Stone Age, South Africa. Proc. Natl. Acad. Sci. 106, 9590–9594.

Wadley, L., Sievers, C., Bamford, M., Goldberg, P., Berna, F., Miller, C., 2011. Middle Stone Age bedding construction and settlement patterns at Sibudu, South Africa. Science 334, 1388–1391.

Wadley, L., Trower, G., Backwell, L., d'Errico, F., 2015. Traditional glue, adhesive and poison used for composite weapons by Ju/'hoan San in Nyae Nyae, Namibia.

Implications for the evolution of hunting equipment in prehistory. PLoS ONE. 10 (10), e0140269.

Waguespack, N., Surovell, T.A., Denoyer, A., Dallow, A., Savage, A., Hyneman, J., et al., 2009. Making a point: wood- *versus* stone-tipped projectiles. Antiquity 83, 786—800.

Walton, J., 1954. South West African rock paintings and the triple-curved bow. Bull. S. Afr. Archaeological Soc. 9 (36), 131—134.

Watt, J.M., Breyer-Brandwijk, M.G., 1962. The Medicinal and Poisonous Plants of Southern and Eastern Africa, Edn 2. Oliver & Boyd, London, Edinburgh, 1457 pp.

Wawrzyniak, S., 2013. Khoisan—the best hunters in the World. Investigationes Linguisticae XXVIII, 92—100.

Wei, C.H., Hartman, H.C., Pfuderer, P., Yang, W.-K., 1974. Purification and characterization of two major toxic proteins from seeds of *Abrus precatorius*. J. Biol. Chem. 249, 3061—3067.

World Health Organization, 2017. WHO Model List of Essential Medicines. World Health Organization. ⟨http://www.who.int/medicines/publications/essentialmedicines/en/⟩ (accessed May 1, 2018).

Widlok, T., 1999. Living on Mangetti: 'Bushman' Autonomy and Namibian Independence. Oxford University Press, Oxford, 291 pp.

Wiessner, P., 1983. Style and social information in Kalahari San projectile points. Am. Antiq. 48, 253—276.

Wikar, H.J., 1779. The journal of Hendrik Jacob Wikar. With an English translation by A. W. van der Horst and the journals of Jacobus Coetsé Jansz: (1760) and Willem van Reenen (1791) with an English translation by Dr. E. E. Mossop. Dr. E.E. Mossop (Editor, with footnotes).: The Van Riebeeck Society, Cape Town 1935.

Wilhem, H.J., 1963. Die !Kung-Bushmanner, 12. Jahrbuch Museum f. Völkerkunde, Leipzig, p. 134.

Wilkins, L., Schoville, B., Brown, K., Chazan, M., 2012. Evidence for early hafted hunting technology. Science 338, 942—946.

Will, K.W., Attygalle, A.B., Herath, K., 2001. New defensive chemical data for ground beetles (Coleoptera: Carabidae): Interpretations in a phylogenetic framework. Biol. J. Linn. Soc. 73 (1), 167—168.

Wilmsen, E.N., 1989. Land Filled With Flies: A Political Economy of the Kalahari. University of Chicago Press, Chicago, IL, 420 pp.

Wink, M., van Wyk, B.-E., 2008. Mind-Altering and Poisonous Plants of the World. Briza, Pretoria.

Wintersteiner, O., Dutcher, J.D., 1943. Curare alkaloids from *Chondodendron tomentosum*. Science 97 (2525), 467—470.

Wong, K., 2014. Rise of the human predator. Sci. Am. April 310 (4), 46.

Woollard, J., 1986. The active chemical components of the Basarwa arrow-poison. Botsw. Notes Rec. 18, 139—141.

Woollard, J.M.R., Fuhrman, A.A., Mosher, H.S., 1984. The Bushman arrow toxin, *Diamphidia* toxin: isolation from pupae of *Diamphidia nigro-ornata*. Toxicon 22 (6), 937—946.

Wrangham, R.W., 1977. Feeding behaviour of chimpanzees in Gombe National Park, Tanzania. In: Clutton-Brock, T.H. (Ed.), Primate Ecology. Studies of Feeding and Ranging Behaviour in Lemurs, Monkeys and Apes. Academic Press, London, pp. 503—538.

Wurz, S., Lombard, M., 2007. 70 000-year-old geometric backed tools from the Howiesons Poort at Klasies River, South Africa: were they used for hunting?. South. Afr. Humanities 19, 1—16.

Wynn, T., Coolidge, F.L., 2007. A Stone-Age meeting of minds. Am. Sci. 96, 44—51.

Yamauchi, T., Abe, F., 1990. Cardiac glycosides and pregnanes from *Adenium obesum* (Studies on the constituents of *Adenium*. I). Chem. Pharm. Bull. (Tokyo) 38 (3), 669–672.

Yellen, J., 1974. The !Kung settlement pattern; an archaeological perspective. Ph.D. Dissertation. Harvard University, 442 pp.

Yellen, J., 1977. Archaeological Approaches to the Present. Academic Press, New York, NY, 259 pp.

Further Reading

Ambrose, S.H., 2010. Coevolution of composite-tool technology, constructive memory, and language implications for the evolution of modern human behavior. Curr. Anthropol. 51 (Suppl. 1), S135–S147.

Backwell, L., d'Errico, F., 2001. Evidence of termite foraging by Swartkrans early hominids. Proc. Natl. Acad. Sci. 98 (4), 1358–1363.

Badenhorst, S., 2015. Intensive hunting during the Iron Age of Southern Africa. Environ. Archaeol. 20, 41–51.

Bradberry, S.M., Dickers, K.J., Rice, P., Griffiths, G.D., Vale, J.A., 2003. Ricin poisoning. Toxicol. Rev. 22 (1), 65–70.

Brandmayr, P., Bonacci, T., Giglio, A., Talarico, F.F., Zetto, T., 2009. The evolution of defence mechanisms in carabid beetles: a review. In: Casellato, S., Burighel, P., Minelli, A. (Eds.), Life and Time: The Evolution of Life and its History. Cleup, Padova, pp. 25–43.

Brooks, A.S., Yellen, J.E. 2010. The role of savanna-woodlands in human ancestry: a review of current research findings from central and eastern Africa. In: Paper Presented at Conference on Africa from Stages 6 to 2, Population Dynamics and Palaeo-Environments—2–4 July, McDonald Institute for Archaeological Research, University of Cambridge.

Campbell, A.C., Lamont, G., 1968. Gcwi Bushmen: some notes on hunting with poisoned arrows. Botsw. Notes Rec. 1, 95–100.

Campbell, A.C., Robbins, L.H., 1993. A decorated bone artifact from the Tsodilo Hills, Botswana: implications for rock art. Botsw. Notes Rec. 5, 19–28.

Chapman, J., 1864. Travels in the Interior of South Africa. Clowes and Sons, London.

Crowell, A.L., Hitchcock, R.K., 1978. Basarwa ambush hunting in Botswana. Botsw. Notes Rec. 10, 37–51.

Delaney-Rivera, C., Plummer, T.W., Hodgson, J.A., Forrest, F., Hertel, F., Oliver, J.S., 2009. Pits and pitfalls: Taxonomic variability and patterning in tooth mark dimensions. J. Archaeol. Sci. 36 (11), 2597–2608.

Fourie, L., 1928. The Bushmen of South West Africa. In: Hahn, C.H.L., Vedder, H., Fourie, L. (Eds.), The Native Tribes of South West Africa. South West African Administration, Cape Town, pp. 79–105.

Hawkes, K., 2016. Ethnoarchaeology and Plio-Pleistocene sites: some lessons from the Hadza. J. Anthropol. Archaeol. 44, 158–165.

Lee, R.B., DeVore, I. (Eds.), 1968. Kalahari Hunter-Gatherers. Studies of the !Kung San and their neighbors. Harvard University Press, Cambridge, 408 pp.

Madibela, O.R., Seitiso, T.K., Thema, T.F., Letso, M., 2007. Effect of traditional processing methods on chemical composition and *in vitro* true dry matter digestibility of the Mophane worm (*Imbrasia belina*). J. Arid Environ. 68, 492–500.

Neuwinger, H.D., 1998. Alkaloids in arrow poison. In: Roberts, M.J., Wink, M. (Eds.), Alkaloids: Biochemistry, Ecology, and Medical Applications. Plenum Press, New York, NY, pp. 45–84.

Philippe, G., Angenot, L., 2005. Recent developments in the field of arrow and dart poisons. J. Ethnopharmacol. 100 (1), 85–91.

Philppe, G., Angenot, L., Tits, M., Frédérich, M., 2004. About the toxicity of some *Strychnos* species and their alkaloids. Toxicon 44, 405–416.

Retief, J.A., 1988. Cultivation of Pachypodium namaquanum. Aloe 25, 6–8.

Schulz, A., Hammar, A., 1897. The New Africa: A Journey up the Chobe and Down the Okovango Rivers, A Record of Exploration and Sport. William Heinemann, London.

Tanaka 1996, The African Studies Monograph is published by Kyoto University, Japan. Available from: < http://jambo.africa.kyoto-u.ac.jp/asm/index.html >.

Tanaka, J., 1996. The world of animals viewed by the San hunter-gatherers in the Kalahari. In: Tanaka, J., Sugawara, K. (Eds.), Social, Ecological, and Linguistic Aspects of the /Gui and //Gana San of the Central Kalahari Game Reserve. African Study Monographs Supplementary Issue 22(1), Kyoto University, Japan, pp. 111–128.

Tobias, P.V. (Ed.), 1978. The Bushmen: San Hunters and Herders of Southern Africa. Human and Rousseau, Cape Town.

Tommaseo-Ponzetta, M., 2005. Insects: food for human evolution. In: Paoletti, M.G. (Ed.), Ecological Implications of Minilivestock. Potential of Insects, Rodents, Frogs and Snails. Science Publishers, Enfield, pp. 141–161.

Turnbull, C.M., 1963. The lesson of the Pygmies. Sci. Am. 208 (1), 28–37.

Turnbull, C., 1965a. The Mbuti pygmies: an ethnographic survey. Anthropol. Pap. Am. Mus. Nat. Hist. 50 (3), 139–282.

Turnbull, C., 1965b. Wayward Servants: The Two Worlds of the African Pygmies. Natural History Press, New York, NY.

Turnbull, C., 1968. The importance of flux in two hunting societies. In: Lee, R.B., DeVore, I. (Eds.), Man the Hunter. Aldine Publishing, Chicago, pp. 132–137.

Wiessner, P., 2002. Hunting, healing, and Hxaro exchange: a long term perspective on !Kung (Ju/'hoansi) large game hunting. Evol. Hum. Behav. 23, 407–436.

3

Toxicology in Ancient Egypt

Gonzalo Sanchez[1] and W. Benson Harer, Jr[2]

[1]Medical Associates Clinic, Pierre, SD, United States [2]Independent Scholar, Seattle, Washington, United States

3.1 INTRODUCTION

The ancient Egyptians were astute observers of natural phenomena. The ability to incorporate these observations into their beliefs about religion, politics, and magic was a major factor in the survival of their culture for so many centuries. *Maat* was their term for the proper order of things in the world.

Poisonous snakes and insects pose a special challenge to incorporate into a proper order of the universe. The solution to incorporating them into their belief system was by being able to control them by magic. The scorpion became the goddess Selket, one of the guardians of the king's sarcophagus. At least 10 gods and goddesses who could be protective were portrayed with serpent heads. The king, for example, converted the cobra to his protector. The horned viper was the hieroglyph word for the masculine designation as well as the letter "f." When the horned viper depiction in a glyph was needed in an inscription in a tomb, there was a perceived risk it could come to life to bite the owner so it might be portrayed with the head severed from the body to be safe. Magic, or "ritual power" was produced to protect the deceased from the most dangerous of all, the serpent Apophis, who threatened to block the king's passage through the dangerous night to rebirth in the afterlife.

3.2 SNAKES AS DESCRIBED IN THE BROOKLYN PAPYRUS

The study of venomous snakebite, its chemistry, mode of action, and the biology of the venom-producing organism, are considered part of the science currently known as toxinology. The Brooklyn Museum Papyri, c. 525–600 CCE, consist of two sections, 47,218.48 and 47,218.85, which describe individual snakes and treatments for snakebites, respectively. These papyri, translated by Sauneron, are our major source of information on snakebites in ancient Egypt. The Brooklyn Papyri originally contained 100 paragraphs; §1–13 are missing, but presumably contained material similar to that found in §14–38. A summary of the information included therein:

Paragraphs 14–37:

1. Snake identification by name, physical characteristics of size, color, shape of the head, neck, bite marks, behavior, reptation (i.e., the manner in which it crawls), aggressiveness, and the snake's state of alertness (some text is missing).
2. Effects of the bite, local symptoms (swelling, bleeding, pain, necrosis, discoloration) and systemic symptoms (fever, vomiting, fainting, loss of strength, loss of consciousness, tetanization, possible seizures, coma).
3. Prognoses and recommendations for initial treatment.

Paragraph 38: Description is of a chameleon, not a snake.
Paragraphs 39–100 include treatments for the snakebites described.

3.2.1 Snake Identification

The identification of snakes from ancient Egyptian descriptions in section 4721,848 of the Brooklyn Papyri is not established with certainty.

fy ḥr dbwy

(Viper with horns)

Sahara horned viper

Cerastes cerastes

Regarding a *Viper with horns*, its color is similar to that of a quail; it has two horns [on] its forehead; its head is broad, his neck is narrow, [his] tail is thick.
(If) the orifice of his bite is broad, the face of the injured swells; (if) its bite is small, one who was bitten becomes inert, but [...] (?)...]. Fever for nine days, (but) he will srvive. It is a manifestation of Horus. Its venom is [out] attracted by making him vomit profusely, and exorcizing[...].

A short stout viver, the head is broad, flat and triangular, the eye is prominent on the side of the head, and has a vertical pupil in bright light. The neck is thin. The body is broad and flattened, tail short. There is a long horn above each eye, this horn consists of a single scale. Ground color, yellowish, biscuit colored with brown blotches along its back.
Bites involve swelling and hemorrhage, and necrosis at bite site. Nausea and vomiting occur. Hematuria. Fatalities are infrequent.

Serge Sauneron. Papyrus du Musee de Brooklyn No 47.218.48 + 85 Premiere Parte. Paragraph 28, p.25.1989   Hieratic text translation.

Partial text from S. Spawls and B. Branch "The Dangerous Snakes of Africa," 1995

FIGURE 3.1   Sahara Horned Viper. Comparative text from the Brooklyn Papyri with a contemporary source. *Photo by Gonzalo M. Sanchez, author.*

In paragraphs 13–37 Sauneron labels his identification of 18 of them as "probable" and his identification of four of them as "possible." Likewise, Nunn and Warrell identify 13 of these and label those as tentative identifications. An example of certainty in snake identification is shown in parallel texts §28 and the contemporary description of *Cerastes cerastes* (Fig. 3.1).

The snake identification in the Brooklyn Papyri has raised an interesting issue with the snake in §15, Apophis, as this snake features most prominently in the mythology of ancient Egypt as the personification of evil. Apophis is represented in the various royal tombs and in the Book of the Dead, with consistent morphological characteristics: a large snake with a small head, large eye and round pupil. The body is dark on top with a distinct light underbelly. Apophis in the Brooklyn Papyri is described as a large snake, reddish-brown in color, with a light underbelly, having four fangs, and a lethal bite. Regarding the identification of the snake in §15, John Nunn remarks: "A positive identification would be valuable, but no Egyptian snake fits the description. The largest and most rapidly fatal snake is the Egyptian cobra." The two large dangerous snakes currently in Egypt are: the Egyptian Cobra and the Desert Black Snake. Although we agree with Nunn that *"no Egyptian snake fits the description"* of the snake in Brooklyn Papyri §15, it is possible that such a snake can be found in Sudan. The desertification shift occurring in Southern Egypt in the last five millennia resulted in its

Papyrus du Musée de Brooklyn No 47.218.48 + 85 Premiere Partie § 15

[Quant] au grand serpent d'Âpopi, il est rouge en totalité*; son ventre est blanc**; il y a quatre crocs dans sa bouche. S'il mord quelqu'un, celui-ci meurt aussiôt.

* a chromatic gamma near red, brown, bronze, reddish-brown

** white, whitish, cream.

The god Atum spearing Apophis - Tomb of Ramses I, Luxor, Egypt.

Dispholidus typus

Boomslang

Rear-fanged snake
Highly variable in coloration, from green to dark brown or black above with a pale belly. Eyes very large, round pupils. Most dangerous of Africa's Colubridae family. Its venom is hemotoxic.

Photo courtesy :
Terry Philip, Curator,
Black Hills Reptile Gardens
Rapid City S.D.

FIGURE 3.2 Brooklyn Papyri serpent Apophis, comparison with Boomslang. Reproduction of the god Atum spearing the serpent Apophis from the tomb of pharaoh Ramsess I in Luxor, Egypt, is from public domain File: Apep 1 jpg, from Wikipedia Commons. *The composite hieroglyph text and legends done by Gonzalo M. Sanchez, author. The Boomslang snake photo in this regarding Apophis is used with permission from his owner Travis Phillip. Permission filed.*

flora and fauna retreating to Sudan (5,2). Four large venomous snakes found today in Sudan are: the Black Mamba, Puff Adder, Black Spitting Cobra, and Boomslang. While no longer found in Egypt, the Puff Adder can be easily identified as the snake in §33. Of these four snakes, only the Boomslang (*Dispholidus typus* in the Colubridae Family) has those physical characteristics that could correspond to the snake in §15, Apophis (Fig. 3.2).

3.2.2 Symptoms of Snakebite

The Brooklyn Papyri and other Egyptian texts demonstrate that the ancient Egyptians recognized the different symptoms and signs of serious snake envenomation according to their source: vipers, or cobra and related snakes (elapids). An example is found in the legend of the god

Ra suffering the effects of viper envenomation after having been struck by a snake (6).

Today we understand these clinical effects are due to the main toxic components of the specific venoms in the various snake families, predominantly hemotoxic in vipers and neurotoxic in elapids.

Viper venom causes tissue damage at the bite site and in its proximity, with changes in red blood cells, defects in coagulation, damage to blood vessels and often to the heart, kidneys, and lungs.

Neurotoxic venoms of elapid snakes (i.e., cobras) cause less local tissue damage, but rapid alterations to the nervous system, secondary cardiac and respiratory failure, and death.

3.2.3 Prognosis

- "He will die immediately": (§15−17). "Death hastens very quickly" (§19).
- "He will live": (§22, 25, 26, 28). "One does not die because of it": (§21, 32, 33, 37).
- "One can save him" (uncertain): (§14, 18, 20, 24, 27, 29, 30, 31, 33, 34).

The prognosis of the number of days the patient could survive is sometimes related to whether or not he vomited and whether magic could be administered.

3.2.4 Treatment

Some treatments were for any snakebite and some were snake-specific.

Bites by the lethal snakes §15−17, 19, received no treatment. The bite by snake §33 received treatment, except when the patient had developed signs of what today we understand to be cranial nerve dysfunction. Magic is considered useful in §17.

Therapeutic measures, whether local or systemic, were primarily symptomatic. In viper bites, local edema, necrosis, weakness, sweating, tremors, bleeding, thirst, fever, and pain were treated. In elapid bites, treatment was intended for paralysis, dyspnea, heart failure, aphasia, and seizures.

For local wounds, the bite site was treated with medications kept in place by bandages, with fumigations and sometimes with incisions or debridement over which medications were applied (§31, 32). No tourniquets were used.

Discussion of the extensive Pharmacopeia included in the Brooklyn Papyri is beyond the scope of this chapter. Because of the extensive use of *Allii Cepae*, the onion, it will be addressed here. Onion/water preparations were applied to the bite from the snake identified as Sekhtef §46 in the treatment section only. This snake is not listed in the extant section on snake identification due to the loss of paragraphs 1–13. It was also used in coating and fumigating the bite of the Black Spitting Cobra §47g, 82b; in compresses for *Echis coloratus* §48c bite; in bites of "any venomous snake" §92, 95b; and particularly with copper filings to the bite of the Persian viper §51c.

Possible effects of the topical use of *Alli Capae* in snakebite:

1. Phospholipase A2
    Snake venom phospholipase A2 (PLA2) toxins are present in almost all venomous species. Having different tissue targets, these toxins, we know today, can be myotoxic, neurotoxic (7), and can lyse the phospholipid membranes of red blood cells. The release of arachidonic acid metabolites produces inflammation. The topical application of an aqueous extract of *bulbus Allii Cepae* (10% in a gel preparation) has been found to inhibit mouse ear edema induced by arachidonic acid. The flavonoids quercetin and kaempferol in onions inhibit phospholipase A, cyclooxygenase, protein kinase, and the release of histamine from leukocytes.
2. Paralyzing the lymphatic pump
    Large toxin molecules from elapid snake venoms need first to be absorbed and transported into lymphatic vessels before entering the bloodstream. The current use of pressure immobilization in elapid snakebites is done for the purpose of impeding lymphatic transport without compromising arterial blood flow, thereby delaying toxins entering the circulation. Slowing of the lymphatic pump by up to 400% by topical use of ointment containing glyceryl trinitrate, which releases nitric oxide, has been demonstrated by Australian investigators lead by Dirk vanHelden. Thus, the release of nitric oxide from S-nitrosoglutathione contained in extracts of aqueous garlic, onion, and leek could explain some benefit by the use of topical applications of aqueous onion in elapid snakebites by the ancient Egyptians. The nitric oxide effect on the lymphatic pump would have been, at least, temporarily beneficial.

Systemic treatments also played a role. From their belief that vomiting would get rid of the poison and that onion would kill the venom of any snake (§42), onion and emetics were often used. Recent research has demonstrated the plausibility of onion as a therapeutic agent. Onion mixed with beer (and other substances), for example, was ingested, and then vomited after a specified period of time (2:§41, 45a, 46b, 47a, 48b, 50a, 57, 59.). Oral administration of an ethanol extract of onions to guinea-pigs inhibited smooth muscle contractions in the trachea and also inhibited histamine, serotonin, and acetylcholine-induced contractions in the ileum.

3.3 SCORPIONS

The Berlin Papyrus: The Berlin Medical Papyrus (c. 1200 BCE) is a general medical document that parallels the medical content of the Ebers Medical papyrus. §78 lists unknown ingredients used for fumigation treatment of a scorpion "bite." Other treatments given for scorpion sting in the Egyptian literature are magic incantations, like the "Spell for conjuring a cat," which involved repelling the poison by magic, along with the use of fabric and incisions to remove the venom.

3.4 TETANUS

Tetanus infection developing as a complication of an open wound is addressed in the Edwin Smith Papyrus Case #7. This document (c. 1650−1550), currently housed at the New York Academy of Medicine, is the second major text related to ancient Egyptian medicine. It is a copy of the older original (date range: 2200−1000 BCE) which has not been found. The Edwin Smith Papyrus, the first comprehensive trauma treatise in the history of medicine, presents a pragmatic approach to the classification and management of head, neck, and upper body injuries and is the source of numerous anatomical and functional concepts of the nervous system. In Case #7 the ancient Egyptians appear to extend the concept of illness originating from penetration of the body (by unknown entity) through an open cranial wound, developing symptoms of tetanus: "because of that wound...". They understood the effects of tetanus and the futility of treatment, even though the nature of this toxin would not be known for millennia.

3.5 PLANT AND MINERAL TOXINS

The ancient Egyptians did not embrace poisoning as a means of execution or assassination as was later done by the Greeks and Romans. A possible exception is the much discussed murder of Ramses III in a conspiracy hatched in his harem. The trial of the perpetrators was documented and indicates that he survived for 2 weeks after the attempt by unspecified means, involving magic. This suggests the use of a slow-acting poison such a ricin found in the seed of the castor plant. They probably were aware of the danger of the castor bean since they understood the use of castor oil as medicine. A magical incantation was considered essential to activate ingredients.

Dosage determines the ranges of therapeutic action and toxicity of drugs. The Ebers Papyrus, dated to 1536 BCE (the ninth year of the reign of Amenophis I) is the most voluminous papyrus document known. It is a medical text that provides over 800 recipes to treat almost any symptom imaginable. All the components were dispensed by volume rather than weight, and treatments could be administered by mouth, anum, vaginal suppository, or by lotion or ointment to the skin or even as smoke by fumigation. A critical element, however, is that they be accompanied by an incantation to assure their efficacy. Therapeutic or toxic effects were dependent on the degree of absorption into the patient's body. Experience would have permitted them to avoid toxic levels of drugs such as wormwood, hyosyamine, and mandrake, but no actual references to toxicity are made.

Alcohol is well understood for its dose-related toxicity. The Egyptians consumed it in both beer and wine. Indeed, beer was a staple of their diet. Their writings berate overindulgence and they were aware that alcohol could be habit forming. Wine was the drink of the elite. Both were used as vehicles for many prescriptions. Their beer probably contained about 5% alcohol and their wine 10%–14%. With that concentration, they probably didn't drink sufficient amounts to raise the blood alcohol to a lethal level. We know of no account of death from alcohol poisoning.

It appears that they understood that some of their therapeutic alkaloids were only soluble in alcohol, and used beer or wine as vehicles for their delivery. A papyrus in Leiden gives us the original recipe for a "Micky Finn," a potion to put an unsuspecting man to sleep—mandrake and hyosyamine steeped in wine. Another example is the Egyptian "lotus," more accurately described as Nymphaea caerulea and Nymphaea alba. Both contain narcotic alkaloids. The latter is concentrated in the flower, but not in the seeds or leaves. The blue lotus figures prominently in Egyptian mythology with its yellow stamen mimicking the sun in the blue sky of the petals. Four active ingredients have been isolated and studied. While the narcotic properties of the lotus are not well-recognized by Egyptologists, there is good evidence that it was understood by the ancients. Two prescriptions in the Ebers papyrus use beer and wine in which it has "spent the night" as the vehicles to carry it. The true Nelumbo lotus, which contains 14 powerful narcotic alkaloids, did not enter Egypt until the Persian conquest in 525 BCE (Fig. 3.3).

In 1934, Dawson proposed that the word Shemshemet could be cannabis. Since then it has figured prominently in literature, but there is absolutely no archeological evidence of cannabis being present in ancient Egypt, even to make rope. There is no evidence it was used as a drug until after the Arab conquest centuries later.

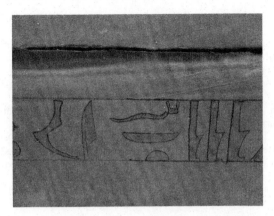

FIGURE 3.3 The horned viper was a hieroglyph indicating the letter F. The ancient Egyptians regarded the *Cerastes cornutus* as so dangerous that when used on many Middle Kingdom coffins it was considered too great a risk to portray intact. They feared the picture could come to life in the after world and bite the mummy. The safest way was to sever the head of the hieroglyphic snake needed in the inscription. They had similar faith in the curative value of their magic. *Photo by W.B. Harer, courtesy of Michael C. Carlos Museum, Emory University.*

Galena or lead sulfide was the black eye paint used for many ophthalmic problems. It probably also was used with soot and other binders to make mascara. Despite its toxicity, the effects from absorption through the skin would be slow to develop. We have no evidence this danger was recognized by the Egyptians.

References

Condrea, E., De Vries, A., Magner, J., 1964. Hemolysis by lysing the phospholipid cell membranes of red blood cells. Specialized section on lipids and related subjects. Biochim. Biophys. Acta 84, 60–73.

Dawson, W.R., 1934. Studies in the Egyptian medical texts. J. Egypt. Archaeol. 20, 185–188.

Ghalioungi, P., 1987. The Ebers Papyrus. Academy of Scientific Research and Technology, Cairo.

Griffith, F.L., Thompson, H., 1974. The Leyden Papyrus. Dover Publications, New York, NY, pp. 148–151.

Grman, M., Misak, A., Cacanyiova, S., Tamascova, Z., Bertova, A., Ondrias, K., 2011. The aqueous garlic, onion and leek extracts release nitric oxide from S-nitrosoglutathione and prolong relaxation of aortic rings. Gen. Physiol. Biophys. 4, 396–402.

Harer, W.B., 1985. Pharmacological and biological properties of the Egyptian lotus. J. Am. Res. Center Egypt 22, 49–54.

Houlihan, P.F., 1995. The Animal World of the Pharaohs. The American University in Cairo Press, Cairo, p. 45.

Kousoulis, P., 2011. Stop, O poison, that I may find your name according to your aspect"; a preliminary study on the ambivalent notion of poison an the demonization of the scorpion's sting in ancient Egypt and abroad. J. Ancient Egypt. Interconnections 3 (3), 15.

Leitz, C., 1996. Die Schlangensprueche in den Pyramidentexten. Orientalia (Nova Series) 65, 382.

Montecucco, C., Gutierrez, G.M., Lomonte, B., 2008. Cellular pathology induced by snake venom phospholipase A2 myotoxins and neurotoxins: common aspects of their mechanisms of action. Cell. Mol. Life Sci. 18, 2897–2912.

Nunn, J.F., 1996. Ancient Egyptian Medicine. University of Oklahoma Press, Norman, OK, pp. 183–186.

Oakes, L., Gahlin, L., 2006. Isis and the Sun God's Secret Name. Ancient Egypt. Anness Publishing, Barnes and Noble, New York, NY, pp. 324–325.

Redford, S., 2002. The Harem Conspiracy. Northern, Illinios Univ Press, Dekalb, IL.

Sanchez, G.M., Meltzer, E.S., 2012. The Edwin Smith Papyrus – Updated Translation of the Trauma Treatise and Modern Medical Commentaries. Lockwood Press, Atlanta, GA, pp. 71–85.

Saul, M.E., Thomas, P.A., Dosen, P.J., Isbister, G.K., O'Leary, M.A., Whyte, I.M., et al., 2011. A pharmacological approach to first aid treatment for snakebite. Nat. Med. 17 (7), 809–811.

Sauneron, S. (1989). Un Traité Égyptien d'Óphiologie. Papyrus du Brooklyn Museum Nos. 47.218.48 et 85. Institut Français d'Archéologie Orientale, Le Caire, pp. 164–165; 48, 148–149; 143; 21.

Shupe, S., 2013. Venomous Snakes of the World. Skyhorse Publishing, New York, NY, pp. 155–156, 165, 194, 198, 203.

Sutherland, S.K., Coulter, A.R., Harris, R.D., 1979. Rationalisation of first-aid measures for elapid snakebite. Lancet 1 (8109), 183–185.

Texeira, C.F., Landucci, E.C., Antunes, E., Chacur, M., Cury, Y., 2003. Inflammatory effects of snake venom myotoxic phospholipases A2. Toxicon 42 (8), 947–962.

Westendorf, W., 1998. Handbuch der Altägyptischen Medizin. Brill, Leiden, Boston, Koln. Berlin, p. 278.

WHO, 1999. Bulbus Allii Cepae. Monographs on Selected Medical Plants. World Health Organization, Geneva, p. 10.

The Death of Cleopatra: Suicide by Snakebite or Poisoned by Her Enemies?

Gregory Tsoucalas[1] and Markos Sgantzos[2]

[1]History of Medicine, Department of Anatomy, School of Medicine, Democritus University of Thrace, Alexandroupolis, Greece [2]Department of Anatomy, Faculty of Medicine, University of Thessaly, Larissa, Greece

4.1 CLEOPATRA'S ANCESTRY AND HISTORICAL BACKGROUND OF THE ERA

Soon after the death of Alexander the Great, his generals ruled a divided empire. One of his favorites, Ptolemy (father of Cleopatra), was assigned to govern Egypt. A supreme dynasty had arisen in this part of the northeast corner of the African continent. Ptolemy had decided to transform the city of Alexandria into the greatest trade, cultural, and religious center of the known world. His plans included the largest harbor of the Mediterranean Sea, majestic palaces, extraordinary public

FIGURE 4.1 On the antique painting in encaustic of Cleopatra, discovered in 1818 by Sartain, John, 1808–1897. Published in 1885 by G. Gebbie & Co. in Philadelphia.

buildings, museums, the wondrous Pharos lighthouse, wide boulevards laid out in a grid, translation centers, libraries, and irrigation and drainage systems. Inside the walls of the famous Library of Alexandria were gathered manuscripts from all over the world to encompass all the known knowledge of the era (Heather, 2010; Dimitsas, 1885).

In this magnificent city, at the prime of its civilization, the great Cleopatra VII (Fig. 4.1) was born in 69 BC. At the time of her birth, Rome was perhaps the only city that could be considered a rival to Alexandria. Rome, though, was vastly superior to the Egyptian metropolis, in terms of the magnitude and extent of its military power, which it used to control its vast territories. Meanwhile, Alexandria ruled over Egypt, and over a few of the neighboring coasts and islands. The fruitfulness of this isolated territory, at the magnificent delta of the river Nile, along with its strategic location, made it a perfect target for the Romans. While the idea of a new Roman province was compelling, Egyptian military and strategic power tempered such thoughts. Political events, though, were in their favor and the bountiful prey, Egypt, was lured by Rome and fell into the trap (Burstein, 2004).

The Roman government at the time was a republic, and the senate had extended powers. The two most powerful men in the state of Rome were Pompey and Caesar. Caesar was in the ascendency in Rome when Ptolemy first petitioned him for an alliance between the two great nations of that era, that is, Rome and Egypt. Meanwhile, Pompey was otherwise occupied in Asia Minor, engaged in a war with Mithridates, the powerful monarch of Pontos, who was at that time challenging Rome. Caesar was deep in debt, and much in need of financial resources, not only for relief from existing public embarrassments but also to enable him to accomplish the assorted political schemes which he was entertaining, and to secure his hold on Rome. Negotiations were difficult and delays long, but it was finally agreed that Caesar would exert his influence to secure an alliance between the Roman people and Ptolemy, on the condition that Ptolemy would pay him 6000 talents (ancient Egyptian currency), an enormous amount and, essentially, a bribe. While this served to establish a formal alliance, it also had a serious impact on the Egyptian people. Subsequently, due to an uproar by the citizens over the imposition of heavy taxes, Ptolemy fled to Rome in search of safety. There, a formal alliance was established. The door was then open for Rome to more directly intervene in Egypt (Burstein, 2004; Seppard, 2006; Roller, 2010).

During the above turmoil, in 58 BC, with Ptolemy XII absent, his daughter, Berenice, sister of Cleopatra, ascended to the throne. She married the king of Syria, Seleucus, whom she reportedly strangled. Her second marriage to the prince Archelaus was more successful. Meanwhile, Ptolemy was plotting with the Romans for his return. Although negotiations were difficult, the Romans had decided to permit Pompey to send troops into Egypt. Gabinius, who governed Syria with his second-in-command, Mark Anthony, was a wild and dissolute young man who had lost his family inheritance and embarked on a brief but successful campaign against Egypt. Archelaus was slaughtered in the final ferocious encounter and Berenice was taken prisoner. Cleopatra was only 15 when the unnatural quarrel between her father, Ptolemy, and her sister, Berenice, was working its way toward a dreadful termination. She was a quiet spectator, neither benefitting nor suffering. Ptolemy recaptured the throne. Mark Anthony and the Roman troops remained in Alexandria to help rule a wounded empire (Burstein, 2004; Maloney, 2010).

In Rome, a civil war between Caesar and Pompey was raging. Cleopatra was the oldest, trusted child, and a princess of great promise, endowed with a sharp intellect and personal charms. In order to retain control over his family legacy, Ptolemy ordered her, before his death, to be married to her brother, Ptolemy XIII, who was only 10 years old at the time. Cleopatra was 18 when she was crowned queen, and rivalry among her ministers inside the royal palace was continuous. Pompey went to Egypt seeking assistance against Caesar but instead was brutally murdered in front of Ptolemy XII, her father, an act that infuriated even Caesar, who subsequently marched against Alexandria (Burstein, 2004; Maloney, 2010).

Sometime later, Cleopatra and her loyal guardian Apollodorus appeared in front of Caesar to plead for her own agenda in the ensuing power struggle with Ptolemy XII. She forthwith gained Caesar's heart. Caesar made it possible for her to gain Egypt's throne, making Rome appear more an ally than a conqueror. Together they fought against her father, Ptolemy XII, and her minister Pothinus, ultimately securing both their deaths. In 47 BC, Cleopatra gave birth to Caesar's son, Caesarion. Some months later, during a visit with her son to Caesar's palace in an island of the Tiber River, she informed him that she was the undeniable Queen of Egypt, mother of a Roman prince (Burstein, 2004; Maloney, 2010).

In 44 BC, Caesar was assassinated. Mark Anthony joined with Caesar's adopted son, Octavian, and Lepdius to form the Second Triumvirate. Mark Anthony had already been attracted to Cleopatra and she went on to fully capture his heart. He spent the winter of 41–40 BC, in Egypt, drunk in love with Cleopatra. The couple began to expand Cleopatra's realm through continuous warfare. Anthony soon sent a message to Rome saying that he would abandon his wife (yes, he had a wife back home) to marry his intoxicating concubine, Cleopatra. Rome, specifically Octavian, furiously turned against the couple, declaring yet another war against the Egyptian Empire. After fierce sea battles and violent slaughters on land in the area of western Greece, Cleopatra and Anthony were defeated and retreated back to Egypt. Cleopatra had supported two great Roman leaders and kept Egypt an independent state for some 20 years, but in the end lost everything (Burstein, 2004; Maloney, 2010).

As Anthony was consumed by his defeat, Cleopatra has a message sent to him saying that she has died. Eager to save herself, her sons, and the crown from the advancing Octavian, she realized that she could neither kill Anthony nor exile him. But she believed that if he could be induced to kill himself for love of her, she could save herself. And, indeed, Anthony attempted suicide by a self-inflicted sword wound. The injury, though, was not immediately fatal. Discovering that his queen was still alive, he managed to reach her chambers and see her once more before he expired. Octavian had entered Alexandria as a conqueror on August 1, 30 BC. There, at the Palace of the Ptolemies, he found the lifeless body of Mark Anthony and met the notoriously charming queen. Cleopatra, with all her subtlety and political foresight, had already backed two losers, first Caesar and then Anthony, to whose downfall she had notably contributed. Now, at the age of 40, she tried to seduce the new Roman conqueror, Octavian, but he was coldly indifferent toward her advances (Burstein, 2004).

Cleopatra understood that she would likely be publicly humiliated as a trophy of Octavian. She remained locked in her mausoleum during her final days, searching for a solution out of this crisis. Two handmaidens and, possibly, a eunuch were with her at the time. Cleopatra ordered a bath and asked for a basket of figs as well, a request that was granted by the guards. Soon after, she sent messengers to Octavian indicating that she wanted to be buried alongside Anthony. Cleopatra was subsequently found dead by Octavian and his men. The specific mysterious circumstances surrounding her actual death continue to intrigue historians even today. Plutarch said, "The messengers came at full speed, and found the guards apprehensive of nothing; but on opening the doors they saw her stone dead, lying upon a bed of gold, set out in all her royal ornaments." Her son Caesarion, possible offspring of Caesar, was later executed, supposedly by order of Octavian (Burstein, 2004).

4.2 CLEOPATRA'S REIGN. HER DOWNFALL AND HER DEATH

The queen of all queens, Cleopatra VII, was enthroned as queen of the Ptolemaic Dynasty and Pharaoh of Egypt, inheriting, apart from the Crown, the great inclination of the Ptolemies toward medicine and their love for science. Despite the fact that she reigned during politically turbulent times, she managed to acquire at least some rudimentary knowledge of medicine. She was aware of gynecological diseases and was conversant in pharmacology and botany, and may have authored several scientific texts. Some essays were found beside her head in her tomb but are no longer extant. She produced her own eye makeup, aromatic oils, remedies for baldness, antiseptics, and beauty products (Caesar and Gardner, 1967; Tsoucalas et al., 2013). She wrote a treatise called the "Cosmetics," which discussed remedies, potions, and ointments. This treatise was mentioned by Galen (third century), Aëtius of Amida in Alexandria (sixth century), Paul of Aegina (seventh century), and John Tzetes (twelfth century) (Skevos, 1902). The longevity of this work's usefulness owes a debt to her knowledge of pharmacology and cosmetics.

During Cleopatra's reign, the city of Alexandria was a center of science. Inside the famous Library of Alexandria were 700,000 treatises and manuscripts, works from mainland Greece, the Aegean, Asia Minor, and Pontos, all gathered in an extraordinary assembly. The Alexandrian medical school was famous for advances and instruction in anatomy, physiology, and pharmacology. The knowledge of potions, poisons, and antidotes was so pervasive during that era in Alexandria that philosophers and physicians believed that production could and should be administered by simple pharmacists, so that the philosophers and physicians could focus on the dosage of the drugs and the patient himself. The Greeks introduced the concept of *theriac*, a versatile, multifunctional drug-antidote to cure all diseases. The theriac was composed of a large number of plant, animal, and mineral substances, mixed with a base of viper venom or blood and blood from animals fed with poisonous plants. Mithridates himself invented a theriac named Mithridatiki. It is more than likely that Roman scholars knew about this ultimate potion and that the Egyptians became familiar with it as well. Cleopatra herself gave poisonous potions in various dosages to slaves to test its toxic limits (Plat, 2004).

Cleopatra was, in fact, a neglected child who wasn't meant to rule. She was well educated and fluent in seven languages, including Egyptian, unlike the rest of the royal family. She lived in a competitive environment, full of conspiracies, intrigues, murders, revolts, civil wars, and infidelity. Her only weapon for survival was to believe in herself. She was forced to marry her young brother, to battle against her family, to deal with the Roman invasion, and to counterplot against her ministers. Over time, she developed a narcissistic personality, a strong character to withstand obstacles, and a complex emotional fabric that could brook no defeat, a difficult mask to uphold (Burstein, 2004; Tsoukalas, 2004). Paying great attention to cosmetics and personal charm, she became an example of the perfect *femme fatale* whose technique for ruling was to control the minds and hearts of the vigorous governors of the era (Burstein, 2004).

The circumstances surrounding her death are still cloaked in mystery. First off, did she commit suicide or was she murdered? Three possible scenarios need to be considered:

1. suicide by poison (possibly hidden somewhere in her mausoleum),
2. suicide by the venom of the Egyptian cobra, usually referred to as an asp, or viper, or
3. poisoning by Octavian and/or his men (Orland et al., 1990).

No written documents were found to divulge the secret. The oldest reference was that of Strabo (64/63 BC–C. 24 AD), who mentioned that her death could have been the result of a toxic poison or the bite of an asp, so that it was not clear whether it was a murder or suicide (Strabo, 1924). Cleopatra's physician, Olympus, came to almost the same conclusion, except to emphasize that if poison was taken, it was as an ointment administered dermally, rather than a potion swallowed, a fact that increases the certainty that Cleopatra was an expert in the field, cognizant of drugs in their different formulations (Orland et al., 1990).

Considering the two indistinct prick marks on her body, Galen theorized that she broke the skin by biting her own arm (Galen, 1964–1965). Thus, several death scene scenarios are:

1. The skin marks were self-inflicted. This is the less likely scenario, though, for a woman so devoted to her own beauty (Plat, 2004). She could, instead, have committed suicide simply by swallowing the proper dose of poison.
2. The marks were made by Mark Anthony when dying in her arms. Realizing Cleopatra's lies and her role in having him take his own life, he could have bitten her in vengeance (Burstein, 2004). This could have happened if we see Mark Anthony as an angry, betrayed, and deceived emperor of the East Roman empire, and a victim of Cleopatra's charms.
3. The marks were made by instruments used by Cleopatra's maids, or her eunuch, or by Octavian's men to apply the poison by penetrating the skin. Because of her revulsion for pain and bodily imperfections, this would have been undertaken by force.
4. The marks were actually made by a snakebite, a less likely scenario.

Cleopatra had a deep understanding of poisons (Tsoucalas et al., 2013), as well as a personal physician trained in the Alexandrian School (Orland et al., 1990), and almost certainly possessed an apothecary of potions, poisons, and antidotes. Along with the poison, Cleopatra might have been able to take a drug to reduce the suffering. She might, as well, have developed a partial immunity to common poisons if she had been regularly self-administering a theriac. If this were the case, she may have required a more potent poison to successfully achieve suicide. Psychologically, Cleopatra saw herself as the woman who had almost magically controlled both Caesar and Mark Anthony, and must have felt that she could do the same with Octavian, albeit without seduction as the outcome. Thus, it is reasonable to assume that she was primed to take the situation in hand and would not have sacrificed her ego by giving in to Octavian.

Cleopatra was permitted by Octavian to conduct Mark Anthony's burial rites. It is possible that when she and her retinue were returning to the royal palace, someone had brought one or more snakes inside the walls. Later, she prepared a feast and asked for a royal bath before sitting down to her meal. She had ordered some figs, delivered in a container of some sort, which presented a second opportunity for snakes to be smuggled into her quarters, unnoticed by the guards (Burstein, 2004). Soon after Cleopatra's death, the tale of its being caused by a snakebite became an accepted myth in Rome (Burstein, 2004; Vergilius, 1825; Horatius, 1912) (Fig. 4.2). But was a eunuch or maid strong enough to carry a basket or water jug filled with large and heavy Royal

FIGURE 4.2 Jean-André Rixens. The death of Cleopatra. Musée des Augustins, Toulouse, 1874.

Cobras (each averaging 3–4 m, or 9.8–13 ft, length and weighing some 6 kg or 13 lbs), and camouflaged by figs? It is difficult to confirm this course of events.

More than one snake might have been needed because two of Cleopatra's handmaidens, Iras and Charmion, were also found dead. Curiously, there were no reports of shouts which might have been caused by pain, nor was swelling of the victim's body noted by the guards or Cleopatra's physician (Orland et al., 1990). The vipers' hemotoxic venom causes subcutaneous and intestinal hemorrhage, resulting in a brutal death. Both cobra and viper venom kill after 2–4 and 6–12 hours, respectively, quite a long time frame, as Strabo points out. Three women, Cleopatra, Iras, and Charmion, had died within a few minutes, from something which left no physical or pathological marks (Guillemain, 2009). Cleopatra, according to the snake poisoning theory, would have been bitten by the serpent, and after the envenomation would have had to be physically and emotionally capable of handing it, if a single snake, to her handmaiden, who, after receiving her own mortal bite, would hand it over, in turn, to the other handmaiden and/or to the eunuch. This scenario does not seem plausible even though a large Egyptian cobra is capable of inflicting a quick death with inconspicuous marks (Burstein, 2004). The fact that only 22%–30% of snakebite victims die after African cobra or viper envenomation (Espinoza, 2001; Davies, 1989) again argues against the likelihood of such a death for Cleopatra and her maids. We may also question where so large a snake or snakes quickly disappeared to, as Octavian and his men supposedly rushed to Cleopatra's windowless and sealed mausoleum. In fact, they arrived so quickly that the second handmaiden was still alive (Plat, 2004). Moreover, Octavian was familiar with Cleopatra's cunning and dexterous movements and had her heavily guarded (Burstein, 2004; Guillemain, 2009). This leads us to the final possible scenario: her murder by the ambitious and resourceful Octavian himself.

Octavian knew that his future would be more secure if he could totally remove Cleopatra and her political influence from Egypt. He understood that even if she were imprisoned, she would still pose a danger either by fomenting an outright revolt or by charming the next Roman ruler of the Egyptian province. If Octavian had chosen a public lynching or other humiliation, he would have diminished the nobility of Cleopatra as a mighty and royal queen, and he would have been seen as a strict, vindictive tyrant. Thus, for Octavian one way in particular would have held a great appeal: an injection of one of various poisons which would provide a quick and relatively painless death (Burstein, 2004). The Roman legions, during their campaigns, traveled with their personal physicians (Davies, 1989). A mixture of poisons, say of hemlock, opium, and aconite, could induce a deep sleep resulting in coma and death (Mihailidis, 2010). Octavian and his men might have injected Cleopatra with an instrument which would make it appear that she had been bitten by a snake. Cleopatra was a revered queen for most Egyptians. Death from the sacred Egyptian Royal Cobra might have been deemed almost romantic. Octavian had ample time to eliminate all evidence and fabricate a convincing story of whatever death scene he chose. He had both the motive, that is, to permanently remove the political threat of Cleopatra, and the know-how and time to murder her. The subsequent murder of Caesarion, the politically threatening offspring of Cleopatra and Julius Caesar, further strengthens the plausibility that she was murdered (Plat, 2004).

4.3 EPILOGUE

Cleopatra VII, Pharaoh of the Egyptians (Tsoucalas et al., 2013), ruler of the Egyptian empire, whisperer of powerful men, charming, well-educated, and a strong woman, died relatively young. Although we may never know for certain, the preponderance of evidence seems to suggest that rather than committing suicide with an asp, she was murdered. A proponent of Egyptian independence, for which she fought but which battle she lost, she became a victim of politics. Octavian quite possibly had her killed with a poisonous concoction, and to avoid turmoil in the streets of Egypt, presented the world with a trumped-up story about her suicide. After Cleopatra's demise, Octavian became known as Pharaoh. He was given unprecedented powers and named Augustus by the Senate. The Roman Republic became the Roman Empire, and Octavian was its first Emperor.

References

Burstein, S.M., 2004. The Reign of Cleopatra. Greenwood Press, London.

Caesar, G.J., Gardner, J.F., 1967. Caesar, the Civil War. Penguin Books, New York, NY.

Davies, R., 1989. Service in the Roman Army. Edinburgh University Press, Edinburgh.

Dimitsas, M., 1885. History of Alexandria. Palamidis, Athens.

Espinoza, R., 2001. In relation to Cleopatra and snake bites. Rev. Med. Chil. 129 (10), 1222–1226.

Galen. De theriaca ad pisonem [Kunh, CG, Trans.]. In: Claudii Galeni opera omnia. Hildesheim: G. Olms; 1964–1965.

Guillemain, B., 2009. Mort de Cléopâtre. Hist. Sci. Med. 43 (4), 369–373.

Heather, P., 2010. The Great Library of Alexandria? Library Philosophy and Practice, San Diego, CA.

Horatius, F.Q., 1912. Odes, Book I. Dimitrakos, Athens.

Maloney, W., 2010. The death of Cleopatra, a medical analysis of the theory of suicide by *Naja haje*. WebmedCentral Toxicol 8, WMC00502. Available from: http://www.webmedcentral.com/article_view/502.

Mihailidis, C., 2010. How Cleopatra Died? Eleyterotypia Tegopoulos, Athens.

Orland, R.M., Orland, F.J., Orland, P.T., 1990. Psychiatric assessment of Cleopatra: a challenging evaluation. Psychopathology 23 (3), 169–175.

Plat, I.M., 2004. Women Writers of Ancient Greece and Rome. Anthony Rowe Ltd, Chippenham.

Roller, D.R., 2010. Cleopatra, a Biography. Oxford University Press, New York, NY.

Seppard, S., 2006. Pharsalus 48 BC, Caesar and Pompey, Clash of the Titans. Osprey Publishing Ltd, Oxford.

Z. Skevos, War Kleopatra von Aegypten ein Arzt? [Was Cleopatra of Egypt a doctor?]. In: Archives internationales pour l'Histoire de la Médecine, Haarlem 1902.

Strabo, 1924. In: Jones, H.L. (Ed.), The geography of Strabo. Harvard University Press–William Heinemann Ltd, London.

Tsoucalas, G., Kousoulis, A.A., Poulakou-Rebelakou, E., Karamanou, M., Papagrigoriou-Theodoridou, M., Androutsos, G., 2013. Queen Cleopatra and the other "Cleopatras": their medical legacy. J. Med. Biogr. .

Tsoukalas, I., 2004. Paediatrics from Homer Until Today. Science Press, Thessaloniki–Skopelos.

Vergilius, P.M., 1825. Aeneis, Book VIII. Delalain, Paris.

Kohl Use in Antiquity: Effects on the Eye

Zafar A. Mahmood, Iqbal Azhar and S.W. Ahmed

Department of Pharmacognosy, Faculty of Pharmacy and Pharmaceutical
Sciences, University of Karachi, Karachi, Pakistan

5.1 INTRODUCTION

Exploring the history of ancient civilizations is still relevant today, helping us learn valuable lessons regarding both their achievements and their failures. Some of the oldest civilizations had a surprisingly sophisticated knowledge of science, but how they reached such a level of understanding remains a mystery. The history of ophthalmology and ophthalmic products provides one example of diverse classical preparations still used in alternative systems of medicines in different parts of the world. The use of kohl throughout history illustrates that many ancient civilizations had sound knowledge of science. Over the centuries, kohl was used under different names and both men and women of all socioeconomic levels made use of it.

Comprehensive details relating to the eye and its diseases, including eye treatments in ancient Egypt, have been amply documented in various scientific reviews, highlighting the application and role of lead-based eye preparations, kohl (Anderson, 1997; Tapsoba et al., 2010). However, no specific reviews or research articles have yet been published authenticating lead poisoning in ancient Egypt or in any other ancient civilization. If we assume that use of kohl was pervasive in ancient Egypt, can we conclude that everyone in Egypt was a victim of lead poisoning? Certainly not. Scientists still need more data to explore the benefits and toxicity of kohl. The questions most people ask today relate to its constituents, dating, mechanism, and effect; that is, what kind of materials were used to manufacture kohl, when and where was it made, how does it work and what effects does its application produce in the eyes.

The exact composition of kohl has long been an important topic and a matter of dispute within the scientific community. German and French scientists have played a prominent role in authenticating the exact chemical nature of kohl through chemical analysis, electron microscopy, and X-ray diffraction (Tapsoba et al., 2010; Fischer, 1892; Walter et al., 1999; Martinetto et al., 2000; Habibullah et al., 2010). During Egyptian rule, galena was known by the name *mestem* or *stim*; the latter word is identical to the Greek, *stimmi* or *stibi*, and to the Latin *stibium*, meaning antimony (Fischer, 1892). Therefore, some authors have misinterpreted or, rather, mixed these words and reported antimony as the active ingredient, instead of lead sulfide. The controversy regarding the chemical composition of kohl was finally resolved after the publication of Professor V. X. Fischer's research article (Fischer, 1892) in which he analyzed 30 samples of ancient Egyptian eye preparations (kohls) obtained from Fayum (Egypt) and demonstrated that galena was indeed the chief constituent. In view of this clear evidence, there can be no doubt that the word *stibium* referred primarily to galena (lead sulfide) and not to antimony (Lucas and Harris, 2012; Mahmood et al., 2009). Another analytical report, this one by French scientists (Walter et al., 1999), confirmed the presence of galena and other lead salts after analyzing a huge number of samples (dating from between 2000 and 1200 BC) that were preserved in their original containers. The crystallographic and chemical analysis indicated the presence of galena (PbS), along with some quantity of cerussite ($PbCO_3$) and two synthetic products, laurionite ($PbOHCl$) and phosgenite ($Pb_2Cl_2CO_3$), reflecting the Egyptians' extensive knowledge of "wet" chemistry.

Galena is found near the Red Sea, Aswan, and the Eastern Desert at Gebel-el-Zeit, and Gebel Rasas, also known as Lead Mountain (Pauline, 2007). Fig. 5.1 pictures the Egyptian, Sinai, and Arab Peninsula and Galena's possible trade route into Egypt (North Africa) and the Middle East (Catherine, 2005). No evidence of antimony mines was found in

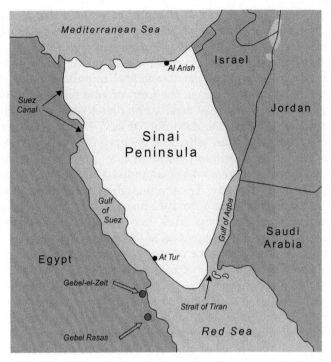

FIGURE 5.1 Showing location of Jabel Rasas and Gebel-el-Zeit on the Red sea coast.

Egypt, Sinai, Saudi Arabia, or Iran. Instead, antimony mines have been identified in Macedonia, Turkey, and Armenia. Apparently, the galena used to prepare kohl in both the Egyptian and the Arabian Peninsula was brought from the two large mines located in Egypt. There is little chance that antimony brought from Macedonia, Turkey, or Armenia was used by the Egyptians and Arabs to manufacture kohl. This further supports the earlier statement that antimony was mistaken for galena and that galena was the actual ingredient of kohl, which its manufacturers still use today.

Studies of the medicinal properties of natural substances used during medieval and Ottoman times also show that galena was used to cure eye diseases (Lev, 2002). Evidence relating to the composition of kohl is cited in *The Encyclopaedia of Islam* (Bosworth et al., 1986) and in *Medieval Islamic Civilization—An Encyclopaedia* (Meri, 2006). The various literatures of the time report differing opinions as to the relative benefits and toxicity of kohl. The scientific community posits that an intellectual war arose between two schools of thoughts. The first proposed that since lead is toxic, it was likely hazardous for the human body and should not be used even in eye preparations, regardless of the type of lead (organic or inorganic), and its physicochemical behavior and route of application. The second school of thought maintained that lead toxicity relates to organic lead (such as tetraethyl lead or tetramethyl lead) or some soluble inorganic lead salts. Moreover, it was believed that the inorganic lead in galena was an insoluble lead salt; it had been used for thousands of years owing to its biomedical importance and was not toxic when applied to the eyes in the form of kohl because of its physicochemical nature and mode of action. The first school of thought based its premises primarily on the detection of lead sulfide in various kohl preparations collected from different geographical regions. As additional support for their perspective, members of this school stated that if kohl were misused, there could be an indirect relationship between its application and lead toxicity. Thus, although some studies on children have concluded that kohl may produce toxicity, they have not been controlled studies, nor have they taken into account the environmental, nutritional, and other relevant factors of the region and people.

In contrast, the second school of thought contends that there is a scientific basis for the application of kohl, a centuries-old preparation, and that it certainly has biomedical importance when applied to the eyes. In support of this belief, a number of controlled research trials conducted on both humans and animals have been published suggesting that kohl applied in the eyes does not increase blood lead levels, nor do the studies show that kohl produces any toxicity. There are some reports (Grant and Schuman, 1993) of minute conjunctival abrasion when lead sulfide (galena) is applied in the form of an eye preparation (kohl/surma), possibly due to substandard products having larger particle size, but no toxic injury. An extensive literature search done to investigate the issue of kohl's toxicity has been reported in a review article (Mahmood et al., 2009). It should be noted that, as a protective agent, the lead in kohl promotes the production of nitric oxide (diatomic free radicle), which is known to boost the immune system's response to infection.

Kohl has been closely associated with almost all human civilizations. Its use dates back to the Bronze Age (3500–1100 BC), and it is even mentioned in the Old Testament (see Kings II 9:30 and Ezekiel 23:40, particularly the reference to "painted eyes"). The word "kohl" is Arabic in origin; Arab oculists called it Kahal (Sweha, 1982). It was accepted by people of many ancient civilizations, including the Sumerians (3500–1950 BC in Iraq), Egyptians (3050–30 BC), Greeks (1550–100 BC), Romans (753 BC–AD 476), Chinese (2100 BC–1911 AD), Japanese (1800–1500 BC), Phoenicians (1200–146 BC in Lebanon), Persians (569–330 BC), Indians (1500 BC), and Muslims (AD 641). Its use continued right through to the Coptic Period (the phase of Christian Egyptian culture) which lasted from the end of the Roman Period (the end of the 3rd century ad) to the coming of Islam ad 641. Kohl is indeed one of the most ancient ophthalmological preparations known to humans.

The effects of kohl on the eyes have been reviewed by many research workers during the last 50 years. The classical views of two schools of thoughts have been described above. For a more detailed explanation, we have examined two of the most important effects of kohl: the protective effect of kohl against UV radiation from the sun and its antimicrobial action for both therapeutic and prophylactic purposes.

5.2 PROTECTIVE EFFECT AGAINST UV RADIATION

The UV absorptive property of galena in the deserts of the Sinai-Egyptian and Arabian Peninsula has been highlighted by many researchers. Galena's black, shiny particles have been reported to screen the eyes from the brilliance and reflection of sunlight and thus protect the eyes from the harmful effect of the sun's UV rays and the flies in the deserts (Heather, 1981; Cohen, 1999; Kathy, 2001; Cartwright-Jones, 2005). A series of reviews and studies also document galena's solar absorption properties (Nir et al., 1992; Pop et al., 1997), thus supporting the ancient civilization's application of kohl to protect their eyes from sun, especially in the deserts of the Sinai-Egyptian and Arab Peninsula. Scientific evidence regarding the absorption and transmittance rate of sunrays by lead sulfide is available and can be used to correlate the solar protective property of kohl when applied to the eyes (Nir et al., 1992; Pop et al., 1997). The light absorption spectrum of a thin film of lead sulfide prepared on indium tin oxide is reported to be high and low in transmittance in the UV band, which further increases with deposition voltage (Li-Yun et al., 2008). This implies that lead sulfide's thin film will have higher absorption and lower transmittance in the UV light band. Therefore, when kohl (which is made up of lead sulfide) is applied to the eyes as a thin film, it should react similarly, thus absorbing the sun's UV light and protecting the eyes from its harmful effects.

Lead sulfide has been reported to be an important direct narrow-gap semiconductor material with an energy band gap of -0.4 eV at 300K and a relatively large excitation Bohr radius of 18 nm. These properties also make lead sulfide suitable for infrared (IR) detection applications (Gadenne et al., 1989). These findings offer reasonable justification to conclude that kohl containing lead sulfide as a major ingredient has a natural protective effect against the sun's glare when applied to the eyes in the form of kohl and thus support claims and uses reported elsewhere. The role of other ingredients of kohl was also investigated. Some interesting formulations reflecting the benefits of these ingredients for the eyes are reported elsewhere in the literature. For example, zinc oxide was probably used in kohl because of its powerful natural sunblock property (Mitchnick et al., 1999), and it may enhance the protective capacity of galena against the glare of the sun. Interestingly, zinc oxide is a modern sunscreen ingredient. Neem (*Azadirachta indica*) is very well known worldwide for its astringent and antibacterial properties (Almas, 1999; Linda, 2001). Like silver leaf, neem also possesses antiviral activity (Badam et al., 1999).

5.3 ANTIMICROBIAL ACTION AND BIOMEDICAL IMPORTANCE

Although the antimicrobial action of kohl was known for centuries, inasmuch as the Egyptians used it to protect against eye infection, recently a more scientific approach has been launched to establish kohl's antimicrobial action. French researchers have reported that the heavy kohl-based eye makeup that the ancient Egyptians used for centuries may actually have had some medical benefits. At low dose, the specially made lead compounds actually boost the immune system by stimulating production of nitric oxide (Tapsoba et al., 2010). During the last 30 years, nitric oxide has been recognized as an extremely versatile agent in the immune system (Bogdan, 2001). At the time of its discovery (1985–1990), nitric oxide was simply defined as a product of macrophage activated by cytokines. With the passage of time, however, its role has now been broadened to include antimicrobial (DeGroote et al., 1999; Nathan and Shiloh, 2000; Bogdan et al., 2000), antitumor (Nathan, 1992; Bogdan and Mayer, 2000; Xie et al., 1996; Pervin et al., 2001; Bauer et al., 2001), and antiinflammatory−immunosuppressive activity (Bogdan, 2001; Bogdan and Mayer, 2000; Henson et al., 1999; Spiecker et al., 1998; Dalton et al., 2000; Bobe et al., 1999; Tarrant et al., 1999; Shi et al., 2001; Allione et al., 1999; Angulo et al., 2000). Nitric oxide is reported to possess broad-spectrum antibacterial activity. This property is based primarily on two reactive by-products, peroxynitrite ($ONOO^-$) and dinitrogen trioxide (N_2O_3). Nitric oxide is quite effective on both Gram-positive and Gram-negative bacteria, including methicillin-resistant *Staphylococcus aureus* (DeGroote et al., 1999). It is now well established that nitric oxide (NO) is an endogenous cell-signaling molecule of fundamental importance in physiology, and so it has become the subject of considerable scientific interest in recent years. There is evidence that certain diseases are related to a deficiency in the production of nitric oxide. This creates the possibility of developing new drug treatments that can donate nitric oxide when the body cannot generate sufficient amounts to permit normal biological functions. NO and other molecules involved in NO-mediated signaling are present in ocular tissues. Studies have shown that topical or systemic administration of classic NO-donors (nitroglycerine, isosorbide dinitrate) in patients reduce intraocular pressure (IOP), supporting a role for nitric oxide in regulating IOP (Kotikoski et al., 2002; Schuman et al., 1994; Nathanson, 1992; Wizemann and Wizemann, 1980). This finding is of particular interest in the potential treatment of glaucoma, which is often associated with an increased IOP and can lead to blindness if not treated.

The antimicrobial activity of nitric oxide has been shown to take place through several mechanism. This activity may be the result of DNA mutation; inhibition of DNA repair and synthesis or inhibition of protein synthesis; alteration of proteins by S-nitrosylation; adinosine diphosphate (ADP)-ribosylation or tyrosine nitration or inactivation of enzymes by disruption of Fe−S clusters, zinc fingers, or heme groups; or by peroxidation of membrane lipids (Bogdan, 2001). Unfortunately, nitric oxide cannot as yet be used as a drug due to its very short half-life (only a couple of seconds). However, scientists are working on delivering this chemical through nanotechnology. Nitric oxide is soluble in both aqueous and lipid media and thus readily diffuses through the cytoplasm and plasma membranes. Fig. 5.2 is based on studies by French scientists and others, and reports on heavy metal ions, such as Pb^{2+} as well as the release of nitric oxide and its effect on eyes. Amazingly, many scientists now believe that the ancient Egyptians were aware of the beneficial effects of galena-based kohl.

After the publication of the paper by the French scientists (Tapsoba et al., 2010), a number of web-based reviews and one review article published opposing opinions. It is of course very difficult to unequivocally demonstrate that the ancient Egyptians were familiar with the role of NO, when kohl is applied to the eyes therapeutically, but several papers (Anderson, 1997; Walter et al., 1999; Martinetto et al., 2000; Habibullah et al., 2010; Lucas and Harris, 2012; Mahmood et al., 2009) have indicated that the Egyptians were quite aware of various diseases and their treatment. Even today, some pharmaceutical companies manufacture ophthalmic products based on nitric oxide activity.

According to ancient Egyptian manuscripts, lead-based eye preparations were essential remedies for treating eye illness. This conclusion seems astonishing to us today when we consider the well-recognized toxicity of lead salts. French scientists, using microelectrodes, obtained new insight into the biochemical interactions between lead(II) ions and cells, which support the ancient medical use of insoluble or sparingly soluble lead compounds. In Tapsoba's study (Tapsoba et al., 2010), it was reported that a submicromolar concentration of Pb^{2+} ions is sufficient to elicit a specific oxidative stress response of keratinocytes, leading to overproduction of NO. Based on NO's established biological role in stimulating nonspecific immunological defenses, it can be concluded that lead-based eye preparations (kohl) were manufactured and used in ancient Egypt to prevent and treat eye illnesses by supporting the immune system (Tapsoba et al., 2010).

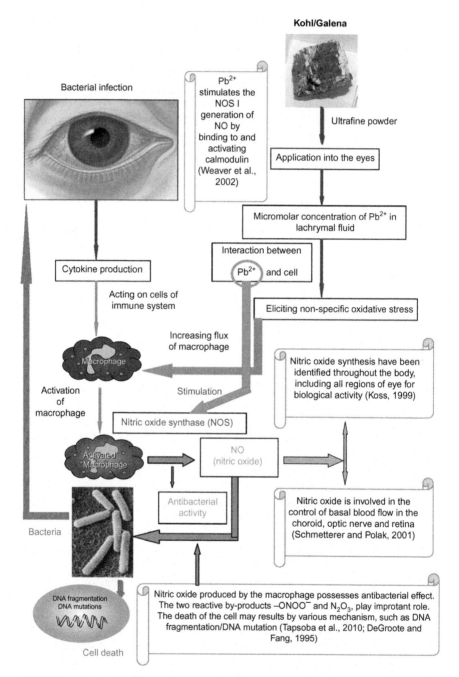

FIGURE 5.2 Role Pb^{2+} from Kohl in the biosynthesis of nitric oxide for antibacterial and other effects.

References

Allione, A., Bernabei, P., Bosticardo, M., Ariotti, S., Forni, G., Novelli, F., 1999. Nitric oxide suppresses human T lymphocyte proliferation through IFN-gamma-dependent and IFN-gamma-independent induction of apoptosis. J. Immunol. 163, 4182–4191.

Almas, K., 1999. The antimicrobial effects of extract of *Azadirachta indica* (neem) and *Salvadora persica* (arak) chewing sticks. Indian J. Dent. Res. 10 (1), 23–26.

Anderson, S.R., 1997. History of Ophthalmology, The eye and its diseases in ancient Egypt. Acta Ophthalmol. Scand. 75, 338–344.

Angulo, I., Federico, G., José, F.G., Domingo, G., Angeles, M.F., Fresno, M., 2000. Nitric oxide producing CD11b$^+$ Ly-6G(Gr-1)$^+$CD31(ER-MP12)$^+$ cells in the spleen of cyclophosphamide-treated mice: implications for T cell responses in immunosuppressed mice. Blood. 95, 212–220.

Badam, L., Joshi, S.P., Bedekar, S.S., 1999. In vitro antiviral activity of neem (*Azadirachta indica* A. Juss) leaf extract against group B coxsackieviruses. J. Commun. Dis. 31 (2), 79–90.

Bauer, P.M., Fukuto, J.M., Pegg, A.E., Ignarro, L.J., 2001. Nitric oxide inhibits ornithine decarboxylase via S-nitrosylation of cysteine 360 in the active site of the enzyme. J. Biol. Chem. 276, 34458–34464.

Bobe, P., Karim, B., Danièle, G., Paule, O., Linda, L., Roger, H., 1999. Nitric oxide mediation of active immunosuppression associated with graft-versus-host reaction. Blood. 94, 1028–1037.

Bogdan, C., 2001. Nitric oxide and the immune response. Nat. Immunol. 2 (10), 907–916.

Bogdan, C., Mayer, B., 2000. The function of nitric oxide in the immune system. In: Mayer, B. (Ed.), Handbook of experimental pharmacology. Nitric Oxide. Springer, Heidelberg, pp. 443–492.

Bogdan, C., Rollinghoff, M., Diefenbach, A., 2000. Reactive oxygen and reactive nitrogen intermediates in innate and specific immunity. Curr. Opin. Immunol. 12, 64–76.

Bosworth CE, Donzel EV, Lewis B, Pellat C. The Encyclopedia of Islam. Prepared under the patronage of the International Union of Academics. 1986;5:356–57.

Cartwright-Jones, C., 2005. Introduction to Harquus: Part 2: Kohl as traditional women's adornment in North Africa and the Middle East. TapDancing Lizard Publications, Stow, Ohio.

Catherine, C.J., 2005. Kohl as traditional women's adornment in North Africa and Middle East. Introduction to Harquus: Part 2. Kohl, Stow, Ohio, pp. 1–9.

Cohen M. Cosmetics and perfumes, Egypt. 10,000 BCE 1999.

Dalton, D.K., Haynes, L., Chu, C.Q., Swain, S.L., Wittmer, S., 2000. Interferon-γ eliminates responding CD4 T cells during mycobacterial infection by inducing apoptosis of activated CD4 T cells. J. Exp. Med. 192, 117–122.

DeGroote, M.A., Fang, F.C., Fang, F.C., 1999. Antimicrobial properties of nitric oxide. In: Fang, F.C. (Ed.), Nitric Oxide and Infection. Kluwer Academic/Plenum Publishers, New York, NY, pp. 231–261.

Fischer, V.X., 1892. The chemical composition of ancient Egyptian eye preparations (English Translation) Arch. Pharm. 230, 9–38.

Gadenne, P., Yagil, Y., Deutscher, G., 1989. Transmittance and reflection in-situ measurements of semicontinuous gold film during deposition. J. Appl. Phys. 66, 3019.

Grant, W.M., Schuman, J.S., 1993. Toxicology of the Eye, fourth ed. Charles C Thomas, Springfield, IL, pp. 682–685.

Habibullah, P., Mahmood, Z.A., Sualeh, M., Zoha, S.M.S., 2010. Studies on the chemical composition of kohl stone by X-ray defractometer. Pak. J. Pharm. Sci. 23 (1), 48–52.

Heather, C.R., 1981. Art of Arabian Costume—A Saudi Arabian Profile. Information on Kohl Application. Arabesque Commercial, Saudi Arabia.

Henson, S.E., Nichols, T.C., Holers, V.M., Karp, D.R., 1999. The ectoenzyme γ-glutamyl transpeptidase regulates antiproliferative effects of S-nitrosoglutathione on human T and B lymphocytes. J. Immunol. 163, 1845–1852.

Kathy, C., 2001. An A to Z of Places and Things Saudi. Published by Stacey International, London, UK.

Kotikoski, H., Alajuuma, P., Moilanen, E., 2002. Comparison of nitric oxide donors in lowering intraocular pressure in rabbits: role of cyclic GMP. J. Ocul. Pharmacol. Ther. 18, 11–23.

Lev, E., 2002. Reconstructed materia medica of the Medieval and Ottoman al-Sham. J. Ethnopharmacol. 80, 167–179.

Linda, S.R., 2001. Mosby's Handbook of Herbs & Natural Supplements. Mosbey.

Li-Yun, C., Wen, H., Jian-Feng, H., Jian-Peng, W., 2008. Influence of deposition voltage on properties of lead sulfide thin film. Am. Ceram. Soc. Bull. 87 (6), 9101–9104.

Lucas, A., Harris, J., 2012. Ancient Egyptian Materials and Industries. Courier Dover Publications, Mineola, New York, USA.

Mahmood, Z.A., Zoha, S.M.S., Usmanghani, K., Hasan, M.M., Ali, O., Jhan, S., 2009. Kohl (Surma): retrospect and prospect. Pak. J. Pharm. Sci. 22 (1), 107–122.

Martinetto, P., Anne, M., Dooryhee, E., Tsoucaris, G., Walter, P., Creagh, D.C., et al., 2000. A synchrotron X-ray diffraction study of Egyptian cosmetics. In: Creagh, D.C., Bradley, D.A. (Eds.), Radiation in Art and Archeometry. Published by Elsevier, Amsterdam, The Netherlands, pp. 297–316.

Meri, JW., 2006. Cosmetic. In: Medieval Islamic Civilization—An Encyclopedia, vol. 1., Mayflower Books, New York, USA, p. 177.

Mitchnick, M.A., Fairhurst, D., Pinnell, S.R., 1999. Microfine zinc oxide (Z-Cote) as a photostable UVA/UVB sunblock agent. J. Am. Acad. Dermatol. 40, 85–90.

Nathan, C., 1992. Nitric oxide as a secretory product of mammalian cells. FASEB J. 6, 3051–3064.

Nathan, C., Shiloh, M.U., 2000. Reactive oxygen and nitrogen intermediates in the relationship between mammalian hosts and microbial pathogens. Proc. Natl. Acad. Sci. U.S.A. 97, 8841–8848.

Nathanson, J.A., 1992. Nitrovasodilators as a new class of ocular hypotensive agents. J. Pharmacol. Exp. Ther. 260, 956–965.

Nir, A., Tamir, A., Zelnik, N., Iancu, T.C., 1992. Is eye cosmetic a source of lead poisoning? Isr. J. Med. Sci. 28 (7), 417–421.

Pauline WT., 2007. Ancient Egyptian Costume History, Part 6—Ancient Egyptian Make Up and Cosmetics, ⟨www.fashion-era.com⟩.

Pervin, S., Singh, R., Chaudhuri, G., 2001. Nitric oxide-induced cytostasis and cell cycle arrest of a human breast cancer cell line (MDA-MB-231): potential role of cyclin D1. Proc. Natl. Acad. Sci. U.S.A. 98, 3583–3588.

Pop, I., Nascu, C., Lonescu, V., 1997. Structural and optical properties of PbS thin films obtained by chemical deposition. Thin Solid Films. 307, 240–244.

Schuman, J.S., Erickson, K., Nathanson, J.A., 1994. Nitrovasodilator effects on intraocular pressure and outflow facility in monkeys. Exp. Eye. Res. 58, 99–105.

Shi, F.D., Flodström, M., Kim, S.H., Pakala, S., Cleary, M., 2001. Control of the autoimmune response by type 2 nitric oxide synthase. J. Immunol. 167, 3000–3006.

Spiecker, M., Darius, H., Kaboth, K., Hübner, F., Liao, J.K., 1998. Differential regulation of endothelial cell adhesion molecule expression by nitric oxide donors and antioxidants. J. Leukocyte Biol. 63, 732–739.

Sweha, F., 1982. Kohl along history in medicine and cosmetics. Hist. Sci. Med. 17 (2), 182–183.

Tapsoba, I., Arbault, S., Walter, P., Amatore, C., 2010. Finding out Egyptian God's secret using analytical chemistry: biomedical properties of Egyptian black makeup relealed by amperometry at single cells. Anal. Chem. 82, 457–460.

Tarrant, T.K., Silver, P.B., Wahlsten, J.L., Rizzo, L.V., Chan, C.C., Wiggert, B., 1999. Interleukin-12 protects from a Th1-mediated autoimmune disease, experimental autoimmune uveitis, through a mechanism involving IFN-γ, nitric oxide and apoptosis. J. Exp. Med. 189, 219–230.

Walter, P., Martinetto, P., Tsoucaris, G., Brniaux, R., Lefebvre, M.A., Richard, G., 1999. Making make-up in ancient Egypt. Nature. 397, 483–484.

Wizemann, A.J.S., Wizemann, V., 1980. Organic nitrate therapy in glaucoma. Am. J. Ophthalmol. 90, 106–109.

Xie, K., Dong, Z., Fidler, I.J., 1996. Activation of nitric oxide gene for inhibition of cancer metastasis. J. Leukocyte Biol. 797, 797–803.

6

Nicander, *Thêriaka*, and *Alexipharmaka*: Venoms, Poisons, and Literature

Alain Touwaide[1,2]

[1]Brody Botanical Center, The Huntington Library, Art Collection, and Botanical Gardens, San Marino, CA, United States [2]Institute for the Preservation of Medical Traditions, Washington, DC, United States

The *Thêriaka* and *Alexipharmaka*, by the otherwise not-well-known ancient Greek author Nicander of Colophon, are probably the most paradoxical works in the history of toxicology (Gow and Schofield, 1953). They have been transmitted by more than 40 manuscripts—one of which dates as far back as the 10th century and is richly illustrated—with multiple editions, translations, and commentaries by Renaissance physicians and scholars. The first of these was printed as early 1499 by the famous scholarly printer Aldo Manuzio (1449–1515) of Roman origin established in Venice, at the very beginning of Greek scholarly printing, and repeated printed editions by modern Western classicists, including innumerable erudite publications, aimed to decipher and interpret the text, sometimes cryptic, of these two works.

The *Thêriaka* and *Alexipharmaka* are two poems of 958 and 630 verses, respectively. The former is devoted to venoms and the latter to other poisons. A word of explanation about the use of these words is in order. *Venom* usually refers to a poisonous substance requiring an animal delivery system (for example, snake venoms via biting). On the other hand, *poison* more generally refers to a substance causing harm typically by ingestion, although other routes of exposure, such as absorption through the skin are also possible. Thus, plants such as poison ivy, minerals such as cinnabar (due to its mercury content), and even parts of animals, such as fugu fish, may correctly be called *poisons*.

The verses of Nicander's two poems reproduce the meter of the *Iliad* and the *Odyssey*, which consist of hexameters (or dodecasyllable verses). Whatever the exact period of the composition of the *Thêriaka* and *Alexipharmaka* (possibly the later 2nd century BCE, but this is a matter of debate, as we shall see), such verses gave an archaic tone to Nicander's poems. It certainly sounded familiar to the Greek readers of Nicander in antiquity since the *Iliad* and the *Odyssey* formed the backbone of ancient education and every educated Greek memorized them at an early age. Nevertheless, the use of such verses conveyed a sense of solemnity, majesty, and power—if not of inexorability and fatality—that strongly contrasted with the content of the two poems and the description of cases of envenomation and poisoning.

6.1 THE *THÊRIAKA*

The *Thêriaka* is about venomous animals, the effects of their venoms on humans, and the treatment of envenomation. The title refers to wild animals (*thêr* in ancient Greek) and is an adjective in neuter, plural, functioning as a substantive meaning "the things about wild animals." The term *thêr* was normally used in classical Greek to refer to a wide variety of wild animals, from bears to wolves. Common to all of them was their possible noxiousness to humans. This notion justifies the use of the term *thêr* in the field of toxicology, but with the understanding that the possible harm arose from venom of the animals in question rather than the carnivorous qualities of creatures such as bears and wolves. Indeed, the toxicology of the time concerned itself with rather small animals such as scorpions, spiders, and insects.

The subjects of the *Thêriaka* are not limited to such obviously dangerous animals as those just mentioned (to which snakes of all kinds must be added). They also include other animals like shrews, lizards, and fishes. The common denominator in all such animals is the presence (real or supposed) of both an anatomical structure that makes it possible to deliver a substance by transcutaneous injection, and a liquid that becomes harmful when injected into the human physiological system.

The anatomical structures functioning as needles may be the fangs of snakes, spiders, and shrews; the stings of scorpions, stingrays, and bees; or the acerate, needlelike dorsal fins of fishes. These structures allow for the delivery of substances whose effects can range from irritant to lethal (Fig. 6.1).

In some cases, there was no such poisonous substance (for example the shrew). But most probably a pathogen of unknown nature, but certainly infectious, vehicled by the animal contaminated by its environment and possibly poor sanitary conditions.

In this context, *Thêriaka* are the venomous animals identified as a group in the natural world. This group is not defined as a genus in terms of modern taxonomy, but rather by its anatomical structure, which allows for the injection of venom in most cases and of a pathogenic substance in others. In spite of this overarching definition, most of the *Thêriaka*—no less than 350 verses, or 40% of the whole work—is devoted to snakes.

The composition of the *Thêriaka* is clear. After the brief narration of two myths accounting for the creation of snakes and scorpions in the Greek legends accounting for the world's history (verses 8—12 and 13—20, respectively), Nicander briefly mentions the activities of humans in the countryside that expose them to the risk of envenomation

FIGURE 6.1 A venomous snake described by Nicander (Paris, Bibliothèque nationale de France, manuscript supplementum graecum 647, 10th century).

(verses 31–34). This is a pretext for him to introduce a tableau and to evoke country life in the manner of Greek literature, particularly of the Hellenistic period, best illustrated by the bucolic *Idylls* of the Alexandrine poet Theocritus (3rd century BC). We thus see farmers collecting grain, spending the night in the open air and looking at the sky and its constellations, or relaxing and sipping a cup of wine. Nicander then moves to the prevention of envenomation, mainly through fumigations with odoriferous substances that are supposed to repel snakes (verses 35–79), when farmers and lumberjacks were spending the night in nature, and body creams with the same objective (verses 80–114), not only for farmers and other workers in the fields and the wild, but also for people enjoying a walk in nature.

Once these preambles are finished, Nicander examines several cases of snake envenomation. Again, he opens this section with some tableaux, taking their substance from the description of the annual life cycle of snakes, their sloughing and breeding seasons (including the classical myth of female vipers killing their mates after coupling), and their supposed higher irritability (and hence the increased noxiousness of their venom) during scorching summer heat (verses 115–144). He then studies 15 snakes; for each one, he describes its habitat, body, and any other ethological characteristic, followed by a precise enumeration of the symptoms caused by envenomation (verses 145–492). As an example of Nicander's way of presenting symptoms, below are two different cases, including a description of the anatomical structure of the jaw:

> Consider now the murderous asp … in the violence of its wrath it fastened death upon wayfarers who meet it. It has four fangs, their underside hollow, hooked, and long, rooted in its jaws, containing poison, and at their base a covering of membranes hides them. Thence it belches forth poison unassuageable on a body. Be they no friends of mine whose heads these monsters assail. For no bite appears on the flesh, no deadly swelling with inflammation, but the man dies without pain, and a slumberous lethargy brings life's end **Thêriaka, *vv. 157, 180–189.***
>
> You would do well to mark the various forms of the viper … two fangs in his upper jaw, as they spit poison, leave their mark upon the skin, but of the female always more than two, for the lays hold with her whole mouth, and you can easily observe that the jaws have opened wide about the flesh.
>
> And from the wound she makes there oozes a discharge like oil or, it may be, bloody or colorless, while the skin around starts up into a painful lump, often greenish, now crimson, or again of livid aspect. At other times it engenders a mass of fluid, and about the wound small pimples like slight blisters rise flabbily from the skin, which looks scorched. And all around spread ulcers, some at a distance, others by the wound, emitting a dark blue poison; and over the whole body the piercing bane eats its way with its acute inflammation; and in the throat and about the uvula retchings following fast upon one another convulse the victim. The body is oppressed also with failures of sense in every part, and forthwith in the limbs and loins is seated a burdening, dangerous weakness, and heavy darkness settles in the head. Meantime the sufferer at one moment has his throat parched with dry

thirst, often too he is seized with cold from the finger-tips, while all over his frame an eruption with wintry rage lies heavy upon him. And again a man often turns yellow all over his body and vomits up the bile that lies upon his stomach, while a moist sweat, colder than the falling snow, envelops his limbs. In some cases his color is that of sombre lead, in others his hue is murky, or again it is like flowers of copper **Thêriaka**, *vv. 209, 231–257.*

After this zoological-pathological section, Nicander devotes no less than 220 verses (493–713) to the treatment of the 15 cases of envenomation that he has just presented. He cites the ingredients for 23 remedies and explains how to prepare them. Strangely enough, he does not specify for which exact snake envenomation each remedy is used, whereas in the description of the symptoms following the different cases of envenomation, he makes clear the difference between hemolytic and neurotoxic venoms.

Demonstrating that his major focus is on snakes, Nicander analyzes all the other venomous animals in just 120 verses (715–836): spiders (715–768); scorpions (769–804); some insects (805–814); the shrew (815–816), which is supposed to be venomous but probably just propagates contagious diseases; lizards (817–821); and fishes (822–836).

The poem ends with a second section on therapeutics (verses 837–956). In no particular order, Nicander describes first and at great length a remedy made of 82 substances (verses 837–914). He then moves to emergency measures (verses 915–933), which include sucking the venom from the wound or applying leeches with the same purpose, applying cups and cauterization to aspire or burn the venom and close the wound, applying a tourniquet to prevent the diffusion of the venom in the cardiovascular system, and using irritant botanicals aimed at provoking hyperemia to the same effect. The poems conclude with the description of a panacea made of 24 drugs.

6.2 THE ALEXIPHARMAKA

This second poem is substantially shorter than the *Thêriaka*, with just 630 verses. Its topic has not always been correctly understood. In the literature, it is not infrequent to find statements that the *Alexipharmaka* deals with the antidotes to venoms and poisons. Actually, the poem is devoted to poisons and poisoning and their treatment, and it represents the second part of the diptych that always constituted toxicology in antiquity, with the venoms on one side and the poisons on the other.

The meaning of the word *Alexipharmaka* is still unclear. It is manifestly composed of two elements, the second of which is *pharmakon*. In spite of its apparently obvious interpretation referring to a *remedy*, this term was used in antiquity to refer to both a remedy and a poison. Its exact meaning was defined by using a determinant [such as *esthlon* (salutary) or *kakon* (deadly)] that helped clarify its fundamental ambiguity. This contradictory use indicates that the overarching meaning of *pharmakon* is neither "medicine" nor "poison," but rather "a substance introduced into the body (or applied to it) in order to provoke a modification." As for the first component of *Alexipharmaka*, *alex-*, it is of uncertain origin and meaning, although it clearly conveys the meaning of "treating" or "healing." The association of the two elements, *alex-* and *pharmakon*, may signify both the treatment (=*alex-*) of the deadly substances introduced into, or in contact with, the human body, or healing medicines (=*pharmakon*). Used as the title of this poem, it refers to the treatment of the finest effects of substances introduced into, or making contact with, the human body. This is all the more true because, if the *Alexipharmaka* is defined in that way, it completes the *Thériaka* and constitutes the second half of the ancient toxicological discourse. In this way, the study of toxicology has two elements: venoms (injected by animals) and poisons (taken orally or through contact with the skin). The two parts make a coherent set of works on toxicology.

The structure of the *Alexipharmaka* is much more directly perceptible than that of the *Thériaka*. It deals with 23 poisons from the three natural kingdoms (vegetable, mineral, and animal) and devotes a section to each of them, with three elements: the description of the solution containing the poison, that of the symptoms following its absorption per os, and finally, a list of remedies and their preparation.

These poisons are the following (cited below in the order in which they appear in the text):

Aconite;
Gypsum;
White lead;
Blister-beetles;
Coriander;
Hemlock;
Arrow-poison (unidentified);
Meadow-saffron;
Chamaeleon-thistle;
Bull's blood;
Buprestis;
Clotted milk;
Thorn-apple;
Pharicum (unidentified);
Henbane;
Opium;
Sea-hare;

FIGURE 6.2 Two poisonous animals mentioned by Nicander (Paris, Bibliothèque nationale de France, manuscript supplementum graecum 647, 10th century).

Leeches (ingested);
Fungi (poisonous);
Salamander (mucous secretion ingested) (Fig. 6.2);
Toads (idem);
Litharge;
Yew.

6.3 THE NICANDREAN QUESTION

The Homeric style of the two poems is not limited to their form (i.e., each consists of 12 syllables); they both imply the epic mode, with its feeling of uniformity, grandeur, and, at the same time, tragedy. In this view, the description of the symptoms that follow the envenomation or the poisoning is not strictly medical in nature—even though it uses medical terminology, sometimes specific and sometimes not. Rather, it transformed toxicology into a grandiose epic battle between humans and death. The victims become heroes, fighting, with all their strength and energy, a desperate battle against a treacherous enemy, often hidden in the depth and the dark and attacking without a break, in spite of the repeated administration of remedies.

Just as the Homeric epic poem sometimes moves from the battlefield to the tents of the warriors on the plain of Troy and details the daily life of the soldiers, Nicander's narrative of the fight between humans and venoms/poisons is interrupted by descriptions of homes, with their cozy interiors, and of the quiet life of the countryside. For example, the variegated color of the skin of a snake is compared to a rug; the many leaves of a plant used for therapeutic purposes become a forest; and the inflammation of the body recalls the sun of summertime, together with the need for water and refreshment.

Delving deeper into these associations of ideas, the fight of humans against the destructive forces of venoms and poisons is equated with the primordial forces that shaped the cosmos; the narrative assumes a grandiose dimension with a desperate tone that inspires a feeling of tragedy.

This comparison of the report of clinical cases to the Homeric epic poem with its multilayered interpretations, of which we have just highlighted some, poses the question of the exact nature of the *Thêriaka* and *Alexipharmaka* and of the identity of Nicander. Should the two poems be considered as medical works, or just literary fantasies on a medical theme? And what kind of background did their author have? Was he a physician with a perfect knowledge of toxicology, or a writer and a poet whose genius consisted of imbuing the toxicological discourse with an epic feeling? The toxicological knowledge—or the lack thereof—displayed in the two poems has rarely been taken into consideration in the polarized debate generated by these questions. Nor have the scanty elements known about Nicander's life been studied.

6.4 ANCIENT TOXICOLOGY

The knowledge of venoms and poisons in the ancient Greek world dates as far back as memory can go. In Greek mythology, Telegonus, son of Odysseus and the nymph Circe, accidentally killed his father with a spear topped with the venomous spine of a stingray, believed in Antiquity to be venomous, whereas it is not and hurts by the violence with which it is introduced into the body (sometimes deeply into the organs) and killing in some cases by hitting the heart or vital organs, just as an arrow. Although toxicology was not formalized as a discipline in its own right before the 1st century BCE/CE, it gradually developed into a coherent body of data organized according to a well-defined principle: the specificity of the action of the venoms and poisons, characterized by a selectivity for a determined physiological system in the human body. Such specificity required counteraction by means of equally specific therapies. As a result, when toxicology reached its peak sometime at the beginning of the Common Era, its discourse was organized according to this principle—that is, in a tripartite division: the description of the animals responsible for the envenomation or of the substances causing the poisoning for identification purposes and, on this basis, administration of the specific therapy; the description of their effect for the same reason, in case the lethal agent could not be seen or was no longer present; and the remedies.

The substance and the composition of the *Thêriaka* and *Alexipharmaka* do not demonstrate a full mastery of this principle or of the toxicological discourse. In the *Thêriaka*, the effects of hemolytic and neurotoxic venoms are distinguished very well and the text does not cause any confusion. However, no such distinction appears in the enumeration of the therapies that are grouped after the description of all the cases of envenomation. In the *Alexipharmaka*, the description of the substances and their effects instead are directly followed by a list of the therapies for such a substance. Besides the differences in treatment between the *Thêriaka* and the *Alexipharmaka*, the lack of internal coherence in the *Thêriaka* does not support the hypothesis of Nicander as being a physician who was well informed and aware of the latest advances in toxicology. Rather, he seems to be a literate man who applied his art to create a paradoxical oeuvre.

6.5 VENOMS, POISONS, AND ART

Although his work is cited in some Greek and Latin literature, not much is known about Nicander. One of Nicander's works is dedicated to a king of Pergamon, usually identified as Attalus III, who reigned from 138 to 133 BCE. Attalus is reputed to have had a keen interest in toxicology, particularly in poisonous plants. He is credited with testing their effects via experiments on convicted criminals.

However unacceptable this might seem nowadays, it was not exceptional at the time, as other kings in that period did similar work. Besides the king of Macedonia, Antigonus II Gonatas (reigned 277–239 BCE), Antiochus III, king of Seleucia (reigned 222–187 BCE), and Ptolemy IV, king of Egypt (reigned 221–205 BCE), there was the king of Pontos, Mithridates VI Eupator (reigned 120–63 BCE), who distinguished himself by experiments of all kinds, including some that he performed on himself.

These kings give credibility to the activity of Attalus III of Pergamon and to the presence of a literate person at the royal court who devoted some of his works to the interest of his king. By writing these works in Homeric verse, the poet transformed (or maybe *transmuted* is a better word) the activities of his sovereign from a dubious activity into a literary expression of a drama of ancient daily life, particularly envenomation.

In perfect harmony with the artistic taste of his time and also in total continuity with the centuries-long Greek educational tradition, Nicander wrote the *Thêriaka* and *Alexipharmaka* in a Homeric way as a pathetic fight against death. His works are a graphical, almost visual representation of blood and gore, of violence and destruction, and of death and putrefied bodies. They echo the dramatic feeling of the Hellenistic era best concretized in the grandiloquent statuary of the Pergamon Altar. Simultaneously, the Homeric inspiration permeates the works with an epic sense of dignity as they compare the victims of poisons to the warriors whose heroic deeds shaped the Greek identity in history.

More than a diptych on toxicology, Nicander's *Thêriaka* and *Alexipharmaka* indicate the seductive power of venoms and poisons on

FIGURE 6.3 Safely entering a garden after a fumigation aimed to expel snakes as per Nicander's recommendation (Paris, Bibliothèque nationale de France, manuscript supplementum graecum 647, 10th century).

the human imagination, however tragic this literary theme might be. As such, they have a place in the history of toxicology as an interesting source of information, however implicit and imperfect (Fig. 6.3).

Reference

Gow, A.S., Schofield, A.F., 1953. Nicander, the poems and poetical fragments. Edition with introduction, translation and notes. In: Gow, A.S., Schofield, A.F. (Eds.), Nicander, the Poems and Poetical Fragments. Edition with Introduction, Translation and Notes. Cambridge University Press, Cambridge.

The Case Against Socrates and His Execution

Okan Arihan[1], Seda K. Arihan[2] and Alain Touwaide[3,4]

[1]Department of Physiology, Faculty of Medicine, Hacettepe University, Ankara, Turkey [2]Department of Anthropology, Faculty of Letters, Van Yuzuncu Yil University, Van, Turkey [3]Brody Botanical Center, The Huntington Library, Art Collections, and Botanical Gardens, San Marino, CA, United States [4]Institute for the Preservation of Medical Traditions, Washington, DC, United States

7.1 INTRODUCTION

The Greek philosopher Socrates (469–399 BC) is famous in the history of Western thinking for constantly asking questions and pressing his interlocutors in inquisitive dialogues that were immortalized by his disciple Plato (428/7–348/7 BC). In 399 BC, he was sentenced to death by the senate of Athens because he supposedly corrupted the Athenian youth and did not respect the gods.

According to historiographical tradition, Socrates was given a poisonous draught whose ingredients have been discussed by modern authors. Whereas it has been traditionally considered to be hemlock, some thought it was hemlock and opium, and others believed it to have been other ingredients.

Hemlock (*kôneion* in ancient Greek) is now identified as *Conium maculatum* L. It is among the most toxic plants of the *Apiaceae* family and is commonly found in the flora of the Mediterranean region and Greece (Boissier, 1872). Although its toxicity had been known since antiquity, it was not studied scientifically until 1760, when it was discussed by the Austrian physician Anton von Stoerck (1731–1803). His (von Stoerck, 1760) work on the subject was first published in Latin under the title *Libellus, quo demonstratur: cicutam non solum usu interno tutissime exhiberi, sed et esse simul remedium valde utile in multis morbis, qui hucusque curatu impossibiles dicebantur* and translated the same year into English as *An essay on the medicinal nature of hemlock: in which its extraordinary virtue and efficacy, as well internally as externally used, in the cure of cancers. . . are demonstrated*.

The main biologically active component of hemlock is the alkaloid coniine, which was isolated in 1827 by the German chemist August Louis Giseke (Giseke, 1827). Its structure was later clarified by August Wilhelm Hoffmann (1818–92) (Hoffmann, 1881). Hemlock has frequently been cited in literature on poisonings since antiquity; studies of the bioactivity and pharmacological properties of coniine were initiated as early as 1898 (Moore and Row, 1898).

The mechanism of the action of coniine was clarified more recently. Its action on the nervous system is by the nicotinic acetylcholine receptors (Forsyth et al., 1996). It stimulates the ganglions, and this stimulation is followed by a ganglionic blockade; finally, asphyxia occurs after the blockage of the phrenic nerves, which are responsible for breathing (Bowman and Sanghvi, 1963).

Coniine was once a popular molecule in medical research. However, the antinociceptive properties of coniine were not studied until 2009 (Arihan et al., 2009).

7.2 HISTORICAL LITERATURE

Although much literature has been devoted to Socrates' last days, the textual evidence brought to light so far has been scanty.

To compensate for this lacuna, we have identified the following texts as sources of primary information for the analysis of Socrates' death (we list them in alphabetical order according to each ancient author's name):

- Aelian (2nd/3rd century AD) (Bowie, 2002), author of the vast collection of data on animals entitled *Natura animalium* (*On the characteristics of animals*) (Schofield, 1971).
- Andocides (440–after 392/1 BC) (Furley, 2002), Attic orator, author of discourses including *On the peace with Sparta* (Maidment, 1982).
- Aristophanes (d. in the 380s BC) (Nesselrath, 2002), author of comedies, among which the *Ranae* (*Frogs*) (Rogers, 1979).
- The so-called *Corpus Hippocraticum* (Jones et al., 1923–2010), that is, a series of 60 + treatises produced between the late 5th century BC and the 2nd century AD and attributed to Hippocrates (460–between 375 and 351 BC) (Potter, 2005).
- Diodorus of Sicily (1st century BC) (Meister, 2005), a historian who wrote a *Bibliothêkê* (*Library*) in 40 books covering the whole history of the world from its beginning up to the middle of the 1st century BC (Oldfather, 1963).
- Diogenes Laertius (mid-3rd century AD) (Runia, 2004) composed a series of biographies of philosophers (*Vitae philosophorum—Lives of the Philosophers*), including Socrates (Hicks, 1980).
- Dioscorides (1st century AD) (Touwaide, 2007), compiler of *De materia medica*, the largest encyclopedia of antiquity on the natural substances (of plant, animal, and mineral origin) used for the preparation of medicines (Wellmann, 1906–1914; Beck, 2005).
- The treatise *On poison* attributed to Dioscorides (Touwaide, 1981), although probably not written by him, and added to *De materia medica* at a certain point in time, at any rate before the 9th century AD (Touwaide, 1983; Touwaide and Garzya, 1990).
- Galen of Pergamum (AD 129–after 216 [?]) (Nutton, 2004) abundantly wrote on all medical topics. Of interest here are three of his treatises:
  - *De simplicium medicamentorum temperamentis ac facultatibus* (*On the mixtures and properties of simple medicines*) (Kühn, 1826)
  - *De morborum causis* (*On the causes of diseases*) (Kühn, 1824; Grant, 2000)
  - *De alimentorum facultatibus* (*On the properties of food*) (Kühn, 1824; Grant, 2000)

- Nicander of Colophon (3rd/2nd century BC) (Fantuzzi, 2006), author of the most ancient Greek works currently known on poisons and venoms (Gow and Schofield, 1953).
- Plato (428/7–348/7 BC) (Szlezák, 2007), the disciple of Socrates and the author of several philosophical dialogues, among which *Euthyphron, Crito,* and *Phaedo* are of particular interest here (Fowler, 1914; Fowler, 1925).
- Pliny (AD 23/4–79) (Sallmann, 2007), a contemporary of Dioscorides (above) and the author of the *Historia Naturalis (Natural History),* which is an all-encompassing encyclopedia of the natural historical knowledge of his time (Rackmann, 1938–1962).
- Plutarch of Chaeronea (ca. AD 45 and before AD 125) (Baltes, 2007). A prolific writer, he compiled a set of biographies in which he compared Greeks and Romans who had an important role in history, the so-called *Vitae parallelae (Parallel Lives)* (Perrin, 1969).
- Seneca (d. AD 65) (Dingel, 2008), a Roman philosopher who wrote a series of moral essays (*Moralia*) on different topics such as *De providentia* (Basore, 1970) and exchanging *Letters (Epistulae)* with his friend Lucilius in which he discussed all manners of moral topics (Gummere, 1971).
- Theophrastus (371/0–287/6 BC), a disciple of Aristotle (384–322 BC) (Harmon, 2009) and the author of *Historia plantarum (Inquiry into Plants)* (Hort, 1916–1926), considered to be the foundational work of botany.
- Xenophon of Athens (ca. 430–354 BC) (Schütrumpf, 2010), the author, among others, of the *Hellenica* (or *Greek History*) (Brownson, 1968).

On the basis of these sources, we have collected the following data:

1. We have one main account of Socrates' death by a contemporary, based on a personal autoptic participation in the event, in the *Phaedo* (117a–118) by Plato.
2. The term used by Plato to identify the substance absorbed by Socrates is generic: *poison* (*farmakon* in ancient Greek). This is also the case in Seneca's *Letter to Lucilius* (*Letter* 104: *venenum*). In his essay *On Providence*, Seneca uses the generic term *potio* (*draught*) (3.12), which does not contain any notion of toxicity. Nevertheless, Diodorus of Sicily and Diogenes Laertius both explicitly mention hemlock (*kôneion* in classical Greek) (Diodorus, 14.37.7; Diogenes, 2.5 [= *Life of Socrates*], §35).
3. Passages in ancient nonscientific literature confirm that hemlock was used in Athens as a lethal agent during Socrates' lifetime and in the following century (passages below are listed in the chronological order of the cases they report):

a. 405 BC: Aristophanes, in the *Frogs*, mentions hemlock as the quickest way to commit suicide (124). Also, he says that many ladies took hemlock because they could not stand "the shame and sin" of certain situations (1051).

b. 404 BC: Xenophon, in the *Greek History*, 2.3.56, reports a death due to hemlock.

c. 404/3 BC: Andocides, in *On the peace with Sparta*, 20, mentions that, during the oligarchy of the 30 (i.e., during the year 404/3 BC), "many citizens [died] by the hemlock-cup."

d. Between 399 and 387: Plato mentions hemlock in *Lysis* (219e), in the scene where he discusses a son and his father, with the son having drunk hemlock and the father doing all he can to save him. This dialogue was written after Socrates' death (399) and before Plato's first trip to Sicily in 387 BC.

e. 318 BC: Diodorus of Sicily (18.64—67) and Plutarch (Life of Phocion 31—37) reports that the Athenian orator, politician, and general Phocion (402/1—318 BC) was sentenced to death and forced to ingest hemlock.

4. The action of the poison ingested by Socrates is described as follows by Plato (*Phaedo* 117b—e):

> the man ... was to administer the poison, which he brought with him in a cup ready for use. And when Socrates saw him, he said: "Well, my good man, you know about these things: what must I do?" "Nothing," he replied, "except drink the poison and walk about till your legs feel heavy and the poison will take effect of itself." ... He walked about and, when he said his legs were heavy, lay down on his back, for such was the advice of the attendant. The man who had administered the poison laid his hands on him and after a while examined his feet and legs, then pinched his foot hard and asked if he felt it. He said "No"; Then after that, his thighs; and passing upwards in this way he showed us that he was cold and rigid. And again he touched him and said that when it reached his heart, he would be gone. The chill had now reached the region about the groin, and uncovering his face, which had been covered, he said—and these were his last words—"Crito, we owe a cock to Aesculapius. Pay it and do not neglect it." "That," said Crito, "shall be done; but see if you have anything else to say." To this question he made no reply, but after awhile he moved; the attendant uncovered him; his eyes were fixed.

7.3 HEMLOCK IN ANCIENT SCIENTIFIC LITERATURE

Botanical information on hemlock can be found in the following works (in chronological order; to find relevant information we needed to expand the chronological frame, using works posterior to Socrates' time);

- Theophrastus, *Inquiry into plants*
- Dioscorides, *De materia medica*
- Pliny, *Natural History*

In this body of literature, hemlock is described as follows:

- Theophrastus, *Inquiry into plants*:
    - "Next of the woods themselves and of stems generally some are fleshy, as in ... hemlock ..." (1.5.3).

    - "Of the others some to a certain extent resemble ferula, that is, in having a hollow stem; for instance ... hemlock ..." (6.29).

    - "Mountain-celery (parsley) exhibits even greater differences; its leaf is like that of hemlock ..." (7.6.4).

    - "Hemlock is best about Susa and in the coldest spots. Most of these plants (that is, the medicinal plants) occur also in Laconia, for this too is a land rich in medicinal plants" (9.15.8).

- Dioscorides, *De materia medica*, 4.78. The text reads as follows:
    - "Hemlock: it sends up a large stem, knotty like fennel, leaves look like those of giant fennel, but narrower and oppressive in scent; at the top it has side-shoots and umbels of whitish flowers; it has seeds like of anise but whiter, and a root that is hollow and not deep."

- Pliny, *Natural History*:
    - "... the stem is smooth, and jointed like a reed, of a dark colour, often more than two cubits high, and branchy at the top; the leaves resemble those of coriander, but are more tender, and of a strong smell; the seed is coarser than that of anise, the root hollow and of no use ..." (25.151).

Extant medicopharmaceutical literature contemporary to Socrates does not contain much relevant information on the action of hemlock on human physiology. Consequently, we expanded the chronological frame by including in our inquiry works of subsequent periods up to the second and early 3rd century AD. Relevant data are the following (in chronological order of the works):

- The *Corpus Hippocraticum*:
    - The fruit of hemlock is ground with wine for a plaster to treat fistulae (6.458.16).
    - Hemlock is administered as a draught with water as an emmenagogue after childbirth (7.356.13).
    - Administration of a mixture of sulfur, asphalt, hemlock, or myrrh with honey as an anal injection to treat *hysteria* (8.278.9).
    - Fumigation with hemlock, or myrrh or incense, together with perfume, to treat red discharges (8.378.10).
    - Application of a mixture made of nightshade and hemlock previously boiled together to treat red discharges (8.380.11).
    - Fumigation with hemlock leaves to treat a female sterility (8.432.18).

- Theophrastus, *Inquiry into plants*:
  - "... in hemlock the juice is stronger and it causes an easier and speedier death even when administered in a quite small pill; and it is also more effective ..." (9.8.3).

- Nicander, *Alexipharmaka*:
  - "Take note too of the noxious draught which is hemlock, for this drink assuredly looses disaster upon the head bringing the darkness of night: the eyes roll, and men roam the streets with tottering steps and crawling upon their hands; a terrible choking blocks the lower throat and the narrow passage of the windpipe; the extremities grow cold; and in the limbs the stout arteries are contracted; for a short while the victim draws breath like one swooning, and his spirit beholds Hades." (186–194).

- Dioscorides, *De materia medica*:
  - "This one, too, belongs to the plants that are deadly, killing by chilling through and through." (4.78).

- Pliny, *Natural History*:
  - "Hemlock too is poisonous, a plant with a bad name because the Athenians made it their instrument of capital punishment, but its uses for many purposes must not be passed by. It is a poisonous seed, but the stem is eaten by many both as a salad and when cooked in a saucepan.... The seed and leaves have chilling quality, and it is this that causes death; the body begins to grow cold at the extremities ..." (25.150).

- Seneca, *On Providence*:
  - "... his blood grew cold, and, as the chill spread, gradually beating of his pulses stopped ..." (3.12).

- Galen:
  - *On the mixtures and properties of simple medicines*:
    - "Everybody knows that hemlock is of the outmost cooling property." (7.10.67).

  - *On the causes of diseases*:

    - "Among the cooling drugs are ... hemlock, the last of which is lethal because of its violent chilling action. We designate the blockage as the third and worst cause of cooling diseases, since it engenders torpor, lethargy and apoplexy." (7.13.17–14.6).

- Aelian, *On the characteristics of animals*:
  - "... if a man drinks hemlock, he dies from the congealing and chilling of his blood ..." (4.23).

- Pseudo-Dioscorides, *On poisons*:
  - "In a draught, hemlock provokes a loss of vision at such point to be blind, gasping, a loss of consciousness and cold in the extremities; at the end, the victims gasp as they are asphyxied because breath stops." (11).

Representations of hemlock can be found in several Byzantine manuscripts containing the Greek text of Dioscorides, *De materia medica*. We have considered only the most ancient ones (cited below in chronological order), since the most recent are copies of one or another of them:

- Vienna, National Library of Austria, *medicus graecus* 1 (sixth cent.), f. 187 verso.
- Paris, National Library of France, *graecus* 2179 (ninth cent.), f. 106 recto.
- New York, Pierpont Morgan Library, M 652 (tenth cent.), 76 recto.
- Mt Athos, Megistis Lavras, Ω 75 (11th cent.), f. 61 recto.

No particularly significant information on hemlock and its uses was found in nontechnical literature beyond those mentioned above, under 1. c), apart from the belief that some animal species were immune to the effect of hemlock:

- Starlings, according to Galen (On *properties of food* 6.567.14)
- Swans, according to Aelian (3.7)
- Hogs, according to the same (4.23)

A particular point is the question of the preparation of hemlock for its medicinal use, on which we have some information coming from technical and nontechnical literature (in chronological order according to author).

- Theophrastus, in the *Inquiry into Plants* (9.16.8–9), details the way hemlock has to be prepared to be used as a lethal agent:

> "Thrasyas of Mantinea has discovered, as he said, a poison which produces an easy and painless end; he used the juices of hemlock, poppy, and other such herbs, so compounded as to make a dose of conveniently small size ... he used to gather his hemlock, not just anywhere, but at Susa or some other cold and shady spot ... he also used to compound many other poisons, using many ingredients ... the people of Kos formerly did not use hemlock in the way described, but just shredded it up for use, as did other people; but now they first strip off the outside and take off the husk, since this is what causes the difficulty, as it is not easily assimilated; they then bruise it in the mortar, and, after putting it through a fine sieve, sprinkle it on water and so drink it; and then death is made swift and easy."

- Seneca, in his essay *On Providence* (3.12), specified that the draught taken by Socrates had been *mixed publicly* (*publice mixtam*), a piece of information implying that there were several ingredients. However, he did not detail the ingredients.
- In the *Life of Phocio* by Plutarch, we read the following (36.2):
  - "... seeing the executioner bruising the hemlock ..."

The plant identified in ancient Greek by means of the word *kôneion* is unanimously identified by the post-Linnean authors of botanico-historical works as *C. maculatum* L. from the *Florae Graecae Prodromus*

(1806–13) by the Englishman John Sibthorp (1756–98) (Sibthorp, 1806–1813) to, recently, Suzanne Amigues (Amigues, 2006).

The post-Linnean flora of Greece confirm that hemlock (=C. *maculatum* L.) is present in the area corresponding to modern Greece. According to Sibthorp, in the *Florae Graecae Prodromus* (1806–13) (Sibthorp, 1806–1813), it was common in the area of Athens. According to Edmond Boissier (1872), author of a *Flora orientalis*, it can be found in all of Europe except Lapponia, North Africa, Abyssinia, and Syria. According to his *Conspectus Florae Graecae*, de Halacsy found it in 1901 (de Halacsy, 1901) in Attica and the region of Athens. Boissier, in the *Supplementum* of his *Flora* published in 1908, added the Attica (Boissier, 1908). More recently, Hegi described it in the *Flora of Mittel-Europa* as omnipresent in Europe, Asia, and North Africa, while the *Flora Europaea* (Flora Europaea, 1968) describes it as typical of almost all Europe except the extreme north (Hegi, 1925).

A comparison of ancient textual data, representations in manuscripts, dry specimens from herbaria, and descriptions in modern flora confirms the following:

- The ancient Greek word *kôneion* corresponds to the taxon currently identified as *C. maculatum* L.
- The plant grows in the area that comprised the ancient Greek world and probably grew there in antiquity.

In modern literature, the most relevant analysis from our viewpoint here is provided by the German toxicologist Louis Lewin (1850–1929) (Lewin, 1920). Capitalizing on the work of his predecessors, he listed and paraphrased in his magisterial work *Die Gifte in der Weltgeschichte* (*Poisons in World History*) (1920) the several historical cases of death provoked by means of hemlock in ancient Greece, specifically in Athens. After he commented on Theophrastus' passage, cited above, about the mixtures prepared by Thrasyas of Mantinea, he stressed that, in Athens, hemlock only was used (without, thus, the addition of any other substance, toxic or not).

7.4 MODERN PHARMACOLOGICAL ANALYSIS

The silent and peaceful death of Socrates raises the question of whether hemlock has an analgesic effect. Until 2009, no information was available in the scientific literature. Dayan (2009) stated that "nociceptive responses were not affected" by poison hemlock. A pharmacological study conducted by Arihan et al. contradicted this statement (Arihan et al., 2009), and proved 2000 years later that Galen's affirmation was pharmacologically sound. In this study, authors found that coniine, the active molecule of poison hemlock, shows analgesic activity at doses of 10 mg/kg and more prominently at 20 mg/kg in two well-established tests, namely the hotplate and writhing tests. Beyond this dose, lethality was observed. Additional information about potentiation was also presented in this publication, as morphine and coniine potentiate their analgesic activity when administered together (Arihan et al., 2009).

7.5 TOWARD A RENEWED INTERPRETATION

On the basis of unanimous ancient documentation (be it related to Socrates' death or not, and be it technical or literary), C. maculatum L. was known in antiquity. All sources provide clear evidence, though scant and fragmentary, of the use of C. maculatum L. as an agent for the execution of humans condemned to death. The plant is considered to have a *cooling* property (just like sleep and death, which are characterized by a reduction of or the loss of the active principle of life). The symptoms of the internal absorption of *kôneion* are precisely described as a gradual loss of sensory perception with increasing asphyxia, gasping, loss of consciousness, and, finally, respiratory arrest.

The clinical signs in the report of Socrates' death in Plato's *Phaedo* indicate a paralysis proceeding from the feet to the abdomen and then through the chest. Death was caused by the cessation of respiration due to the paralysis of the diaphragm muscles.

It has to be noted that none of the signs considered to be typical of the absorption of opium in ancient medical and toxicological literature appears in the clinical description of the death of Socrates by Plato or in the works of the other individuals whom we have mentioned. This confirms the findings of the ancient texts, in which no mention of opium is made, as Lewin already noticed.

Some elements of the description of death by absorption of hemlock (be it the death of Socrates or of other individuals) require further consideration, principally the repeated mention in the texts above (particularly those of Seneca, Galen, and Aelian) of a cold gradually spreading through the body, and rigidity. These phenomena suggest a loss of peripheral sensation in the lower extremities (Dayan, 2009), since Socrates could not feel the mechanical stimulus by the man who gave the poison. Cold should probably be interpreted on the basis of subsequent literature as a loss of painful stimulus, that is, analgesic activity.

However, some symptoms typical of the ingestion of hemlock (vomiting, dizziness, or loss of cognitive functions) do not appear, either in Plato's report (Dayan, 2009) or in some of the other cases. The fact that Socrates covered his face with a sheet and no longer communicated with his pupils could explain why such typical symptoms were not mentioned: most of these symptoms (particularly dizziness and loss of cognitive functions) would not have been visible and thus would not have been noticed by Socrates' pupils, who were gathered around him. On the other hand, these symptoms (except vomiting) are clearly mentioned in Nicander's *Alexipharmaca* and in the treatise *On poison* ascribed to Dioscorides.

The study by Arihan et al. constitutes the first assessment of the analgesic activity of coniine on organisms. The analgesic activity of coniine was significant in doses close to lethal, and it may have been supplemented by a strong analgesic agent such as morphine in the case of the mixture prepared by Thrasyas of Mantinea and reported by Theophrastus.

The pharmacological assessment of the activity of coniine and opium suggests different analgesic mechanisms. Morphine is an agonist for central opiate receptors for analgesic activity, whereas coniine stimulates the nicotinic receptors (Bowman and Sanghvi, 1963; Guyton and Hall, 2001). Analgesic activity was possibly enhanced due to their different modes of action. Theoretically, coniine and opium combine well for a painless execution, as Thrasyas of Mantinea probably already understood in the 4th century BC.

7.6 CONCLUSION

According to our results, the report of Socrates' death made by Plato seems realistic. Since coniine exerts antinociception more strongly at doses close to lethal, Socrates probably felt no pain after he ingested the poisonous mixture made of hemlock. In addition, the statement by Theophrastus about hemlock and opium as ingredients of a toxic potion composed by Thrasyas of Mantinea may also be true, since these two agents complement each other in antinociception.

References

Amigues, S., 2006. Théophraste, Recherches sur les plantes. Tome V: Livre IX. Belles Lettres, Paris.

Arihan, O., Boz, M., İskit, A.B., İlhan, M., 2009. Antinociceptive activity of coniine in mice. J. Ethnopharmacol. 125 (2), 274–278.

Baltes, M., 2007. Plutarchus [2]. Brill, Leiden and Boston, MA, pp. 410–423.

Basore, J.W., 1970. Seneca in Ten Volumes, I Moral Essays in Three Volumes. I. Harvard University Press/W. Heinemann, Cambridge, MA/London.

Beck, L.Y., 2005. Pedanius Dioscorides of Anazarbus, De materia medica. Translated. Olms-Weidmann. Hildesheim, Zürich/New York, NY.

Boissier, E., 1872. Flora Orientalis Sive Enumeratio Plantarum in Oriente a Graecia et Aegypto ad Indiae Fines Hucusque Observatarum. H. Georg, Geneva and Basel.

Boissier, E., 1908. Supplementum Florae Graecae. W. Engelmann, Leipzig.

Bowie, E., 2002. Aelianus [2]. Brill, Leiden and Boston, pp. 200–201.

Bowman, W.C., Sanghvi, I.S., 1963. Pharmacological actions of Hemlock (*Conium maculatum*) alkaloids. J. Pharm. Pharmacol. 15, 1–25.

Brownson, C.L., 1968. Xenophon in Seven Volumes. I. Hellenica, Books I–IV, With an English Translation. Harvard University Press/W. Heinemann, Cambridge, MA/London.

Dayan, A.D., 2009. What killed Socrates? Toxicological considerations and questions. Postgrad. Med. J. 85, 34–37.

de Halacsy, E., 1901. W. Engelmann, Leipzig.

Dingel, J., 2008. Seneca [2]. Brill, Leiden and Boston, pp. 271–278.

Fantuzzi, M., 2006. Nicander [4] of Colophon. Brill, Leiden and Boston, pp. 706–708.

Flora Europaea, 1968. University Press, Cambridge.

Forsyth, C.S., Speth, R.C., Wecker, L., Galey, F.D., Frank, A.A., 1996. Comparison of nicotinic receptor binding and biotransformation of coniine in the rat and chick. Toxicol. Lett. 89, 175–183.

Fowler, H.N., 1914. Plato in Twelve Volumes. I. Euthyphro, Apology, Crito, Phaedo, Phaedrus, with an English translation. Harvard University Press/W. Heinemann, Cambridge, MA/London.

Fowler, H.N., 1925. Plato in Twelve Volumes. III. Lysis, Symposium, Gorgias, with an English Translation. Harvard University Press/W. Heinemann, Cambridge, MA/London.

Furley, W.D., 2002. Andocides. Brill, Leiden and Boston, pp. 677–678.

Giseke, A.L., 1827. Ueber das wirksame Princip des Schierlings *Conium maculatum*. Arch. Apoth.-Ver. 20, 97–111.

Gow, A.D., Schofield, A.F., 1953. Nicander, the Poems and Poetical Fragments. Edited With an Introduction, Translation and Notes. Cambridge University Press, Cambridge.

Grant, M., 2000. Routledge, London.

Gummere, R.M., 1971. Seneca in Ten Volumes. VI Ad Lucilium Epistulae Morales With an English Translation. Harvard University Press/W. Heinemann, Cambridge, MA/London.

Guyton, A., Hall, J., 2001. Textbook of Medical Physiology. W.B. Saunders Company, Philadelphia, PA.

Harmon, R., 2009. Theophrastus. Brill, Leiden and Boston, pp. 508–517.

Hegi, G., 1925. Illustrierte Flora Von Mittel Europa, V.2. J.F. Lehmanns, Munich.

Hicks, R.D., 1980. Diogenes Laertius, Lives of Eminent Philosophers With an English Translation, in Two Volumes. Harvard University Press/W. Heinemann, Cambridge, MA/London.

Hoffmann, W., 1881. Zur Kenntniss der Coniin-Gruppe. Ber. Dtsch. Chem. Ges 18 (5–23), 109–131.

Hort, A., 1916–1926. Theophrastus, Enquiry into Plants and Minor Works on Odours and Weather Signs With an English Translation, in Two Volumes. Harvard University Press/W. Heinemann, Cambridge, MA/London.

Jones, W.H.S., Potter, P., Withington, E.T., 1923–2010. Hippocrates With an English Translation. Harvard University Press/W. Heinemann, Cambridge, MA/London.

Kühn, K.G., 1824. Galeni Opera Omnia, Volumen VII. Knobloch, Leipzig.

Kühn, K.G., 1826. Claudii Galeni Opera Omnia, Volumen XII. Knobloch, Leipzig.

Lewin, L., 1920. Die Gifte in der Weltgeschichte. Toxikologische, allgemeinverständliche Untersuchungen der historischen Quellen. J. Springer, Berlin.

Maidment, K.J., 1982. Minor Attic Orators in Two Volumes. I Antiphon, Andocides, With an English Translation. Harvard University Press/W. Heinemann, Cambridge, MA/London.

Meister, K., 2005. Diodorus [18] Siculus. Brill, Leiden and Boston, pp. 444–445.

Moore, B., Row, R., 1898. A comparison of the physiological actions and chemical constitution of piperidine, coniine and nicotine. J. Physiol. 22, 273–295.

Nesselrath, H.-G., 2002. Aristophanes [3]. Brill, Leiden and Boston, pp. 1125–1132.

Nutton, V., 2004. Galen of pergamum. Brill, Leiden and Boston, pp. 654–661.

Oldfather, C.H., 1963. Diodorus of Sicily, with an English Translation, in Twelve Volumes, VI. Books XIV–XV.19. Harvard University Press/W. Heinemann, Cambridge, MA/London.

Perrin, B., 1969. Plutarchus' Lives, with an English Translation, in Eleven Volumes, VIII. Harvard University Press/W. Heinemann, Cambridge, MA/London.

Potter, P., 2005. Hippocrates [6] of Cos. Brill, Leiden and Boston, pp. 354–363.

Rackmann, H., 1938–1962. Pliny, Natural History, with an English Translation. Harvard University Press, Cambridge, MA/London.

Rogers, B.J., 1979. Aristophanes in Three Volumes. II The Peace, The Birds, The Frogs. With an English Translation. Harvard University Press/W. Heinemann, Cambridge, MA/London.

Runia, D.T., 2004. Diogenes [17] Laertius. Brill, Leiden and Boston, pp. 452–455.

Sallmann, K., 2007. Plinius [1]. Brill, Leiden and Boston, pp. 383–390.

Schofield, A.F., 1971. Aelian, on the Characteristics of Animals, with an English Translation, in Three Volumes. I Books I–V. Harvard University Press/W. Heinemann, Cambridge, MA/London.

Schütrumpf, E.E., 2010. Xenophon. Brill, Leiden and Boston, pp. 824–833.

Sibthorp, J., 1806–1813. Florae Graecae Prodromus: Sive Plantarum Omnium Enumeratio Quas in Provinciis Aut Insulis Graeciae Invenit. Typis Richardi Taylor, veneunt apud J. White, London.

Szlezák, T.A., 2007. Plato [1]. Brill, Leiden and Boston, pp. 338–352.

Touwaide, A., 1981. Les deux traités toxicologiques attribués à Dioscoride. La tradition manuscrite du texte grec, édition critique et traduction, vol. 5. PhD Thesis. Louvain-la-Neuve.

Touwaide, A., 1983. L'authenticité et l'origine des deux traités de toxicologie attribués à Dioscoride. I. Historique de la question. II. Apport de l'histoire du texte. Janus 38, 1–53.

Touwaide, A., 2007. Pedanius [1] Dioscorides. Brill, Leiden and Boston, pp. 670–672.

Touwaide A., Garzya A., 1990. Les deux traités de toxicologie attribués à Dioscoride: Tradition manuscrite, établissement du texte et critique d'authenticité. In: Garzya, A. (Ed.), Tradizione e ecdotica dei testi medici tardo-antichi e bizantini. Atti del Convegno internazionale, Anacapri, 29–31 ottobre 1990. D'Auria, Naples, 1992, pp. 291–339.

von Stoerck, A., 1760. (English translation: An essay on the medicinal nature of hemlock: in which its extraordinary virtue and efficacy, as well internally as externally used, in the cure of cancers... are demonstrated. London) Libellus, quo demonstratur: cicutam non solum usu interno tutissime exhiberi, sed et esse simul remedium valde utile in multis morbis, qui hucusque curatu impossibiles dicebantur. J.J. Trattner, Vindobonae.

Wellmann, M., 1906–1914. Pedanii Dioscuridis Anazarbei, De materia medica libri quinque, 3 vols. Weidmann, Berlin.

8

Murder, Execution, and Suicide in Ancient Greece and Rome

Alain Touwaide[1,2]

[1]Brody Botanical Center, The Huntington Library, Art Collections, and Botanical Gardens, San Marino, CA, United States [2]Institute for the Preservation of Medical Traditions, Washington, DC, United States

OUTLINE

The political, scientific, cultural, and artistic life of the ancient world may be highlighted by such remarkable individuals as Athens' political leader Pericles (c. 495–429 BCE), the Greek philosopher Aristotle (384–322 BCE), the historian of Greek wars Thucydides (460–395 rBCE), and Phidias (490–430 BCE) the sculptor of the magnificent Athena statue at the Parthenon, in Greece, and the founder of the Roman Empire and the first Emperor Augustus (63 BCE; *reign.* 27 BCE–15 CE), the orator Cicero (106–43 BCE), and the naturalist Pliny (23/4–79 CE), as well as by Pompeii's many fresco painters in the Roman world. However brilliant the civilization of Classical Antiquity might have been as these names and works of art suggest, it also seems to have been punctuated by treacherous murders, summary executions, and self-inflicted death for many reasons ranging from lovesickness to desperation and shame. Many illustrious individuals have been reported to have committed suicide: for example, the Athenian orator Demosthenes (384–322 BCE); the Egyptian Queen Cleopatra (69–30 BCE); even the philosopher Aristotle according to the ancient historian and philosopher Diogenes Laertius (180–240 CE) and the 5th- or 6th-century Byzantine lexicographer and historian of literature Hesychius;

and the Carthaginian general Hannibal (247/6–183 BCE) defeated at Zama (Tunisia) in 202 BCE.

The case of the Athenian politician and general, Themistocles, (c. 525–c. 459) is revealing. According to historical accounts, he committed suicide by drinking bull's blood, which in antiquity was believed to be highly toxic (and it probably was due to toxins such as botulinum, anthrax, or others resident in cattle). As the story goes, after he defeated the Persian fleet attacking Athens in the Bay of Salamis in 480 BCE, Themistocles was banished from Athens for political reasons and escaped to Persia. There, King Artaxerxes I (*reign.* 465–424/3 BCE) ordered him to lead a military operation against Greece. Themistocles refused and reportedly committed suicide rather than betray his country. Closer examination of the story reveals, however, that following Themistocles's death (which probably was due to natural causes) its details were deliberately misinterpreted to save the reputation of an esteemed, victorious general turned traitor. To preserve the memory of his sacrifice, a statue was erected depicting Themistocles slaughtering a bull for sacrifice, thereby perpetuating the legend that he committed suicide rather than attack his native land.

Although suicide, murder, and execution were certainly a reality in ancient life, they were not necessarily as frequent occurrences as popular legend, political propaganda, or other ill-intentioned maneuvers would have us believe. At any rate, the use of lethal substances may be questioned. Often documentary evidence is anecdotal and cannot be corroborated for lack of supplementary and independent sources, or for presenting contradictory or implausible scenarios, as Cleopatra's death suggests. A suicide carried out through a cobra's organized biting seems to be more speculative than likely. Ancient medical literature credited cobra's venom with causing instant paralysis, leading to immediate death (Fig. 8.1). A suicide with no suffering was ideal and would have certainly been the supposedly self-indulgent Cleopatra's choice. Nevertheless, it is highly improbable that a cobra could have been brought into her apartments without having been noticed. Cleopatra and Egypt were very important to the Roman Empire because of Egypt's rich resources, especially its abundant agricultural production, which Rome needed to produce; the necessary bread to feed its people. Indeed, Cleopatra herself was viewed as a precious political commodity. At the same time, because of her liaison with Marc Antony (83–30 BCE), who had been defeated by Octavius at Actium (on the West coast of Northern Greece) in 31 BCE, the future Emperor Augustus, Cleopatra became politically undesirable. Under these conditions, Cleopatra was kept under strict control in her palace where she could easily be assassinated. Her assassination was extremely risky, for it could well have provoked mob action and have led to Rome's loss of control over Egypt. Crediting her with suicide—for

FIGURE 8.1   Domenico Riccio (1516–1567), Cleopatra's Death (1552) (Fondazione CariCesena, Cesena, Italy).

reasons of desperation, political calculation, or any other reason—was the ideal coverup and so we will probably never know exactly how she died.

Murder was more often committed by stabbing than by poisoning. The Roman general and politician Caesar (100–44 BCE) is the best example. He was stabbed in plain light, in the Curia, by a group of conspirators, including his adoptive son ("Tu quoque, fili mi"—"You too, child" or "Tu quoque, Brute"—"You too, Brutus," according to the last words pronounced by Caesar, if we are to believe the historiographical tradition) (Fig. 8.2). In many other cases, however, toxic substances were used. In the ancient Greek world, sovereigns who succeeded Alexander the Great (356–323 BCE) and divided his empire showed a particular interest in poisoning. Antigonus Gonatas (c. 320–239 BCE), king of Macedonia in central Greece, Antiochus III (c. 242–187 BCE), king of Seleucia in Asia Minor, and Ptolemaeus IV Philopator, king of Egypt (c. 244–205 BCE), were among such kings, together with Attalus III Philometer Evergetes, king of Pergamum (*reign*. 138–133 BCE), to whom Nicander might have dedicated one of his two poems on toxicological matters. According to historical sources, Attalus cultivated medicinal and toxic plants, including henbane, hellebore, hemlock, and aconite. He is credited with testing their toxic properties on individuals who had been sentenced to death. The purpose was not only to prepare poisons and to identify their lethal doses, but also—if not primarily—to make use of them, particularly in a time when political rivalry was intense and coups d'état were not rare. The most famous among these Hellenistic kings who were manipulators of poisons is without doubt Mithradates VI Eupator, king of Pontus (133; *reign*. 120–63 BCE), originally a small kingdom on the south shore of the Black Sea. An

FIGURE 8.2   Vincenzo Camuccini (1771–1844), Caesar's assassination (1798) (Museo Nazionale di Capodimonte, Napoli, Italy).

ambitious and unscrupulous politician, Mithradates had assassinated several members of his court and even members of his family (including his own mother) to preserve his throne. Furthermore, he fiercely opposed the Roman conquest of Asia Minor and inflicted severe defeat on Roman troops, expanding his kingdom from a small area in the north of Asia Minor to almost all Asia Minor. Correctly fearing that he might be poisoned—whether by his own entourage or by the Romans— he absorbed increasing doses of all the poisons he knew to gradually acquire immunity. He was so successful that, when he was eventually captured by the Romans in 63 BCE and he wanted to kill himself in order to escape the clutches of his captors, he consumed a poison that he always carried with him but did not die. He had no other recourse than to ask his slaves to run him through with a sword.

Poisons were no less common in Rome. As early as 449 BCE, the *Lex duodecim tabellarum* (Law of the Twelve Tables) prohibited poisons, thus clearly implying that they had been in use. The law probably did not have much effect, as a new one needed to be promulgated in 81 BCE: the famous *Lex Cornelia de sicariis et veneficis* (Cornelia Law on Assassins and Poisoners). Again, this new legislation did not necessarily prevent poisoning. Indeed, half a century later, the Latin poet Tibullus (c. 55–19 BCE) wrote an elegy in which he recalled a sickness during which he thought he would die. Imploring mercy from Persephone, the Queen of the Underworld, he proclaimed that he did not poison anybody. Murder by poison even took place within the imperial palace, as is shown by the case of Claudius (10; emp. 41, d. 54 CE). Claudius was known to be fond of mushrooms, and so it was that he was served with an abundant plate of mushrooms, supposedly of *Boletus*, but they might have been mixed with a poison. Whatever the toxic agent used, Claudius did die. It was his second wife, Agrippina (15–59 CE), who initiated plans to get Claudius out of the way so that her son Nero (37; emp. 54; d. 68) would become emperor. Claudius's poisoning was orchestrated by Agrippina in collusion with a woman named Locusta, who was supposedly from Gaul, had mastered the art of poisons, and had been sentenced to jail for poisoning. Taken out of prison, she helped engineer Claudius's murder. Ironically enough, she also prepared the poison that Nero himself would later use to kill himself. As for Locusta herself, she was executed under Emperor Galba (24; emp. 68–69), who, in turn, was killed by mutinous soldiers.

With rare exception, the exact nature of the poisons used for murder—whether they were prepared by an expert such as Locusta or by less competent assassins—is not known. The art of compounding poisons seems to have flourished during the 1st century BCE. Interestingly enough, this period also saw an unprecedented development of compound medicines. The origin of this therapeutic strategy is often attributed to Mithradates. However, a compound medicine appeared earlier in Nicander's *Theriaca*, which suggests that such a practice must have predated Mithradates. Whatever the case, the formula for the medicine mentioned in Nicander's *Theriaca* was refined by the Cretan Andromachus, who was Nero's personal physician. It thus seems that, during the 1st century CE and particularly during Nero's reign (54–68 CE), a substantial amount of research was done on compounding—and also administering—medicines and possibly poisons by mixing several substances, medicinal or toxic, respectively, together.

If the nature of poisons is unknown in many cases, it might be because poison had a special off-limits status in the ancient world, apart from the evident secrecy of criminal poisoning. Its manipulation was often attributed to individuals who were on the margins of society. Locusta is exemplary from that viewpoint. Women had no rights under Roman law but acquired status only by being married; Locusta was of foreign origin and consequently had even fewer legal rights than most Roman women. At any rate, she lost any possibility of acquiring an official status once she was condemned for her poisonings. Locusta's case was not unusual. Long before she appeared on the scene, the archetype of the individual at the margins of society who was a user of poisons had arrived in the person of the legendary Medea. A granddaughter of the sun god Helios and of the magician Circe, Medea was the daughter of the king of Colchis, living on the Black Sea, at the eastern edge of the Greek world, at the border of Scythia. She used her singular talents to manipulate poisons in various ways. She first laid eyes on Jason when he arrived at Colchis from Greece during his expedition to conquer the Golden Fleece; Medea instantly fell in love with him. But being the daughter of the local king, she could not even think of approaching this foreigner, and so she tried to commit suicide. She got hold of a poison defined only by its black color (actually the color of the death it was supposed to provoke), which, however, she did not drink because she was too frightened to do so. An additional deterrent to her taking the poison was her sister's admonition that by killing herself she would destroy a promising life. After Jason conquered the Golden Fleece and returned victorious to Medea's father, the two married and they had two children. Later, they moved to Corinth and, after 10 years, Jason fell in love with the daughter of Corinth's king. Plotting her revenge, Medea poisoned a tunic by dyeing it with a substance described with many details in literature, but more fictional than realistic. She offered this tunic to her rival who upon slipping it on died in severe pain. As further, complete revenge against Jason, immediately afterward Medea killed their children, while she also provoked a fire that destroyed all the royal palace of Corinth. Medea's legend was more paradigmatic than historical and was told to future generations to provide ethical guidelines. Indeed, the story aimed to show that poisoning out of jealousy was the typical act of irrational people, as Medea's murder of her own children demonstrated, whereas committing suicide out of lovesickness was more understandable. All the same, it was to be avoided since, for the youthful lover, it meant the end of a life that had just begun. In this view, the exact nature of the poison did not matter; poisoning was a concept rather than a particular tool.

Toxic substances had a special status in historical sources at the edge of reality and imagination. According to Theophrastus (372/0–c. 288/6 BCE), a pupil of Aristotle and the compiler of a work on botany (*Historia plantarum—Inquiries into plants*) considered to have laid down the basis of botany in Western science, skilled poisoners were able to prepare aconite in such a way that it would kill in a fixed time: 2 months, 3 months, and even 1 or 2 years—but this is likely a fiction. Nonetheless, in this same work, Theophrastus displays a genuine grasp of pharmacology in his description of how a contemporary of Aristotle, Thrasyas of Mantinea, mixed hemlock and poppy, thereby producing a poison that would kill without pain.

Although cases of poisoning have been attested to in written documents, their frequency should not be overstated. The same can probably also be said about execution. The execution of Socrates with hemlock in 399 BCE should not lead to the conclusion that poisons were common instruments of execution. Indeed, techniques other than poisoning were used. Seneca (c. 1–64 CE), who had been Nero's tutor, was condemned to death (actually, he was forced to commit suicide) for his supposed participation in a conspiracy against the emperor (Fig. 8.3). Rather than

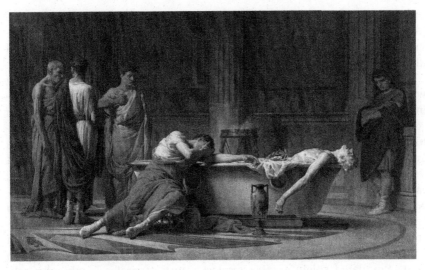

FIGURE 8.3 Manuel Dominguez Sánchez (1840–1906), Seneca's suicide (ca. 1871) (Museo del Prado, Madrid, Spain).

take poison, Seneca chose to open his veins (Fig. 8.3). Nevertheless, hemlock was said to be available in Massalia (present-day Marseilles) in a safe box for those who were condemned, as well as for others who chose to end their lives.

Information is no more specific for suicide than it is for murder. Again, we can look to mythology for precedents. One myth tells the story of Antheia, the wife of Proetus, the king of Tyrins, in continental Greece. The two had been married for many years and had several children. However, one day when the beautiful Bellerophon paid them a visit, Antheia fell in love with him. When Bellerophon spurned her advances, Antheia got her revenge by accusing him of attempting to rape her. Proetus thereupon expelled Bellerophon, who later returned to Tyrins after he was proven innocent of Antheia's charge. Upon his arrival, Antheia committed suicide through a means that is not specified in the literature. Contrary to Medea's account, suicide committed out of lovesickness was accepted: Antheia was not in her prime and so was unlike those whose young life would have ended before really getting started.

In historical times, many individuals, famous or not, committed suicide by poison. One of these individuals was the Carthaginian general, Hannibal. A fierce enemy of the Romans, Hannibal was defeated by them in 202 BCE. About to be captured by the Romans in 195, he fled to Asia Minor and was granted refuge first by the court of the poisoner king of Seleucia, Antiochus III, and subsequently by the king of

Bithynia, Prusias I (c. 230—182 BCE). When the Romans asked Prusias to extradite Hannibal, the general took poison from a ring he was wearing; the nature of this poison is not known, being referred to in ancient literature only by the imprecise term *venenum*.

There is an abundant body of ancient literature on poisons, starting with Nicander in Asia Minor, possibly in the 2nd century BCE, proceeding to Galen (129—after [?] 216 CE) in Rome, and later to Oribasius (4th century), Aetius (6th century) in Byzantium, and Paul of Egina (7th century) in Alexandria. Their works reflect a precise knowledge of poisons and their actions. These works are fundamentally medical in nature and devoted to the treatment of poisoning, whether criminal or accidental, rather than being guides to the art of poisoning, although this knowledge of poisons could very well have been used by nonspecialists with criminal intentions. It is much more likely that knowledge of toxic substances to be used for suicide, assassination, and execution—apart from some notorious exceptions such as Socrates' death and Plato's description of it—circulated solely among a select group of professionals who sold their secrets, were hired as professional killers, or were state executioners. At the same time, common people likely possessed a rudimentary knowledge of poisons and their deadly use; women were even more likely to have such knowledge, as the use of abortifacients suggests. Such knowledge probably never made its way into learned medical literature, particularly because the authors of this literature were physicians who having sworn the Hippocratic oath pledged that they would not administer poisons.

Further Reading

Amberger-Lahrmann, M., Schmähl, D., 1988b. Gifte. Geschichte der Toxikologie. Springer, Berlin.

Arihan, O., Karaoz Arihan, S., Touwaide, A., 2014. The Case Against Socrates and His Execution. Elsevier, Amsterdam, pp. 69—82.

Cilliers, L., Retief, F., 2014. Poisons, Poisoners, and Poisoning in Ancient Rome. Elsevier, Amsterdam, pp. 127—137.

de Maleissye, J., 1991. Histoire du Poison. François Bourin, Paris.

Golden, C.L., 2005. The Role of Poison in Roman Society. Ph.D. Thesis., University of North Carolina, Chapel Hill.

Lewin, L., 1920. Die Gifte in der Weltgeschichte. Toxikologische, Allgemeinverständliche Untersuchungen der Historischen Quellen. Julius Springer, Berlin.

Macinnis, P., 2004. Poisons. A History from Hemlock to Botox. MJF Books, New York, NY.

Martinetz, Lohs, K., 1987. Poison: Sorcery and Science, Friend and Foe. Edition Leipzig, Leipzig.

Mayor, A., 2014a. Mithridates of Pontus and His Universal Antidote. Elsevier, Amsterdam, pp. 21—33.

Mayor, A., 2014b. Alexander the Great: A Questionable Death. Elsevier, Amsterdam, pp. 52—59.

Pichon-Vendueil, E., 1914. Etude sur les pharmaques et venins de l'antiquité. Poisons de guerre, de chasse, de justice et de suicide des anciens peuples de l'Europe (Scythes, Hellènes, Italiotes, Celtes, Germains et Ibères). Ph.D. Thesis. University of Bordeaux, Bordeaux.

Touwaide, A., 2014. Harmful Botanicals. Elsevier, Amsterdam, pp. 60−68.

Tsoucalas, G., Sgantzos, M., 2014. The Death of Cleopatra: Suicide by Snakebite or Poisoned by Her Enemies? Elsevier, Amsterdam, pp. 11−20.

Vuillet-A-Cilles, C., 1986. Les empoisonnements dans l'antiquité chez les peuples indo-européens. Ph.D. Thesis. University of Besançon, Besançon.

The Oracle at Delphi: The Pythia and the *Pneuma*, Intoxicating Gas Finds, and Hypotheses

Jelle Z. de Boer[†]

Department of Earth Science, Wesleyan University, Middletown, CT,
United States

Delphi's sanctuary lies cradled on the southern flank of Mount Parnassus and overlooks the Pleistos river valley. Archeological remains suggest that the oracular site dates back to the Mycenaean period when Gaia (Mother Earth) was worshiped. By the 8th century bce the cult of Apollo (god of prophesy) had been established and Delphi developed into a Panhellenic cradle of religion that also regulated politics, relationships between city states, and public and private affairs. Its zenith, when pilgrims from across the Mediterranean world came for advice, lasted more than a thousand years, from c. 700 bce until 361 ce when Oribasius, sent to Delphi by Emperor Julian, reported that the well-built hall had fallen and the speaking waters were stilled. A few years later

[†]Deceased.

Eunapius wrote, "To all intelligent men the end of the temples . . . was painful indeed" (Philostratus (the Athenian), Eunapius, MCMXXII).

Ancient literary texts with information about the oracle's workings include testimonies by a philosopher (Plato), historians (Diodorus and Pliny), poets (Aeschylus and Cicero), geographers (Pausanias and Strabo), and the famous essayist/biographer Plutarch, who served as high priest at Delphi. Strabo (64 bce–25 ce) contributes:

They say that the seat of the oracle is a cave that is hollowed out deep down in the earth, with a rather narrow mouth, from which rises a pneuma (gas, vapor or breath) that inspires a divine frenzy; and that over the cleft is placed a high tripod, mounting which the Pythian priestess receives the pneuma and then utters oracles in both verse and prose (Strabo, 1927).

Plutarch (46–120 ce) left several accounts of the activities at the oracle and describes the relationship between Apollo and the Pythia as that of a musician and his instrument. He tells us that the *pneuma* was emitted "as from a spring" in the *adyton* (reserved and restricted inner sanctum) and scented like sweet perfume. Priests and consultants could on some occasions detect the aroma of the *pneuma* in the antechamber where they waited for the Pythia's responses. Plutarch mentions that during his tenure, emissions had become weaker and more irregular. He suggests that either the vital essence had run out or that heavy rains in the mountains had diluted it. Earthquakes, he proposes, might have partially blocked the mouth of the cave, thus causing reduced flow. His hypotheses indicate a remarkable understanding of the hydrology and geology of the area! (Plutarch, 1936).

John Collier, Britain, 1850–1934, (*Priestess of Delphi*, 1891, London, oil on canvas, 160.0 × 80.0 cm, Gift of the Rt. Honourable, the Earl of Kintore 1893, Art Gallery of South Australia, Adelaide) (Fig. 9.1).

Pausanias (c. 110–180 ce) writes about seeing a spring (the Kerna?) emerge on the slope above the temple and had heard that it plunged underground and emerged in the *adyton*, "where its waters made the priestesses prophetic" (Pausanias, 1935).

Oracular sessions were held on Apollo's Day, the seventh day after each full moon in spring, summer, and fall. The oracle did not operate in the winter months. Pilgrims would travel for weeks, mostly by boat, disembark at Kirrha, and climb the steep path to the oracle. Once there, they had to participate in several ritual procedures and wait their turn. If they missed sessions, or were passed over by more important patrons, they had to wait a full month or longer. During that time they could, and no doubt did, exchange information on the socioeconomic and political conditions in their kingdoms, cities, or towns. This was valuable material for the priests, who became well aware of the goings-on in their Mediterranean world. One wonders whether such intelligence may

FIGURE 9.1 Priestess of Delphi.

have colored their interpretations of the Pythia's responses during mantic services.

The Pythias were women from the village selected by the priesthood of the temple. Early in the morning on Apollo's Day, priests would lead the chosen priestess, who had been fasting for days, from her secluded home to the Kastalia spring. After purification, she would slowly and ceremoniously ascend the Sacred Way to the temple. Before entering she was offered a cup of holy water, obtained at the prophetic spring.

After a few more ceremonies she would enter the temple and descend into its *cella* where the small *adyton* was located. After being seated on the tripod and given a few minutes to inhale the *pneuma*, the Pythia would go into a trance and the head priest would relay questions submitted by the visitors, who remained in a separate antechamber. During her trance, the Pythia spoke in altered voice and at times chanted her responses, on occasion indulging in wordplay, which the priests would decipher for the enquirer(s) to interpret.

Reconstructions of the *adyton* by Courby and Amandry suggest an asymmetric structure, vis-à-vis that of the temple that occupied a small space in the *cella* against its southeastern wall, below the site where the sixth column of the internal colonnade is believed to have been purposefully omitted during (re)construction in the 4h century bce. The *adyton* was presumably covered by a low ceiling and measured about 2.90 m × 5.60 m. Inside stood a tripod and a beehive-shaped *omphalos* (the umbilicus of the earth) (Courby, 1927).

Since the volume of the *pneuma* that rose into the *adyton* presumably was small, there must have been a device that allowed for it to be concentrated. In September 1913, in the late stages of the long excavation process, a small *omphalos*-shaped relic appeared mysteriously inside the temple. French archeologists dismissed it as a modern base for a cross of the type commonly found on Christian shrines that had somehow slid inside the *cella*. Others, including Holland and the author, believe it to be genuine. This relic (28.7 cm high) is similar in shape and size to *omphaloi* depicted on bas-reliefs, coins, and ceramic wares from both the Classical and Hellenistic periods. It contains a square hole (3 cm × 4 cm wide) carved from top to bottom, widening downwards. Holland proposes that this funnel served to convey the inspiring fumes to the Pythia. In the month between successive mantic sessions, a simple cork would have allowed for as much as 300 mL of gas to accumulate inside this conduit and much more in the space carved out of its foot block (Bousquet, 1951).

The design of the Apollo temple is unique in all of Greece. First of all, it was constructed on a slope with abundant evidence for instability as the result of rock falls and landslides. Second, unlike most Greek temples, Apollo's realm had a recessed earthen *cella*. And third, a spring emerged inside as testified by the presence of thick travertine crusts and construction of a drainage tunnel leading through the southeastern retaining wall to the Fountain of the Muses. Clearly its design was based on a special need: the protection of a holy spring (Fig. 9.2).

The orientation and setting of the temple has to do with the presence of a major, active fault, which is part of the many fractures that developed along the northern flank of the Corinth rift zone that cuts an east—west swath through central Greece. It is one of the most seismically

FIGURE 9.2   Temple of Apollo at Delphi.

active and rapidly extending areas in the world. Tectonic slip planes of the Delphi fault are exposed both east and west of the oracle site and their connection indicates that it passes at depth below the temple. With a tectonic recurrence periodicity of several centuries, segments of the crustal block south of this fault have slipped down a few meters at a time, causing earthquakes. Evidence of at least three historic ruptures can be recognized on the exposed fault planes. They might be related to destructive seismic events that occurred in the 6th century bce (leading to construction of the Cyclopean wall), in 373 bce, and in 83 bce.

The Delphi fault intersects the Parnassus—Ghiona tectonic unit of the Hellinides, a massive complex of westward-thrusting limestone/dolomite slabs of the Cretaceous and Paleogene ages. The former were deposited in a shallow subtropical ocean and contain strata enriched in bituminous matter. In addition to this major east—west fault, swarms of older northwest- and northeast-trending fractures cut through the area. Groundwater rising along a concave segment of the Delphi fault emerges at the Kerna spring and continues downslope inside the northwest-trending Kerna fracture zone. Its presence inside the sanctuary is clearly shown by the linear arrangement of four extinct springs with travertine crusts.

In the late 1990s, an American team composed of a geologist (author), archeologist (Hale), and geochemist (Chanton) carried out an interdisciplinary study of the oracle site and Delphi region. The study included detailed analyses of the fault/fracture systems and the chemistry of the groundwater and travertine deposits. Sampling included travertine from inside the temple and thick crusts that had adhered onto the massive Ischegaon wall constructed after an earlier temple crumbled during the 373 bce seismic events. These samples were found to contain small quantities of both methane and ethane. The highest concentrations were encountered in samples collected inside and near the southeastern wall of the temple. Commonly only very small volumes of gas are captured inside porous travertine deposits around active springs. The presence of hydrocarbon gases in the old crusts, however small, therefore indicates that significant volumes surfaced when this rock was formed (de Boer, 1992).

Both ethane and methane were also encountered in water from the Kerna spring, as expected, in larger volumes. The Kerna water furthermore included trace amounts (7 nL/L) of ethylene. These discoveries led the researchers to conclude that hydrocarbon gases had emerged with groundwater in the *adyton* and that the gases likely represented the intoxicating vapors mentioned by the ancient sources (Hale et al., 2003).

The volume of water and gases that commonly surfaces along active faults varies significantly as a function of fault topography and recurring local as well as regional seismicity in the Gulf of Corinth rift zone. Brecciation during rupture and/or strong shaking opens fault zones and frequently increases their permeability. In March 1981, the author observed a major upwelling of gas in the Gulf of Alkyonides following rupture and seismic aftershock activity along a fault similar to the one at Delphi. Over time, newly opened spaces are slowly choked by the formation of calcareous crusts, which reduce or can even stop groundwater flow until seismic reactivation occurs. The present-day reduced flow of the Kerna spring is probably related to additional problems. Widespread deforestation and drainage of a large wetland on the southern flank of Mount Parnassus increased regional runoff and reduced groundwater penetration into bedrock.

In 2006, an Italian—Greek team of scientists (Etiope, Papatheodorou, Christodoulou, Geraga, and Favali) reappraised the natural gas occurrences at the oracle site. Their surveys included gas fluxes from soils, gas in groundwater, and isotopic analyses of the travertine deposits. They confirmed the earlier discovery of the hydrocarbon gases and concluded that methane, ethane, and carbon dioxide had been released from a thermogenic (catagenic) hydrocarbon source at depth, most likely bituminous enriched strata in the limestone complex. They found the highest methane flux in soils around the Kerna spring (145 mg/m^2 day). The soils in the temple *cella* contained 65 mg/m^2 day. Kerna spring water yielded concentrations of 73.2 nmol/L methane and 2.2 nmol/L ethane. They found that the travertine crusts near the Kerna spring were formed mainly by oxidation of hydrocarbons to carbon dioxide, a process that also greatly facilitated the formation of thick travertine crusts elsewhere in the sanctuary.

The Italian–Greek team thus confirmed the surfacing of light hydrocarbon gases but did not believe that sufficient ethylene for inducing trancelike states in Pythias could have been produced in the deep carbonate rocks below Delphi, since that specific environment is not prone to biogenic production of ethylene. However, an abiogenic transformation of ethane to ethylene under thermal conditions generated by frictional heating in the fault zone during seismic rupture cannot be excluded. The team posited that the gas-linked neurotoxic effects could have resulted from oxygen depletion due to carbon dioxide/methane inhalation in the enclosed and nonventilated *adyton*. They also proposed that benzene, an aromatic hydrocarbon commonly dissolved in groundwater springs, might have been the culprit. Short-term exposure to benzene, however, leads to shortness of breath, diminished mental alertness, and impaired vision. It also causes emotional instability. No such effects are reported to have commonly occurred in Delphi's long history (Etiope et al., 2006).

During his long tenure at Delphi, Plutarch described only one instance in which a Pythia was not responding properly. He wrote, "She was like a laboring ship and was filled with a mighty and baleful spirit. Finally she became hysterical and with a frightful shriek rushed toward the exit" (Plutarch, 1936).

Scientists working at substance abuse centers have reported that all three gases—ethylene, ethane, and methane—can potentially produce an altered mental state. In present-day inhalant abuse, light hydrocarbon gases, due to their intoxicating properties, remain the primary sought-after substances. Ethylene, an aliphatic hydrocarbon gas with a sweet odor, is the most likely candidate for inducing a trance. A major advantage of ethylene is its lack of respiratory and cardiovascular depressing effects. Ethane would be nearly as potent. A mixture of the two would produce a significant altered state (Spiller et al., 2002a).

Ancient accounts have provided relatively clear descriptions of the Pythia in trance, namely that of a woman who willingly entered the *adyton*, remained conscious, and was able to respond to questions with visions described in tones and patterns of altered speech. Recovery was said to be rapid, and other than an out-of-body experience—the feeling of being possessed by the god Apollo—the Pythia would not remember events that had occurred while in trance. These effects are similar to those described by 20th- and 21st-century toxicologists for patients under mild ethylene-induced anesthesia and thus add another level of congruency to reconcile contemporary science with classical wisdom concerning the Pythia and the *pneuma* (Spiller et al., 2002b).

When French archeologists began to excavate ancient Delphi in the 1890s, they first had to remove the modern settlement of Kastri, which had been built on top of the ruins. After exhaustive labor they unearthed the temple foundations but did not find a clear trace of the *adyton*. Furthermore, finding no evidence for a cave—and detecting no gases—they concluded that the ancient references must be erroneous. Unfortunately, their assumptions invaded principal textbooks, such as those by Parke and Fontenrose, and became dogmatic among scholars for much of the 20th century (Oppe, 1904).

In 2007, classicists Foster and Lehoux published a criticism of the Delphic ethylene intoxication hypothesis and concluded that the Pythias' behavior could not be accounted for by ethylene intoxication, because insufficient volumes rose into the *adyton*. No substantiation for this statement, other than referring to the hypothesis of Ethiope and colleagues, was provided. Continuing in its century-old vein, their criticism no longer centers on the possibility of vapors having risen in the inner sanctum, but on their volume: "Όσο πιο πολύ αλλάζουν τα πράγματα, τόσο παραμένουν ίδια!" (The more things change, the more they stay the same!) (Foster and Lehoux, 2007).

The ancient belief in intoxicating gaseous emissions at the site of the Delphi oracle is not a myth. An unusual, but by no means unique, combination of faults, bituminous limestone, and rising groundwater worked together to bring volatile hydrocarbon gases to the *adyton*. Contemporary geologic research has thus reaffirmed many if not most aspects of the ancient sources.

References

Bousquet, J., 1951. Observations sur l'omphalos archaique de Delphes. Bulletin de Correspondance Hellenique 75, 210–230; Holland, L.B., 1951. The mantic mechanism at Delphi. Am. J. Archaeol. 37, 201–214; Zeilinga de Boer, J., 1951. Delphi's small omphalos: an enigma. Syll Class 18 (2007), 81–104.

Courby, F., 1927. Fouilles de Delphes, tome II, Topographie et Architecture. E. de Boccardm, Paris; Amandry, P., 1927. La mantique Apollonienne A Delphes; Essai sur le functionnement de l'oracle. E. de Boccard, Paris; Amandry, P., 1927. Recherches sur la cella du temple de Delphes. E. de Boccard, Paris, pp. 215–230.

de Boer, J.Z., 1992. Dilational fractures in the Corinthian and Evian rift zones. Ann. Tectonicae 6, 41–61; de Boer, J.Z., Hale, J.R., 1992. The geological origins of the oracle in Delphi, Greece. Geological Society of London, Special Publication, London, pp. 399–412; de Boer, J.Z., Hale, J.R., Chanton, J., 1992. New evidence for the geological origins of the ancient Delphic oracle (Greece). Geology 29 (8), 707–710.

Etiope, G., Papatheodorou, G., Christodoulou, G., Geraga, M., Favali, P., 2006. The geological links of the ancient Delphic oracle (Greece): a reappraisal of natural gas occurrences and origin. Geology 10, 821–824.

Foster, J., Lehoux, D., 2007. The Delphic oracle and the ethylene-intoxication hypothesis. Clin. Toxicol. 45 (1), 85–89; Lehoux, D., 2007. Drugs and the Delphic oracle. Class World 10 (1), 41–56; Etiope, G., Papatheodorou, G., Christodoulou, G., Geraga, M.,

Favali, P., 2007. The geological links of the ancient Delphic oracle (Greece): a reappraisal of natural gas occurrences and origin. Geology 10, 821–824.

Hale, J.R., de Boer, J. Zeilinga, Chanton, J.P., Spiller, H.A., 2003. Questioning the Delphic oracle. Sci. Am. 8, 67–73.

Oppe, A.P., 1904. The chasm at Delphi. J. Hell Stud. 24, 214–240; Parke, H.W., 1904. A History of the Delphic Oracle. Blackwell, Oxford, UK; Parke, H.W., Wormell, D.E.W., 1904. The Delphic Oracle. Blackwell, Oxford, UK; Fontenrose, J., 1904. The Delphic Oracle, Its Responses and Operations. University of California Press, Berkeley, CA.

Pausanias, 1935. Pausanias: Description of Greece (W. H. S. Jones, Trans.). Harvard University Press, Cambridge, MA.

Philostratus (the Athenian), Eunapius, MCMXXII. Philostratus and Eunapius: The Lives of the Sophists, 427. London/New York, NY: William Heinemann, G. Putnam's Sons.

Plutarch, 1936. De defectu oraculorum, Moralia, Vol. 499–500. Harvard University Press, Cambridge, MA.

Spiller, H.A., Hale, J.R., de Boer, J.Z., 2002a. The Delphic oracle: a multidisciplinary defense of the gaseous vent theory. Clin. Toxicol. 40 (2), 189–196; Dorsch, R.R., De Rocco, A.G., 2002a. A generalized hydrate mechanism of gaseous anesthesia, 1. Theory. Physiol Chem Phys. 5, 209–223.

Spiller, H.A., Hale, J.R., de Boer, J.Z., 2002b. The Delphic oracle: a multidisciplinary defense of the gaseous vent theory. Clin. Toxicol. 40 (2), 189–196; Broad, W.J., 2002b. The oracle. The lost secrets and hidden message of ancient Delphi. The Penguin Press, New York, NY.

Strabo, 1927. Geography, 9.3. Putnam's Sons, New York, NY.

Further Reading

Broad, W.J., 2006. The Oracle: The Lost Secrets and Hidden Message of Ancient Delphi. The Penguin Press, New York, NY.

Lipsey, R., 2001. Have You Been to Delphi? State University of New York Press, Albany, NY.

10

Alexander the Great: A Questionable Death

Adrienne Mayor

Classics Department and Program in History and Philosophy of Science,
Stanford University, Stanford, CA, United States

On his way back to Greece after spectacular conquests from Persia to India, Alexander the Great (b. 356 BC) died in Babylon in the Palace of Nebuchadnezzar II (in modern Iraq) on June 10 or 11, 323 BC. During one of many all-night drinking parties in Babylon, Alexander's companions had heard him cry out in pain, complaining of a "sudden, stabbing agony in the liver." The young conqueror of the known world was put to bed with severe abdominal pains and a very high fever. Over the next 12—14 days, his condition declined and he became gravely ill. The ancient sources report that Alexander suffered restlessness, weakness, extreme thirst, loss of consciousness, possible convulsions, and great pain. Partial paralysis set in—he was only able to move his eyes, head, and fingers with difficulty. He lost the ability to speak and fell into a deathlike state.

Toxicology in Antiquity
DOI: https://doi.org/10.1016/B978-0-12-815339-0.00010-X
151

10.1 ALEXANDER'S LAST DAYS

Several ancient Greek and Roman historians described Alexander's last days. They had access to many contemporary texts that no longer survive, including a mysterious source called the "Royal Diaries" or "Journal" (Badian, 1968; Romm, 2011). We know that five men close to Alexander wrote accounts of his death: Alexander's bodyguard and friend Ptolemy, his admiral Nearchus, his secretary Eumenes, his chamberlain Chares, and his military engineer Aristobulus. Unfortunately, their memoirs are all lost except for fragmentary quotations preserved by later historians, including Diodorus Siculus (1st century BC); Plutarch (about 100 AD); Pliny and Quintus Curtius Rufus (both 1st century AD); Arrian, Pausanias, and Justin (2nd century AD); Aelian (about 200 AD); and the so-called *History* or *Romance of Alexander* (dating to about 250 AD; several manuscript versions exist).

According to Diodorus (17.117), at the last banquet he attended in Babylon, Alexander "drank much unmixed [strong] wine... and finally gulped down a huge beaker. Instantly he shrieked aloud as if struck by a violent blow." His attendants "conducted him to bed, and his physicians were summoned," but Alexander continued to suffer in great pain. Justin (12.13−15) adds more details: "Taking up a cup, he had drunk half of it when he suddenly uttered a groan, as if he had been pierced by a spear; he was carried half-conscious from the banquet. The torture was so excruciating that he called for a sword to put an end to it. The pain upon being touched by his attendants was if he were covered with wounds.... On the sixth day, he could no longer speak."

The historian Arrian (7.25−26) affirmed that Alexander was unable to speak near the end: "Lying speechless as his men filed by, he struggled to raise his head, and in his eyes there was a look of recognition for each individual as he passed." Justin (12.17) had also noted that Alexander was speechless but was able to move his hands: "Unable to speak, he took his ring from his finger, and gave it to Perdiccas [Alexander's trusted general]."

Alexander's biographer Plutarch (*Alexander* 75−77) denied the stabbing pain, describing only "a raging fever" marked by "violent thirst." Plutarch said "he drank a lot of wine, upon which he fell into delirium." Plutarch also described restlessness, loss of appetite, high fever, and inability to speak.

According to the 3rd-century AD *Greek Alexander Romance* (3.30−31), which contains both historical facts and fantasy about Alexander's life and death, after his cupbearer Iolaus served him a beaker of wine late in the evening, Alexander began showing signs of malaise, getting up and pacing around the room. "He again sat down [and] with his hands trembling, complained that it was as if a heavy yoke were upon his neck. When he stood again to drink... he shouted with pain as if pierced in the liver with an arrow." The *Romance* continues: "[R]acked with pain," Alexander's condition deteriorated, and he "could not speak because his tongue was so swollen." He suffered convulsions, delirium, hallucinations, and bouts of unconsciousness. "Throughout the night the king would writhe and shake upon his bed, then he would become still. At other times he would ramble with meaningless words, appearing to speak with spirits in the bedchamber" (Stoneman, 1991).

About 13 days after falling ill at the banquet, Alexander was pronounced dead, on the afternoon of June 11, 323 BC, just before his 33rd birthday. His body was placed in a coffin in a storeroom. The Egyptian and Chaldean embalmers who arrived on June 16 noted that Alexander's body was strangely preserved, even in Babylon's hot climate (Plutarch *Alexander* 77; Curtius 9.19). This effect has been taken by modern historical detectives to indicate that some sort of poison may have preserved the corpse or that a profound coma was mistaken for death (Fig. 10.1).

Many ancient Greek and Roman writers speculated on the true cause of Alexander's death, which remains an unsolved mystery. Several ancient historians reported that rumors of poisoning circulated soon after the untimely death of Alexander (Bosworth, 1971; Bosworth, 2010; Lane Fox, 2004). His closest friends suspected a legendary poison gathered from the Styx River waterfall near Nonacris in Arcadia (north central Peloponnese, Greece), a substance reputed to be so corrosive it could only be contained in the hoof of a horse. The Styx River was thought to be an entrance to the Underworld in classical antiquity and its waters were believed to be toxic.

THE DEATH OF ALEXANDER

FIGURE 10.1   "Death of Alexander," painting by Karl Theodor von Piloty, 1886.

The ancient historians were divided on whether Alexander died of natural causes or was murdered by poison. Justin (12.13–14) and Pliny (*Natural History* 30.53) accepted the poison conspiracy, as did Pseudo-Plutarch (*Lives of the Ten Orators* 56; *Moralia* 849F). Pausanias (8.17.18), Arrian (7.27), and Curtius (10.10.14) were neutral. Plutarch (*Alexander* 75–77) was skeptical. Diodorus (17.117.5–118.1–2) was cautious, pointing out that after Alexander's bitter enemies, Antipater and Cassander, took over Alexander's empire, "many historians did not dare to write about the drug" or the plot. Moreover, he noted that Alexander's mother Olympias sought revenge on Cassander and others because she believed that they had poisoned her son (Diodorus 19.11.8). Olympias was then murdered by Cassander. Diodorus's intimations were more strongly expressed by Justin (12.13), who wrote, "The conspiracy... was suppressed by the power of Alexander's successors." In the Middle Ages and Renaissance, murder by poisoning was generally accepted as the cause of Alexander's death; Voltaire, among others, accepted the poisoning plot.

10.2 MODERN THEORIES OF NATURAL CAUSES

Many modern theories propose an array of natural causes for Alexander's slow death. These retrodiagnoses include alcohol poisoning from extremely heavy drinking, septicemia from infected old wounds that Alexander received on previous campaigns, pancreatitis, malaria, typhoid, West Nile fever, porphyria, schistosomiasis, accidentally harmful treatment by the royal physicians, or a combination of causes. In 2009, John Atkinson, Elsie Truter, and Etienne Truter produced a valuable summary and full bibliography of Alexander's last days, with the reported symptoms, a timeline based on the ancient sources, and an appendix of proposed causes of death and their merits and drawbacks (Atkinson et al., 2009).

Before falling ill in Babylon, Alexander had traveled across India, Pakistan, Iran, and Iraq. Hindu doctors accompanied his army in India and Pakistan; they provided Alexander with exotic natural medicinal plants, venoms, and minerals. Significantly, Alexander was known to have treated himself with unknown, powerful drugs—in one of these experiments he nearly died. One might speculate that some experimental self-treatment was the cause of his death (Mayor, 2010).

It also may be significant that in the 2 weeks before his death Alexander had been sailing for 3 days in the Great Swamp of Babylonia, the marshy Tigris–Euphrates delta (Diodorus 17.116.5–7). This raises the possibility of mosquito-borne disease; many of the symptoms fit the ancient descriptions. West Nile fever has been suggested (Marr and

Calisher, 2003), but that is a recently evolved disease; perhaps an unknown precursor existed in Alexander's era. Malaria was first proposed by a French physician in 1878 and has been promoted by several recent historians (Engels, 1978; Borza, 2010; Thompson, 2005). Malaria is an attractive candidate, but the distinctive recurrent fever curve in malaria caused by the *Plasmodium* parasite was absent from the ancient reports. Typhoid has also been suggested (Oldach et al., 1998; Cunha, 2004). A complication of typhoid is ascending paralysis, which could have caused Alexander to appear dead for several days before he died. But if any of these infectious diseases was the culprit, it seems likely that other people in Babylon would have been struck with similar symptoms. According to the detailed historical accounts, Alexander was the only one to fall ill, a fact that could point to poisoning.

10.3 MODERN THEORIES OF POISONING

Many modern arguments have been made for accidental or deliberate poisoning. In one theory, for example, Alexander's doctors treated him with a fatal dose of medicine in an attempt to counteract poisoning. Leo Schep, a toxicologist, considered what sort of drug might have been administered if Alexander was thought to have ingested poison; in 2009, he concluded that hellebore was the likely culprit (Schep et al., 2009). In 2014, Schep's team suggested that if Alexander had been deliberately poisoned, it would have been by white hellebore (*Veratrum album*) fermented into wine (Schep et al., 2014). Arguments against hellebore include the following facts: hellebore is intensely bitter; there is no evidence for fermentation into wine in antiquity; and hellebore was notorious for causing violent gastrointestinal symptoms and even death in weak patients. Often prescribed as a purge in antiquity, hellebore had to be used very cautiously. The symptoms of hellebore poisoning were very well known generally and an overdose would have been recognized by Alexander's doctors and probably by his companions. The effects are unmistakable: hellebore induces explosive diarrhea and vomiting. Significantly, neither of these hallmark symptoms was mentioned in any of the ancient sources who described Alexander's death.

If Alexander was poisoned by his enemies, the agent may have been an easily available mineral or plant toxin. In 2004, Paul Doherty concluded that Alexander was poisoned with arsenic, which might have preserved the body from decomposition (Doherty, 2004). Strychnine was first suggested by Milnes (1968), citing Theophrastus (*History of Plants* 9.11.5–6). In 2004, Phillips (2004) suggested an alkaloid plant such as belladonna (deadly nightshade), aconite (monkshood)—both easily available—or strychnine. Since Alexander did not suffer vomiting, which always accompanies alcohol, belladonna, arsenic, and aconite poisoning (as well as hellebore ingestion), Phillips concluded that the agent must have been strychnine (*Strychnos nux-vomica*). Strychnine poisoning appears to match some of the symptoms: paroxysmal contractions of muscles followed by complete relaxation, skin very painful to touch, high fever, sweating, high blood pressure, intense thirst, and lockjaw. *S. nux-vomica* has a very bitter taste, however, making it difficult to hide in a drink. This plant would have had to come to Babylon from southern India, never reached by the Greeks. Moreover, poisoning by this and other plant toxins would have required repeated doses to cause Alexander's slow death, and this would increase the chances of the plot being discovered. It is also notable that Alexander, who was described as pathologically paranoid at this time, apparently did not himself suspect poisoning (Engels, 1978).

10.4 THE STYX RIVER POISON PLOT

Rumors of poisoning began to circulate among some of Alexander's companions soon after his death (Justin 12.14; Plutarch *Alexander* 77.1–3; Diodorus 17.118). Many people, both in Babylon and in Macedon, had the motives and the means. Suspicion fell on his enemy Antipater, the viceroy in Macedonia, and on Antipater's son Cassander, who had recently arrived in Babylon from Greece. Plutarch (*Alexander* 75–77) gives a detailed account of the alleged conspiracy and the special poison from the Styx, mentioned above. Some (such as Arrian 7.24–27; Plutarch *Alexander* 77; Pliny 30.53) claimed that it was Aristotle, Alexander's old friend and tutor, who had provided the Styx poison because he now feared his student. Aristotle was in Athens at the time of Alexander's death. He was said to resent the murder of his nephew by Alexander in 327 BC. Was Aristotle also suspected because he had written about Styx poison in a lost book of natural history? We know that Aristotle's fellow natural philosopher Theophrastus did write about Styx poison.

According to Arrian (7.24–27), "Antipater sent Alexander medicine which had been tampered with and he took it, with fatal results. Aristotle is supposed to have prepared this drug.... Antipater's son Cassander is said to have brought it [to Babylon]. Some accounts declare that he brought it in a mule's hoof, and that it was given to Alexander by Cassander's younger brother Iolaus, who was his cup-bearer.... [O]thers state that Medius, Iolaus' lover, had a hand in it.... [I]t was Medius who invited Alexander to the drinking-party [where Alexander] felt a sharp pain after draining the cup." The *Greek Alexander Romance* (Stoneman, 1991) maintained that the banquet was a conspiracy involving Antipater, Cassander, Iolaus, and others. Düring (1957) gathered and commented on the ancient evidence for Aristotle's involvement, and he suggested that the case was a common *topos* in public debates by later peripatetic philosophers. Plutarch (*Alexander* 75–77) identifies the authority for implicating Aristotle as Hagnothemis, who heard it from Antigonus, a trusted contemporary of Alexander. The Styx poison said to have been prepared by Aristotle was "deadly cold water from the rock cliff near Nonacris, gathered [or distilled] like a delicate dew [or exudation] and stored in an ass's hoof, for all other vessels were corrupted by its icy, penetrating corrosiveness."

According to the *Greek Alexander Romance*, Alexander was killed with a poison that destroyed bronze, glass, and clay, and had to be sealed in a lead jar inside an iron jar. When Alexander tried to induce vomiting to rid himself of the poison, Iolaus gave him a poisoned feather. A 14th-century illustrated manuscript of the *Greek Alexander Romance* contains miniature paintings depicting the poison being transported in a lead *pyxis* (lidded box) from Greece to Babylon, the poison passed to Iolaus the cupbearer, and Alexander drinking from a glass goblet. The motif of a feather coated with poison brings to mind the account of the poisoning of the Roman emperor Claudius, in Tacitus (*Annals* 12.56–58): when poisoned mushrooms prepared by the notorious poisoner Locusta failed to kill Claudius, the emperor called for a feather to induce vomiting but his doctor poisoned the feather. Some of Alexander's symptoms and the course of his illness seem to match ancient Greek myths about water from the Styx—for example, he lost his voice, like the Olympian gods who fell into a coma-like state after drinking from the Styx. This similarity of symptoms may have led his companions and others to assume (or to claim for propaganda purposes) that Styx poison was responsible. Such a sacred *pharmakon* or drug would cast an aura of divinity on Alexander. The notion that their great hero had succumbed to the fabled poison taken from the gods' sacred oath river would have carried symbolic resonance. Combined with ancient traditions associating the Styx with immortality, Alexander's friends could blame a mythic drug worthy of their "semidivine" leader (Fig. 10.2).

The true cause of Alexander's untimely death more than 2300 years ago will probably never be solved with certainty. Even if Alexander *was* poisoned, the story of a strange "icy dew" gathered from the banks of the Styx waterfall, the dismal entrance to the Underworld, seems fantastic. Yet the descriptions of that poison and its source remained consistent in many different sources over many centuries. Although some of the historians and naturalists doubted the reality of a poison plot, not one ancient writer ever cast doubt on the existence of a deadly substance from the Styx River. Possible identifications of this toxin include calicheamicin, a deadly bacterium that thrives on limestone crust

THE DEADLY STYX RIVER OF GREEK MYTH
DID POISON FROM THE STYX KILL ALEXANDER THE GREAT?

Adrienne Mayor Research Scholar, Classics & History of Science, Stanford, mayor@stanford.edu
Antoinette Hayes Toxicologist, Pfizer Research, antoinette.hayes@pfizer.com

Myth: The Styx, River of Hades

At the gloomy River Styx, portal to the Underworld, the gods swore sacred oaths. If they lied, Zeus forced them to drink the water, which "struck them down with an evil coma. They were unable to move, breathe, or speak for one year." (Hesiod, ca 750 BC)

Water from the Styx (identified as the Mavroneri River in Greece, see below) was said to be deadly to mortals. "Its lethal power was first recognized after goats tasted the water and died," wrote Pausanias (ca AD 170). "It dissolves crystal and pottery and corrupts bronze, lead, tin and silver vessels. The only thing able to resist corrosion is the hoof of a mule or horse. I have no actual knowledge that this was the poison that killed Alexander, but I have certainly heard it said."

Charon crossing the Styx, oil on wood, Joachim Patinir

Alexander's Mysterious Death in Babylon, 323 BC

At one of many all-night drinking parties in Babylon (in modern Iraq), Alexander's companions heard him cry out from a "sudden, sword-stabbing agony in the liver." The conqueror of the known world took to bed with abdominal pain and very high fever. He never recovered. Over the next 12 days, he worsened, moving only eyes and hands, unable to speak; paralysis followed by deathlike coma. Alexander was pronounced dead on June 11, just before his 33rd birthday. Various theories for cause of death have been proposed: heavy drinking, septicemia, pancreatitis, malaria, West Nile fever, typhoid—and accidental or deliberate poisoning (hellebore, arsenic, aconite, strychnine).

Death of Alexander the Great, Karl Theodor von Piloty

Legendary Styx Poison Suspected

Alexander's closest friends believed that his enemies (many had motives) murdered him with poison. Some suspected "the icy cold exudation or dew, gathered from the mossy rocks at the high Styx waterfall," carried to Babylon sealed in a mule's hoof. Some of Alexander's symptoms seem to match ancient Greek myths about the Styx—he even lost his voice, like the gods. (Arrian, Plutarch, Diodorus, Justin)

Styx/Mavroneri Falls, Mt Helmos, Achaia, photo by R. Pona

Naturally Occurring Deadly Toxin?

The Styx (now called Mavroneri, "Black Water") originates in the high limestone mountains of Achaia, Peloponnese. "Plunging over jagged mountains, the cold water of the Styx cascades" over a limestone crag to form the highest waterfall in Greece (200 meters, 650 feet). In modern times, locals avoid drinking the water and say it corrupts vessels.

Consistent ancient literary evidence suggests that something in the Styx/Mavroneri gave rise to its poisonous reputation. The stream probably contains corrosive or caustic minerals and other materials. These conspicuous effects, observed in antiquity, made the water undrinkable.

But the Styx may have hosted soil-derived killer bacteria as well. We suggest that the "mythic" poison from the Styx was a powerful bacterium new to modern science: **calicheamicin.** A secondary metabolite of an extremely toxic, gram-positive soil bacterium (*Micromonospora echinospora*), calicheamicin was discovered in the 1980s in caliche, crusty deposits of calcium carbonate that form on limestone.

Naturally occurring calicheamicin is extraordinarily potent—one of the most cytotoxic substances known (cellular lethality greater than that of ricin or cyano morpholinyl anthracycline, Lode et al. 1998). A small amount causes irreparable double-strand breaks in DNA and apoptosis (genetically programmed cell death), and could lead to massive organ failure (liver, heart, kidney, bladder, lung, nervous system), and death. Cytotoxic agents attack rapidly dividing cells (bone marrow, intestinal epithelium, hair follicles, epithelial cells of trachea and esophagus); symptoms include severe pain, weakness, fatigue, swelling of mouth and throat. Death would likely be imminent as there is no antidote once a lethal dose is ingested.

Caliche

Conclusion

Although scientists have not yet looked for calicheamicin in Greece, caliche is common in Greece's arid, limestone-dominated landscape, sometimes encrusting moss/lichen. (Higgins and Higgins 1996; Clendenon 2009) We propose 3 hypotheses:

• Calicheamicin was present in the Styx region in antiquity
• Its poisonous effects were observed in antiquity
• Alexander's death appeared to some to be caused by this poison

We may never know the real cause of Alexander's death. If he was poisoned, the agent may have been an easily available plant such as aconite, rather than calicheamicin scraped from Stygian limestone in Greece and transported to Babylon. Yet, just as Alexander's friends believed his death was fabled Styx poisoning, it's interesting to compare Alexander's symptoms to the deleterious effects of calicheamicin, which is 1,000 to 10,000 times more toxic than any known substance.

The rumor that an "icy cold poison gathered from the rocks of the Styx waterfall" killed Alexander may have had a basis in scientific fact.

For details and references, see our working paper: http://www.princeton.edu/~pswpc/papers/authorMZ/mayor/mayor.html

FIGURE 10.2 Did Poison from the Deadly Styx River of Greek Myth Kill Alexander the Great? Poster by Adrienne Mayor and Antoinette Hayes.

(Mayor and Hayes, 2010). Scientific analysis might 1 day reveal the identity of a potentially deadly poison from the Styx. That would account for the river's nefarious reputation—but the demise of Alexander would still remain a mystery.

References

Atkinson, J., Truter, E., Truter, E., 2009. Alexander's last days: malaria and mind games? Acta Classica 52, 23–46.

Badian, E., 1968. A king's notebooks. Harv. Stud. Classical Philol. 72, 183–204.

Borza, E.N., 2010. Alexander's death: a medical analysis. In: Romm, J. (Ed.), The Landmark Arrian: The Campaigns of Alexander. Pantheon, New York, NY, pp. 404–406.

Bosworth, A.B., 1971. The death of Alexander the Great: rumour and propaganda. Classical Q. 21, 112–136.

Bosworth, A.B., 2010. Alexander's death: the poisoning rumors. In: Romm, J. (Ed.), The Landmark Arrian: The Campaigns of Alexander. Pantheon, New York, NY, pp. 407–410.

Cunha, B., 2004. The death of Alexander the Great: malaria or typhoid fever? Infect. Dis. Clin. North Am. 18, 53–63.

Doherty, P., 2004. The Death of Alexander the Great: what—or Who—Really Killed the Young Conqueror of the Known World? Carroll and Graf, New York, NY.

Düring, I., 1957. Aristotle in the Ancient Biographical Tradition. Almquist, Göteborg.

Engels, D., 1978. A note on Alexander's death. Classical Philol. 73, 224–228.

Lane Fox, R., 2004. Alexander the Great. Penguin, London.

Marr, J.S., Calisher, C.H., 2003. Alexander the Great and West Nile virus encephalitis. Emerg. Infect. Dis. 9, 1599–1603.

Mayor, A., 2010. The Poison King: The Life and Legend of Mithradates, Rome's Deadliest Enemy. Princeton University Press, Princeton, NJ.

Mayor A, Hayes A., 2010. The deadly Styx River of Greek myth: did poison from the Styx kill Alexander the Great? In: Poster, Toxicology History Room, XII International Congress of Toxicology. Barcelona, Spain.

Milnes, R.D., 1968. Alexander the Great. Pegasus, New York, NY.

Oldach, D., Richard, R.E., Borza, E., Benitez, R.M., 1998. A mysterious death. N. Engl. J. Med. 338, 1764–1769.

Phillips, G., 2004. Alexander the Great: Murder in Babylon. HB Virgin Publishing, London.

Romm, J., 2011. Ghost on the Throne: the Death of Alexander the Great and the War for Crown and Empire. Knopf, New York, NY.

Schep, L., Wheatley, P., Hannah, R., 2009. The death of Alexander the Great: Reconsidering poison. In: Wheatley, P., Hannah, R. (Ed.), Alexander & His Successors. Regina Books, Camas, WA, pp. 227–236.

Schep, L.J., Slaughter, R.J., Vale, J.A., Wheatley, P., 2014. Was the death of Alexander the Great due to poisoning? Clin. Toxicol. 52, 72–77.

Stoneman, R., 1991. The Greek Alexander Romance (Pseudo-Callisthenes, Trans.). Penguin, London.

Thompson, J., 2005. Disease, not conflict, ended the reign of Alexander the Great. Independent on Sunday.

11

Mithridates of Pontus and His Universal Antidote

Adrienne Mayor

Classics Department and Program in History and Philosophy of Science, Stanford University, Stanford, CA, United States

Mithridates VI Eupator of Pontus inherited the small, wealthy kingdom of Pontus on the Black Sea (today northeastern Turkey) in 120 BC, after his father was poisoned by enemies. With good reason to believe that his mother, Queen Laodice, intended to poison him in order to control Pontus, the teenaged Mithridates went into hiding for several years. Upon his return he assumed his throne and used poison (likely arsenic) to eliminate several treacherous relatives and rivals. Mithridates, who claimed descent from Persian royalty and Alexander the Great, became the most dangerous and relentless enemy of the late Roman Republic in decades-long conflicts known as the Mithridatic Wars. After his defeat

Toxicology in Antiquity
DOI: https://doi.org/10.1016/B978-0-12-815339-0.00011-1

by Pompey in the Third Mithridatic War, he was forced to commit suicide (63 BC).

Driven by his fears of assassination by poison, Mithridates is acknowledged as the first experimental toxicologist, carrying out proto-scientific experiments with poisons and antidotes (Griffin, 1995). His goal was to create a "universal antidote" to make himself and his friends immune to all poisons and toxins. An erudite scholar in many tongues, he had access to myriad poisons, toxicological traditions, and examples, both mythic and historical, to guide him (Mayor, 2010).

11.1 INFLUENCES

Mithridates was aware that his land was the home of the mythical witch Medea, adept in poisons and magic. Medea was said to tame unquenchable flames from the petroleum pools of Baku on the Caspian Sea, and her potions were said to bestow superhuman powers, death-like sleep, or immunity from fire or sword. She also knew the secrets of deadly "dragon's blood" (a name for the toxic mineral realgar) and all the antidotes for serpent venom, secrets eagerly sought by young Mithridates, future toxicologist (Mayor, 2010).

Mithridates' own grandfather, King Pharnaces I of Pontus, was cred-ited with the discovery of a "panacea" (cure-all, centaury plant, Pliny 25.79). Another probable influence would have been the unusual research first begun by "mad" Attalus III of Pergamon, the last king of a neighbor-ing land taken over by Mithridates during his conquests of Asia Minor. Pergamon, with its great library, active scientific community, and the healing temple of Asclepius, was the center of medical learning. Nicander of Colophon, a Greek physician who wrote the *Theriaca* on venomous creatures, and the *Alexipharmaka* on poisons and antidotes, was a member of Attalus's court (Scarborough, 2007). Ancient and modern historians have assumed that the eccentric King Attalus was insane because he preferred scientific experimentation to governing. As a boy, Mithridates heard the rumors accusing Attalus of poisoning his relatives and foes and mocking him for withdrawing from court life, devoting himself to tending his extensive gardens, and studying botany, pharmacology, and metallurgy. He died in 133 BC, around the time of Mithridates' birth (Griffin, 1995; Totelin, 2004).

But was Attalus really insane? Modern historian Kent Rigsby sug-gests that the king's reputation for murder and madness was perpetu-ated by those who wanted to make Attalus seem an unfit ruler. Pointing out that scientific and philosophical pursuits were typical of several other sophisticated Hellenistic monarchs, Rigsby reasons that that "in reality, Attalus was a scientist and scholar." Attalus may have been eccentric, but his activities seem to constitute scientific research (Rigsby, 1988).

Indeed, the most remarkable significance of Attalus's research for understanding Mithridates has been overlooked by Rigsby and other modern historians. Justin's most damning example (36.4) of Attalus's insanity was the king's obsession with "digging and sowing in his garden" and his bizarre practice of concocting "mixtures of both healthful and beneficial plants and drenching them with the juices of poisonous ones." Attalus presented these concoctions as "special gifts to his friends." Ancient historians (Plutarch *Demetrius* 20.2; Diodorus 34–35.3) tell us that Attalus cultivated toxic plants such as "henbane, hellebore, hemlock, aconite (monkshood), and thorn apple (*Datura*) in his royal gardens and became an expert in their juices and fruits." The celebrated physician from Pergamon, Galen (b. 129 AD), added further information. Galen said that Attalus experimented with antidotes against the venoms of snakes, spiders, scorpions, and toxic sea slugs. Galen (*Antidotes* 1.1) praised Attalus for testing his mixtures only on condemned criminals (Scarborough, 2007).

It can be no coincidence that Mithridates engaged in the very same sorts of activities and experiments as his own grandfather Pharnaces, "mad" King Attalus, and Nicolas of Colophon. According to Justin, Mithridates began his investigations of *pharmaka* (ancient Greek for drug or medicine) as a boy, secretly testing toxins and antidotes on others and himself. Mithridates' celebrated "universal antidote," an alexipharmic panacea or theriac that later came to be known as the *Mithridatium*, was created by mixing minuscule doses of deadly poisons with antidotes (Mayor, 2010).

As king, Mithridates was known to have "amassed detailed knowledge from all his subjects, who covered a substantial part of the world" (Aelian *On Animals* 9.29). His international library of ethnobotanical and toxicological treatises described drugs used by the Druids of Gaul and Mesopotamian doctors, and he could have studied the works of Hindu Ayurvedic ("long-life") practitioners, such as the antidote recipe of Sushruta (c. 550 BC), which boasted 85 ingredients, and the *Mahagandhahasti* theriac of Charaka (300 BC), which had 60. In 88 BC, Mithridates received Marsi envoys from Italy, shamans known for their *pharmaka* based on venoms. Mithridates likely studied the alchemical writings of Democritus of Egypt, drawing on those of King Menes, who had cultivated poisonous and medicinal plants in 3000 BC. Mithridates corresponded with the doctor Zopyrus in Egypt, who shared his "universal remedy" of 20 ingredients. Another scientific colleague was Asclepiades of Bithynia, who founded an influential medical school in Rome. He declined Mithridates' invitation to work in Sinope but dedicated treatises to the king and sent him antidote formulas (Mayor, 2010).

11.2 PHARMACOLOGICAL AND TOXIC RICHES

Natural resources with powerful healthful or noxious characteristics abounded in the Black Sea region. There were many venomous snakes, for example. Mithridates' allies on the steppes, the mounted nomad archers of Scythia, poisoned their arrows with a sophisticated concoction of viper venom and other pathogens. Scythian shamans, called Agari, were experts in antidotes based on venoms, and several Agari joined Mithridates' investigations. Mithridates and other experimenters of his day were well aware of the thin line that divided potentially lethal doses from potentially beneficial amounts of powerful agents. His Agari doctors saved Mithridates' life on the battlefield by using snake venom to stop severe bleeding from a thigh wound, a medical milestone in the use of venom beneficially. Today, tiny amounts of viper venom from the Caucasus are used to staunch uncontrollable hemorrhage and investigators are creating anticancer drugs from venoms.

Mithridates' ally to the east, Armenia, had remote lakes with venomous fish. His birthplace of Pontus boasted its own extraordinary flora and fauna (Cilliers and Retief, 2000). Wild honey, distilled by bees from the nectar of poisonous rhododendrons and oleander so profuse on the Black Sea coast, contained a deadly neurotoxin. Even the flesh of Pontic ducks was poisonous. The ducks thrived on hellebore and other baneful plants, and the bees enjoyed a strange immunity to poison. These mysterious natural facts may have inspired Mithridates to search for ways to protect himself from poisons. Beavers were another prized Pontic product; their testicles were valued for treating fever and boosting immunity and sexual vigor, and in perfumes. Castoreum, from beaver musk glands, does contain salicylic acid, the active ingredient in aspirin, derived from willow bark, the beavers' chief food (Mayor, 2010).

The kingdom of Pontus was blessed with rich sources of *pharmaka*, potent substances and products used in many different technologies and crafts such as metal working, dyes, and pigments; in making medicines, unguents, and perfumes; and as poisons. Mithridates' extensive Black Sea Empire possessed myriad toxic plants, used for beneficial drugs or poisons: henbane, yew, belladonna or deadly nightshade, hemlock, thorn apple, monkshood, hellebore, poppies, fly agaric mushrooms, rhododendron, and oleander, to name a few (Mayor, 2010).

Nefarious, rare substances were mined in Mithridates' lands, from plentiful deposits of gold, silver, copper, iron, rock salt, mercury, sulfur, arsenic, and petroleum, among other rare and potentially dangerous substances (Cilliers and Retief, 2000). Sinope, the capital of Pontus, was the center for processing and exporting Sinopic red earth, realgar, orpiment, and other glittering dark red and yellow crystals surrounded by magical and ominous folklore in antiquity. Known by many different names, these minerals occurred in association with quicksilver (mercury), lead, sulfur, iron ore, cobalt, nickel, and gold excavated in Pontus; Armenia was known for arsenic mines. These mines exhaled vapors so noxious that they were worked by slaves who had been sentenced to death for crimes. One of the most infamous mines of Mithridates' kingdom was Sandarakurgion Dag (Mount Realgar), on the Halys River. According to the ancient geographer Strabo (11.14.9; 12.3.40–41), gangs of 200 slaves at a time labored to hollow out the entire mountain. Mount Realgar Mine was finally abandoned as unprofitable, because it was too expensive to continually replace the slaves as they dropped dead from the toxic fumes.

The ancient terms for these groups of related compounds make them difficult to identify today. *Cinnabar, zinjifrah, vermilion, Sinopic red earth, ruby sulfur, sinople, orpiment, oker, sandaracha, sandyx, zamikh, arsenicum, Armenian calche, realgar, dragon's blood*: these were ancient names for the many forms of toxic ores containing mercury, sulfur, and/or arsenic. Sinopic red earth was used to waterproof ships; many of these costly substances were prized as brilliant pigments, varnishes, and textile dyes; they were also important in alchemy and medicine. Arsenic (from ancient Persian *zamikh*, "yellow orpiment") is an odorless poison undetectable in food or drink—the ideal toxin for murder (toxic minerals known in antiquity, Pliny 33.31.98; 33.32.99–100; 33.36–41; 35.13–15; 34.55–56.178; Theophrastus on Stones 8.48–60). The poison that someone slipped into Mithridates' father's meat or wine and that Mithridates used to get rid of enemies was most likely pure arsenic, produced by heating realgar (*rhaj al ghar*, Arabic, "powder of the mine"), red arsenic sulfate.

11.3 AVOIDING ASSASSINATION BY POISON

Mithridates undertook many precautions against assassination by poison. There were guards in his kitchens as well as royal tasters. He knew that some metals and certain crystals and stones were believed to detect—even neutralize—poison in food or drink. Mithridates would have owned so-called "poison cups," goblets made of electrum, a gold and silver alloy. A vessel of electrum was said to reveal the presence of poison when iridescent colors rippled across the metallic surface with a crackling sound, apparently the result of a chemical reaction (Pliny 25.5–7, 33.23.81, 37.15.55–61). Other objects, such as amber, red coral, and *glossopetra* ("tongue stones"), were reputed to magically deflect poisons. Tongue stones were actually fossilized giant shark teeth taken from limestone deposits. On contact with poison, these objects reportedly would "sweat" or change color. The shark teeth could also be ground into a powder that deactivated poison. In fact, calcium carbonate in fossils does react with arsenic through *chelation*, a chemical process in which the calcium carbonate mops up the arsenic molecules (Zammit-Maempel, 1978).

Mithridates tested the nature of poisons for other reasons besides ensuring his own immunity. He also sought to learn which poisons were best for undetectable assassinations of enemies and which poisons were ideal for suicide. According to several sources, Mithridates carried suicide pills and distributed them to his commanders and friends. Those capsules, concealed in rings, amulets, and the hilts of daggers and swords, obviously would have contained a fast-acting, relatively gentle, lethal poison with no known antidote (Pliny 33.5.15, 33.6.25–26; Plutarch *Pompey* 32).

Mithridates' chief toxicological coinvestigator was the Greek "root-cutter" (botanist) named Krateuas. In the course of their systematic study of the effects of common and rare *pharmaka*, Mithridates discovered a curious phenomenon. By ingesting tiny amounts of arsenic each day, he achieved an immunity to larger, otherwise fatal doses. Apparently he achieved tolerance to arsenic as a youth, since sources tell us that conspirators in the palace failed in their attempts to poison him while he was a boy. As king, Mithridates liked to exhibit his remarkable ability to dine safely on poison-laced meat and wine, fatal to others (Aulus Gellius 17). Such theatrical demonstrations enhanced the king's reputation of invincibility. Mithridates was said to follow the "ethical" approach of King Attalus praised by Galen, by experimenting only on himself and condemned criminals (Scarborough, 2007). In one instance, Mithridates received an envoy with a letter and package from his friend Zopyrus in Alexandria. The letter informed Mithridates that the messenger had been sentenced to death and invited the king to test the accompanying antidote on him (Totelin, 2004). The imagined reactions of the courtiers and foreign dignitaries present at Mithridates' sensational demonstrations of immunity inspired A.E. Housman's poem ("Terence, This Is Stupid Stuff," in *A Shropshire Lad*, 1896): "I tell the tale that I heard told. Mithridates, he died old."

Mithridates' mastery of poisons and his unusually long life are memorialized in the term *mithridatism*, the practice of systematically ingesting small doses of deadly substances to make oneself immune to them. With some toxins, the process can be effective. It is possible to acquire tolerance for levels of arsenic that would kill others, for example. It was observed in antiquity that people of North Africa were less affected by local venomous insects and scorpions (*Aelian On Animals* 5.14; 9.29). Mithridates also understood the little-known fact that snake venom can be safely digested if swallowed—it is only deadly if it enters the bloodstream. The rising incidence of poisoning in the Roman Empire inspired the Roman satirist Juvenal to joke that murder weapons of "cold steel might make a comeback if people would take a hint from old Mithridates and sample the pharmacopeia till they are invulnerable to every drug" (Cilliers and Retief, 2000).

11.4 THE SECRET ANTIDOTE

In antiquity, each natural poison—animal, plant, or mineral—was believed to have a natural antidote. Traditional theriacs normally combined substances that were thought to counter poisons (Pliny 25.2.5–8). Mithridates' basic recipe probably contained some of those common ingredients, such as cinnamon, myrrh, cassia, honey, castor, musk, frankincense, rue, tannin, garlic, Lemnian earth, Chian wine, charcoal, curdled milk, centaury, aristolochia (birthwort), ginger, iris (orris root), rue, *Eupatorium*, rhubarb, hypericum (St. John's wort), saffron, walnuts, figs, parsley, acacia, carrot, cardamom, anise, opium, and other ingredients from the Mediterranean and Black Sea, Arabia, North Africa, Eurasia, and India (Mayor, 2010; Totelin, 2004). Modern science reveals that some of these substances can counteract illness and toxins. For example, the sulfur in garlic neutralizes arsenic in the bloodstream. Charcoal absorbs and filters many different toxins. Garlic, myrrh, cinnamon, and St. John's wort are antibacterial. Recent studies of many common *Mithridatium* ingredients reveal bioactivities in the immune system. Certain plants long used by folk healers in Africa and India neutralize cobra, adder, and viper venoms (Mayor, 2010).

Mithridates' personal theriac or tonic was special because it combined *both* toxic *and* beneficial *pharmaka*. Building on the work begun by Attalus III, Nicander of Colophon, and others, Mithridates recorded the properties of hundreds of poisons and antidotes in experiments on prisoners, associates, and himself. "Through tireless research and every possible experiment," wrote Pliny, Mithridates sought ways to "compel poisons to be helpful remedies." Mithridates and Krateuas, joined by the physician Papias, Persian Magi and Scythian Agari healers, and Timotheus, a specialist in war wounds, tested many health-giving essences compounded with minute amounts of poisons. They created an *electuary*, a paste held together with honey and molded into a large pill. Mithridates reportedly ingested his secret theriac with cold spring water on a daily basis (Pliny 25.6–7; 25.17.37; 25.26.62–63; 25.29.65). Apparently the concoction was harmless and may have promoted his immune system. The ancient sources agree that Mithridates enjoyed robust health into his 70s, at a time when the average lifespan was 45 (Griffin, 1995; Mayor, 2010).

After his death, Mithridates' personal library and archives were taken to Rome and translated into Latin by Lenaeus (95–25 BC). Pliny (25.2.5–8; 25.79–82) studied Mithridates' private papers and concluded, "We know from direct evidence and by report" that Mithridates "was a more accomplished researcher into biology than any man before him. In order to become immune to poison by making his body accustomed to it, he alone devised the plan to drink poison every day, after first taking remedies."

The key principle of Mithridates' theriac was the combination of beneficial drugs and antitoxins with tiny amounts of poisons, the approach followed by Attalus and Hindu doctors. Myriad poisons were known in antiquity, from viper, scorpion, and jellyfish venoms to the deadly sap of yew trees and the crimson crystals of cinnabar (Cilliers and Retief, 2000). Pliny described about 7000 venific substances in his encyclopedia of natural history and he listed numerous plants, some with powerful, even dangerous, bioactive properties that were said to counter them, such as scordion, fly agaric mushrooms, artemesia, centaury, polemonia, and aristolochia. Arsenic—the notorious "powder of succession" in antiquity—would have been the first poison Mithridates sought to defend against. Arsenic interferes with essential proteins for metabolism. In small doses, however, enzymes produced by the liver bind to and inactivate arsenic. Taking small amounts over time causes the liver to produce more enzymes, allowing one to survive a normally lethal dose. Mithridates was essentially investigating whether a similar process might work with plant poisons. Mithridates had observed natural tolerances to poisonous plants in rats, insects, birds, and other creatures. According to Pliny and Aulus Gellius (17), the poison blood of Pontic ducks was included in the *Mithridatium* (Totelin, 2004). It is now known that some species of ducks and other birds do eat poison hemlock without harm. Because they do not excrete the toxic alkaloids, their blood and flesh becomes poisonous without harm to them but dangerous to those who eat them (Mayor, 2010).

What other poisons were included in the original *Mithridatium*? Perhaps toxic honey from Pontus—in tiny amounts it was considered a tonic (Aelian *On Animals* 5.4). Various reptiles—such as toxic skinks, salamanders, and vipers—were also said to be part of Mithridates' recipe, based on the notion that all poisonous creatures must produce antidotes to their own toxins in their bodies. Modern scientific experiments show that nonfatal doses of snake venom can stimulate the immune response and allow humans to withstand up to 10 times the amount of venom that would be fatal without inoculation. A similar process works with some insect stings and a variety of toxins. Surprising scientific studies of a "counterintuitive" process called *hormesis* show that very low doses of certain toxins activate a protective mechanism, so that when a larger dose is encountered, it is not as damaging. According to this new concept—remarkably akin to Mithridates' own hypothesis 2000 years ago—minute doses of poison substances can be analogous to a vaccine (Mayor, 2010).

Could the unusual properties of St. John's wort, *hypericum*, listed in many *Mithridatium* recipes, help to solve the ancient riddle of Mithridates' immunity? Molecular scientists have recently discovered hypericum's remarkable antidote effect, not yet completely understood. The herb activates the liver to produce a potent enzyme that is capable of neutralizing a great many potentially dangerous chemicals—as well as prescribed drugs for various conditions. If St. John's wort was included in Mithridates' antidote, it would have stimulated what could be called a hypervigilant "chemical surveillance system" with the capacity to sense and break down normally fatal doses of many different toxins (Mayor, 2010).

11.5 MITHRIDATIUM'S LEGACY

After Mithridates' death, several imperial doctors in Rome claimed to know the secret *Mithridatium* formula. Poisonings and fears of poisoning were commonplace in the Roman Empire. "If you want to survive to gather rosebuds for another day," commented Juvenal (14.251−55), "find a doctor to prescribe some of the drug that Mithridates invented. Before every meal take a dose of the stuff that saves kings" (Cilliers and Retief, 2000) (Fig. 11.1).

Was it possible that Mithridates' genuine recipe was known by some in Rome? Perhaps Mithridates entrusted the secret to his friend

FIGURE 11.1 Portrait of Mithradates VI Eupator of Pontus, silver tetradrachm coin, 88/87 BC.

Asclepiades, the most famous doctor in Rome (Totelin, 2004). According to Galen (*Opera Omnia* 14; *Antidotes* 2), a doctor named Aelius reportedly prescribed *Mithridatium* for Julius Caesar, who was campaigning in Pontus only 16 years after Mithridates' death (Mayor, 2010).

The discovery of an inscription near the Appian Way from the time of Emperor Augustus (b. 63 BC, the year of Mithridates' death) is intriguing. It describes L. Lutatius Paccius (a non-Roman name) as an "incense-seller from the family of King Mithridates." Was he a freed slave? Was he really a relative of Mithridates? Like other ancient apothecaries, Paccius was probably more than just a purveyor of "incense"—why else would he advertise his relationship to Mithridates? Poisons had been strictly regulated since the dictator Sulla's legislation during the Mithridatic Wars. That may explain why an apothecary might only advertise aromatics for sale publicly. Some members of Mithridates' family and his friends did end up in Italy after his death. The inscription suggests that Paccius may have been one of those claiming to know the original *Mithridatium* recipe and that he sold this legendary "trademark" antidote in Rome. A different Paccius, presumably this man's son, later became very rich from selling a very special secret medicine in Rome. This Paccius the Younger bequeathed the "Paccius family recipe" to the Emperor Tiberius, Augustus's successor, in 14 AD, according to the famous imperial doctor Celsius (Totelin, 2004).

Could that mysterious Paccius "family recipe" have been the basis for the later imperial Roman formulas? Many doctors claimed to have improved Mithridates' original, for example, the version compounded by the imperial doctor Andromachus for the Emperor Nero. Andromachus's *Mithridatium* had 64 ingredients; he replaced minced lizards with venomous snakes and added opium poppy seeds (Griffin, 1995). In 2000, Italian archaeologists made a notable discovery at a villa near Pompeii (79 AD). A large vat contained residue consisting of reptile remains and several medicinal plants, including opium poppy seeds. The archaeologists suggested that the vat might have been used to prepare Andromachus's version of the legendary *Mithridatium* (Mayor, 2010).

Nero died in 68 AD, and every Roman emperor thereafter gulped down daily what his personal doctor insisted was a variation based on Mithridates' original antidote. As the number of "authentic" recipes multiplied, more and more exotic, expensive ingredients were added (Pliny 25.3, 29.8.24–26). A century after Mithridates' death, the physician Celsius mixed 36 ingredients in a concoction that weighed almost 3 pounds, for 6 months' worth of pills to be swallowed with wine. In 170 AD, Galen prescribed a liquid *Mithridatium* for Emperor Marcus Aurelius; Galen had added more opium and a fine vintage wine. The wine improved the flavor and the opium certainly guaranteed that the emperor would drink his medicine every day. Later medieval recipes contained as many as 184 ingredients (Griffin, 1995).

In ancient and medieval Islamic toxicology manuscripts, the Arabic theriac (*tiryaq-i-faruq, mithruditus*) and Persian (*daryaq*) recipes followed Mithridates' concept of combining poisons with antidotes. Averroes, the Spanish—Arabic philosopher—physician (b. 1126) wrote a treatise on *tiryaq*. In a veiled allusion to paranoid despots of his day who were obsessed with poisoning, he warned against the prolonged use of theriac by healthy people—cautioning that it "could actually transform human nature into a kind of poison." In 667 AD, Islamic ambassadors presented the Tang Dynasty emperor of China with a gift of the *Mithridatium* theriac (in Chinese called *tayeqie, diyejia*). Chinese chronicles described it as a dark red lump the size and shape of a pig's gall bladder. Chinese manuscript illustrations show envoys in Persian-style costume offering these red *Mithridatium* pills as tribute to the emperor (Mayor, 2010; Nappi, 2008).

From the Middle Ages on in Europe, medicines labeled *Mithridatium* were eagerly purchased. European laws required apothecaries to openly display all the precious, costly ingredients and to mix up the *Mithridatium* outdoors in public squares. For more than two millennia after the death of Mithridates, kings, queens, and nobles from Charlemagne and Alfred the Great to Henry VIII and Queen Elizabeth I ingested some form of *Mithridatium* every day. The royal and aristocratic theriac was kept in ornate gold and pottery apothecary jars, many of them illustrating scenes from the life of Mithridates. Apothecaries sold cheaper varieties of *Mithridatium* to ordinary people, kept in plainer jars. Mithridates' universal antidote became the most popular and longest-lived prescription in history. A *Mithridatium* was advertised by a pharmacy in Rome as recently as 1984 (Griffin, 1995).

Most of the surviving recipes for theriacs in ancient Latin, Greek, Hebrew, Indian, Arabic, and early Islamic medical writings included a range of plant, animal, and mineral *pharmaka* thought to counteract toxins and disease. Aside from Andromachus's addition of chopped vipers for Nero's antidote, however, most of these theriac recipes did not deliberately include poisons (Dio Cassius 37.13; Celsius *De medicina* 5.23.3). Yet most of the ancient Greek and Latin writers agreed with Pliny that Mithridates achieved immunity to poisons by ingesting deadly substances along with a cocktail of specific or general antidotes. In Pliny's words (25.3–7), Mithridates "thought out the plan of drinking poisons daily, after taking remedies, in order that sheer habit might render the poisons harmless."

We can guess at some of the counteracting drugs that Mithridates was likely to have put in his formula, but his precise method of calibrating minuscule doses of poisons and exactly what they were remains a mystery. He and his team worked in secrecy. His original lost recipe was believed to contain more than 50 ingredients, many of them costly, rare substances from distant lands. The confiscated notes translated after his death in Rome listed only a few commonplace ingredients, with the exception of the blood of Pontic ducks. Pliny (29.8.24–26, 23.77.149) expressed surprise at the lack of any obscure or exotic substances in the Mithridatic notes that he studied. He found one scrap of paper in the king's handwriting that said, "Pound together two dried nuts, two figs, and twenty leaves of rue with a pinch of salt: he who takes this while fasting will be immune to all poison for that day." (Griffin, 1995; Totelin, 2004). As Pliny reasonably commented, however, this mundane recipe should not be taken seriously—it could have been a forgery or hoax, or a deliberate red herring.

So what became of the original Mithridatium formula? Perhaps Pliny saw only notes that the emperor allowed him to see. The papers taken to Rome after his death may have recorded only Mithridates' earliest experiments, which had been superseded by successful, more complex experiments whose records did not survive. Mithridates' genuine archives could have been lost, destroyed, or hidden away during the chaos of the Mithridatic Wars. It is even possible that Mithridates' documents may have been encrypted—ancient alchemists often wrote in codes or obscure languages to keep their work secret. Mithridates certainly possessed the linguistic skills; he knew nearly two dozen languages. Were the ingredients of the compound formula somehow divulged to imperial Roman doctors who inherited Mithridates' papers or Paccius's recipe? Were written versions of the perfected formula destroyed on Mithridates' orders? Or were they confided only to his closest friends and allies, such as King Tigranes II of Armenia who, like his son-in-law Mithridates, enjoyed vigorous health and an extremely long life? Perhaps the formula was destroyed when Callistratus, Mithridates' personal secretary, was murdered by Roman soldiers during the wars while carrying important papers. After his defeat of Mithridates in 63 BC, the Roman commander Pompey burned many official papers—he might well have burned some of Mithridates' toxicological archives. Or—as suggested by the historian of medicine Alain Touwaide—maybe Pompey actually obtained the original recipe but kept it secret within his circle (Totelin, 2004). Finally, the instructions for the *Mithridatium* may never have been written down. Perhaps they were recorded only in Mithridates' prodigious memory.

Without new evidence—such as a verifiable, datable recipe from the 1st century BC preserved in writing, or the discovery of sealed jars of the king's own *Mithridatium* containing identifiable residues of known theriac ingredients, or Mithridates' corpse well-preserved enough to

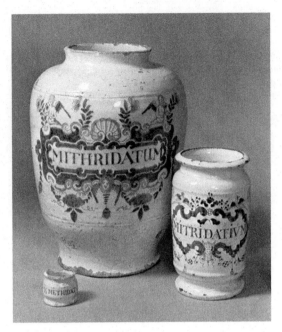

FIGURE 11.2   Mithridatium jars, 16th–17th centuries Europe.

permit hair and bone sampling—we will never know the legendary universal antidote's composition. Yet Mithridates' goal of creating a "universal antidote" lives on. Sergei Popov, a top scientist once employed in the ultrasecret Soviet bioweapons program of the 1980s and 1990s, defected to the United States in 1992. Popov's work now focuses on broad-spectrum biodefenses; in other words, he seeks a "universal" antidote to provide immunity to known biotoxins and "weapons-grade" pathogens. Like the Janus-faced *pharmaka* of the *Mithridatium*, the materials Popov investigates hold the potential for great harm or great good (Mayor, 2010) (Fig. 11.2).

References

Cilliers, L., Retief, F., 2000. Poisons, poisonings and the drug trade in ancient Rome. Akroterion 45, 88–100.

Griffin, J., 1995. Mithridates VI of Pontus, the first experimental toxicologist. Adverse Drug React. Acute Poisoning Rev. 14, 1–6.

Mayor, A., 2010. The Poison King: The Life and Legend of Mithradates, Rome's Deadliest Enemy. Princeton University Press, Princeton, NJ.

Nappi, C., 2008. Bolatu's pharmacy: theriac in early modern China. Early Sci. Med. 14, 737–764.

Rigsby, K., 1988. Provincia Asia. Trans. Am. Philol. Assoc. 118, 123–153.

Scarborough, J., 2007. Attalus III of Pergamon: research toxicologist. In: Paper, 27th Annual Meeting of the Classical Association of South Africa. July 2–5, 2007, Cape Town.

Totelin, L., 2004. Mithradates' antidote—a pharmacological ghost. Early Sci. Med 9, 1–19.

Zammit-Maempel, G., 1978. Handbills extolling the virtues of fossil sharks' teeth. J. Maltese Stud. 7, 211–224.

12

Theriaca Magna: The Glorious Cure-All Remedy

Marianna Karamanou[1] and George Androutsos[2]

[1]Associate Professor of History of Medicine, Medical School, University of Crete, Greece [2]Emeritus Professor of History of Medicine, Medical School, National and Kapodistrian University of Athens, Athens, Greece

OUTLINE

12.1 INTRODUCTION

Once upon a time an ambitious king rivaled Rome's expanding dominions, creating a Black Sea Empire. Claiming descent from Alexander the Great and Persian Royalties, Mithridates VI Eupator (132−63 BC), king of Pontus in Asia Minor, recruited an ethnically diverse army and tried to create an independent East Empire. While Mithridates was still a child, it was said that his mother poisoned his father (Mithridates V) and that after that she become a vicereine. Afraid that his mother would attempt to kill him and living in constant apprehension of being poisoned by his enemies, Mithridates became obsessed with poisons and tried to develop a tolerance by ingesting small amounts of them daily as well as a mixture of antidotes. Experimenting on criminals and slaves, he was one of the first to systematically study poisons in humans. He also created a general antidote, *Mithridatium*, that contained 36 ingredients and was said to protect against any known poison. Defeated by the Roman military leader Pompey (106—48 BC) in 63 BC, Mithridates poisoned his wives and children but no poison could kill him. In the end, he had to ask his Gallic mercenary bodyguard to kill him with a sword. His work on pharmacology and toxicology was translated into Latin by Pompeius Lenaeus, a learned freedman of Pompey, and *Mithridatium* entered the Roman world (Mayor, 2009).

12.2 THERIAC IN ANTIQUITY

In a period in which poisoning was a traditional political weapon and a typical method of royal succession, polyvalent antidotes such as *alexipharmaka* or *theriac* (derived from the Greek *therion*, for wild beast) were highly esteemed. These were supposed to induce immunity according to the principle that like cures like, by gradually introducing small amounts of poison into the organism. Subsequently, several kinds of theriacs were produced but the most celebrated was perhaps that invented by Adromachus the Elder, a Cretan who became *Archiater* (chief physician) to the Emperor Nero (54—68 CE).

Andromachus's concoction, *Galene Theriaca* (tranquility theriac), was an improved version of Mithridates's elixir, containing 65 ingredients with a higher proportion of opiates and minerals and with the original lizard flesh replaced by that of a viper (Watson, 1966). The recipe for *Galene* was written in Greek by Andromachus in the form of elegiac couplets and the prose rendition was quoted by his son Andromachus the Younger:

Pastils of squill 48 drachms; pastils of viper's flesh 24 drachms; pastils of hedychroum and black pepper 24 drachms each; poppy juice 24 drachms; dry roses, water germander, turnip seeds, iris from Illyricum, agaric from Pontus, cinnamon, licorice juice, opobalsam, of each 12 drachms; myrrh, saffron, ginger, rhubarb from Pontus, cinquefoil root, calamint, horehound, parsley, lavender, saussurea lappa, white pepper, long pepper, dittany from Crete, fragrant rush flower, frankincense, turpentine, cassia bark quill, nard from India, of each 6 drachms; germander, ground pine, hypocist, juice of the leaf of white flowered cinnamon, nard from Gaul, gentian root, anise, flesh of lamb from the land of Athamanes, fennel seeds, terra lemnia, roasted copper, cardamon, yellow flag, burnet from Pontus, fruit of balsam-tree, St John's wort, shittah rose, Arabic gum, cardamomum, of each 4 drachms; wild carrot seeds, galbanum, gum from Ferula persica, Hercule's woundwort, bitumen, castoreum, small centaury, small birthwort, of each 2 drachms, of Attic honey 150 drachms; of vetch meal 80 drachms (Galen and Kühn, 1964−5) (Fig. 12.1).

The concoction took 40 days to prepare after which the maturation process began. According to Andromachus, his theriac had more applications than simply as a remedy against venomous bites or an antidote for poisons. It was also effective in the treatment of headaches, poor sight, kidney stones, ulcers, dysentery, and deafness; it could induce the menses and dry up the excess of humors in the body.

However, almost a century later, another theriac appeared that eclipsed all others in fame and popularity to become a universal panacea. It was the theriac prepared by the distinguished Greek physician Galen (c. 130−201) (Fig. 12.2). The basic formula consisted of viper's flesh, opium, honey, and more than 70 other ingredients. Twelve years was considered the proper period to preserve it before use, although the Emperor Marcus Aurelius (121−180 CE) used the mixture within 2 months from its preparation (Prioreschi, 2001). According to Galen, Marcus Aurelius was taking theriac daily, mixing it with wine or water to protect against poisons and to ensure good health status. This habit, which influenced his sleep and resulted in a lack of sleep after it was withdrawn, suggests that he was probably addicted to opium (Africa, 1961). Galen referred to the duration of the drug's efficacy, nowadays known as the expiration date: "Theriac is still effective after thirty years and for all those mild diseases it is recommended even after sixty years. After this span of time, the drug weakens and is no longer effective" (Galen and Kühn, 1965). Moreover, Galen experimented with theriac by performing one of the first randomized control trials in history. In his work *De theriaca ad Pisonem* he writes that he took roosters and divided them in two groups. In group A he administered theriac and in group B he did not. Then he brought both groups into contact with snakes. He observed that the roosters of group B died immediately after the bite, while those in group A survived, proving the therapeutic effectiveness of the preparation. He also mentioned that the above experiment could be used in cases that tested whether theriac was in its natural form or had been adulterated (Galen and Kühn, 1965).

Galen also reported theriac's therapeutic efficacy on patients: "One of the slaves of the Emperor whose duty was to drive away snakes, having been bitten, took for some time draughts of ordinary medicines, but as his skin changed so as to assume the colour of a leek, he came to me

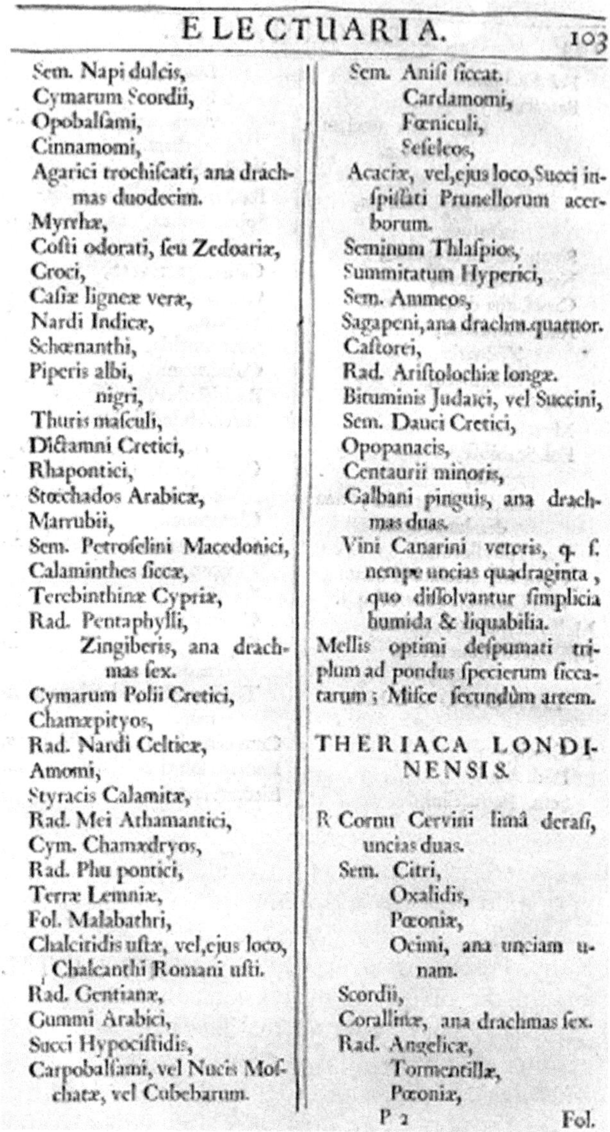

ELECTUARIA. 103

Sem. Napi dulcis,
Cymarum Scordii,
Opobalfami,
Cinnamomi,
Agarici trochifcati, ana drach-
    mas duodecim.
Myrrhæ,
Cofti odorati, feu Zedoariæ,
Croci,
Cafiæ ligneæ veræ,
Nardi Indicæ,
Schœnanthi,
Piperis albi,
    nigri,
Thuris mafculi,
Dictamni Cretici,
Rhapontici,
Stœchados Arabicæ,
Marrubii,
Sem. Petrofelini Macedonici,
Calaminthes ficcæ,
Terebinthinæ Cypriæ,
Rad. Pentaphylli,
    Zingiberis, ana drach-
    mas fex.
Cymarum Polii Cretici,
Chamæpityos,
Rad. Nardi Celticæ,
Amomi,
Styracis Calamitæ,
Rad. Mei Athamantici,
Cym. Chamædryos,
Rad. Phu pontici,
Terræ Lemniæ,
Fol. Malabathri,
Chalciridis uftæ, vel, ejus loco,
    Chalcanthi Romani ufti.
Rad. Gentianæ,
Gummi Arabici,
Succi Hypociftidis,
Carpobalfami, vel Nucis Mof-
    chatæ, vel Cubebarum.

Sem. Anifi ficcat.
    Cardamomi,
    Fœniculi,
    Sefeleos,
Acaciæ, vel, ejus loco, Succi in-
    fpiffati Prunellorum acer-
    borum.
Seminum Thlafpios,
Summitatum Hyperici,
Sem. Ammeos,
Sagapeni, ana drachm. quatuor.
Caftorei,
Rad. Ariftolochiæ longæ.
Bituminis Judaici, vel Succini,
Sem. Dauci Cretici,
Opopanacis,
Centaurii minoris,
Galbani pinguis, ana drach-
    mas duas.
Vini Canarini veteris, q. f.
    nempe uncias quadraginta,
    quo diffolvantur fimplicia
    humida & liquabilia.
Mellis optimi defpumati tri-
plum ad pondus fpecierum ficca-
tarum; Mifce fecundùm artem.

THERIACA LONDI-
    NENSIS.

R Cornu Cervini limâ derafi,
    uncias duas.
Sem. Citri,
    Oxalidis,
    Pæoniæ,
    Ocimi, ana unciam u-
    nam.
Scordii,
Corallinæ, ana drachmas fex.
Rad. Angelicæ,
    Tormentillæ,
    Pæoniæ,
P 2   Fol.

FIGURE 12.1 The prescription of Theriaca Andromachi, from the 1677 edition of Pharmacopea Londinensis. *Source: Wellcome Library London.*

and narrated his accident; after having drunk theriac he recovered quickly his natural colour" (Galen and Kühn, 1965). Galen passed his theriac on under the name of *Theriac of Andromachus* and since then its fame as an antidote has led it to be regarded as a cure-all remedy (Karaberopoulos et al., 2012).

FIGURE 12.2 The eminent Greek physician Galen. *Source: Wellcome Library London.*

12.3 THERIAC IN THE MEDIEVAL PERIOD

During the 6th and 7th centuries, theriac is referred to in several passages of the works of Aetius of Amida (502–575 CE) and Paul of Aegina (625–690 CE), while Arabian medicine, under Galenic influence, adopted theriac's use (Prioreschi, 2001; Adams, 1834). Abulcassis (936–1013 CE) described the preparation of theriac composed of 84 different ingredients, while Maimonides (1135–1204 CE), in his *Treatise on poisons and their antidotes* (1198), noticed the administration of the Great Theriac (Hamarneh and Sonnedecker, 1963; Muntner, 1966). Avicenna (980–1037 CE) provided an explanation for the medicinal action of theriac which was derived from its specific form, the act of combination and not out of the ingredients, claiming that theriac could not only cure cases of poisoning but could also strengths the heart and preserve the health of a perfectly well man who consumed it regularly (Avicenna, 1973). Averroes (1126–1198 CE) held an opposing view: in his treatise on theriac (*Tractatus de tyriaca*), he pointed out that while it was beneficial to the patient as an antidote, it could be dangerous as a regular and repeated medication as it could alter the health status and rendered poisonous. He advised physicians to prescribe it with caution and not as a prophylactic remedy (McVaugh, 1985).

In the 11th century, the translation of the works of Arab physicians renewed Western medicine and theriac is listed in its turn among the remedies recommended in the renowned work *Regimen sanitatis* of the Salernian School: "The raddish, pear, theriac, garlic, rue, all potent poisons will at once undo" (Ordronaux, 1870). In Antidotarium Nicolai (12th century), a Salernian collection of some 150 empirical time-tested recipes, theriac is credited with more far-reaching powers of both a therapeutic and a prophylactic nature: "Theriac is good for the most serious afflictions of the human body as epilepsy, catalepsy, headache, stomach ache, migraine, hoarseness, bronchitis, asthma, jaundice, dropsy, leprosy. It induces menstruation and expels the dead fetus. It is especially good against all poisons, and bites. It strengths the heart, brain, liver and keeps the entire body incorrupt" (Grant, 1974).

In the newly founded Montpellier medical school (1221) theriac stimulated keen interest among medical scholars who attempted to determine its mechanism of action and dosage; their work also raised the question of whether it could be administered to the healthy. Arnaud de Villeneuve (1235–1311 CE), Henri de Mondeville (1260–1316 CE), Bernard de Godron (1285–1318 CE), and others relied more or less on Averroes' doctrines, which stated that theriac should not be taken by healthy individuals (McVaugh, 1972). In his *Chirugia* (1300), Mondeville mentioned, "It must be noted fifthly that according to Averroes in his book de *Tyriaca*, medicines curing poisons are midway between medicines, the body and poisons; in bloody and choleric illnesses theriac is not of value … and it should everywhere and always be administered with the greatest precaution while in fevers the use of theriac is almost wholly to be abandoned" (McVaugh, 1985). Furthermore, physicians and patients seemed to have been aware of the effects of theriac's ingredients and the fatal consequences of overdosing. Bernard de Gordon was called to treat an apothecary who had accidentally received a fatal dose of theriac: "It happened that a certain apothecary had sold a toxic substance and some remained behind the nail of his thumb. When he began to pick his nose with his thumb, his color began to change and he fainted repeatedly. When he took three drams [of the antidote theriac], his condition became more serious. I was called around midnight and I saw that the amount of theriac had been too high. I made him vomit and after he had purged his stomach sufficiently, I gave him one dram of theriac and he was cured" (Demaitre, 1980).

Theriac is mentioned in the 12th century story of *Tristan and Isolt*. Poisoned by a dragon, Tristan receives care from Isolt: "She felt his body and found that he was still alive; she made him drink some theriac and took such care of him that all of his swelling went down and he was cured and restored to his beauty" (Spector, 1973). Thanks to the mystique of legend and magic, theriac usage was widespread in the Middle Ages, notably for curing plague. The 1348 *Consultation de la Faculté de Paris* stressed the importance of theriac's administration for plague prophylaxis (Michon, 1860). In the same way, the notable French surgeon Guy de Chauliac (1300–1368 CE), who witnessed the terrible plague outbreaks of 1358 and 1361, remained healthy thanks to the daily intake of theriac, while several other physicians, among them Johannes Jacobi in his *Regimen contra pestilentiam*, suggested theriac as a cure for and prophylactic remedy against plague (Jacobi, 1371; Byrne, 2012).

12.4 THERIAC IN THE RENAISSANCE

During the Renaissance, theriac became extremely popular and highly prized, taking center stage in the European pharmacopeias, while the ambition to produce a perfect concoction stimulated pharmacists through Europe. The Venetian Republic became the center for the production of the highest-quality theriac known as *Venice treacle*, an official preparation carrying the republic's seal. Its export throughout the rest of Europe provided an important source of revenue (Olmi and Zanca, 1990). Soon an important question rose: Was the prepared theriac as effective as that of the classical age? There had been cases in which theriac was proved to be ineffective and physicians doubted its action. According to Pietro-Andrea Mattioli (1500–1577 CE), many of the ingredients used in classical times were now unavailable, while Bartolomeo Maranta (d. 1571), author of *Della Theriaca ed del Mithridato*, blamed the ignorance or negligence of physicians and pharmacists for the failure to produce an effective theriac (Olmi and Zanca, 1990). Concern for guaranteeing the genuineness of the product led the authorities of several cities to supervise its production (Fig. 12.3). The manufacturing of

FIGURE 12.3 An apothecary publically preparing theriac, under the supervision of a physician, 15th century. *Source: Wellcome Library London.*

theriac in public led to the introduction of strict controls over and standards governing the quality of ingredients, thus stimulating the earliest concepts of medicine regulations.

12.5 CONCLUSION

Kept in ornate porcelain jars, illustrated in scenes from the life of Mithridates, the recipe to produce and to use theriac reached the 19th century as it was included in the 1872 German and 1874 French pharmacopeias, and then progressively disappeared (Olmi and Zanca, 1990) (Fig. 12.4). Was it a universal antidote or just an addicting tranquilizer, thanks to opium? Either way, its impact on humanity lasted more than 2000 years.

FIGURE 12.4 Drug jars for Theriac, 18th century France. *Source: Wellcome Library London.*

References

Adams, F., 1834. The Medical Works of Paulus Aegineta. Welsh, London.

Africa, T.W., 1961. The opium addiction of Marcus Aurelius. J. Hist. Ideas. 22, 97–102.

Avicenna, 1973. The Canon of Medicine of Avicenna [Cameron O, Trans.]. AMS Press Inc, New York, NY.

Byrne, J.P., 2012. Encyclopedia of the Black Death. ABC-CLIO, Denver, CO.

Demaitre, L.E., 1980. Doctor Bernard de Gordon: Professor and Practitioner. Pontifical Institute of Mediaeval Studies, Toronto, ON.

Galen, 1964. De antidotis I. In: Kühn, C.G. (Ed.), Claudii Galeni Opera Omnia. Georg Olms, Hildesheim.

Galen, Kühn, C.G., 1965. De Theriaca ad Pisonem Opera omnia. In: Kühn, C.G. (Ed.), Opera Omnia. Georg Olms, Hildesheim.

Grant, E., 1974. A Source Book in Medieval Science. In: Grant, E. (Ed.), A Source Book in Medieval Science. Harvard University Press, Cambridge, MA.

Hamarneh, S., Sonnedecker, G., 1963. A Pharmaceutical View of Abulcasis al Zahrawi. Brill, Leiden.

Jacobi J., 1371. Preservatio Pestilentie Secundum Magistrum. Geneva.

Karaberopoulos, D., Karamanou, M., Androutsos, G., 2012. The *theriac* in antiquity. Lancet. 379, 1942–1943.

Mayor, A., 2009. The poison king: the life and legend of Mithradates. Rome's Deadliest Enemy. Princeton University Press, Princeton, NJ.

McVaugh, M.R., 1972. Theriac at Montpellier, 1285–1325 (with an edition of the "Questiones de tyriaca" of William of Brescia). Sudhoffs Arch. 56 (2), 113–144.

McVaugh, M.R., 1985. Arnaldi de Villanova opera medica omnia. Publications de la Universitat de Barcelona, Barcelona.

Michon, L.A.J. (Ed.), 1860. Documents inédits sur la grande peste de 1348 (Consultation de la Faculté de Paris, Consultation d'un practicien de Montpellier, Description de Guillaume de Machaut). Baillière, Paris.

Muntner, S., 1966. The medical writings of Moses Maimonides: treatise on poisons and their antidotes. In: Muntner, S. (Ed.), The Medical Writings of Moses Maimonides: Treatise on Poisons and Their Antidotes. J.B. Lippincott Co, Philadelphia, PA.

Olmi, G., Zanca, A., 1990. The prince of all drugs: theriac in pharmacy through ages: ancient drugs. In: Zanca, A. (Ed.), The prince of All Drugs: Theriac in Pharmacy Through Ages: Ancient Drugs. Farmitalia Carlo Erba, Parma, pp. 105–122.

Ordronaux, J., 1870. Regimen Sanitatis Salerni. Scuola Medica Salernitana. Philadelphia, PA: Lippincott.

Prioreschi, P.A., 2001. History of Medicine. Horatius Press, Omaha, NE.

Spector, N.B., 1973. The Romance of Tristan and Isolt. Northwestern University Press, Evanston, IL.

Watson, G., 1966. Theriac and Mithridatum: A Study in Therapeutics. Wellcome Historical Medical Library, London.

13

The Gates to Hell in Antiquity and their Relation to Geogenic CO_2 Emissions

Hardy Pfanz[1], Galip Yüce[2], Walter D'Alessandro[3], Benny Pfanz[4], Yiannis Manetas[5], George Papatheodorou[6] and Antonio Raschi[7]

[1]University of Duisburg-Essen, Department of Applied Botany and Volcano Biology, Essen, Germany [2]Hacettepe University, Faculty of Geological Engineering, Ankara, Turkey [3]INGV, Sezione di Palermo, Palermo, Italy [4]University of Duisburg-Essen, Institute of Applied Botany and Volcano Biology, Essen, Germany [5]Department of Biology, University of Patras, Patras, Greece [6]Laboratory of Marine Geology, University of Patras, Patras, Greece [7]CNR - IBIMET, Florence, Italy

Toxicology in Antiquity
DOI: https://doi.org/10.1016/B978-0-12-815339-0.00013-5

13.1 INTRODUCTION

Friedrich Nietzsche, the classical German philologer and philosopher writes in his "Thus spoke Zarathustra":

> There is an isle in the sea—not far from the Happy Isles of Zarathustra—on which a volcano ever smoketh; of which isle the people, and especially the old women amongst them, say that it is placed as a rock before the gate of the netherworld; but that through the volcano itself the narrow way leadeth downwards which conducteth to this gate. *Friedrich Nietzsche, Thus spoke Zarathustra, XL.*
> *Great events, 1883–1891.*

In Greek mythology, the world is divided into three realms. Zeus rules heaven, Poseidon the sea, and Pluto is the king of the shadows. During all times the living world was separated from the world of the dead, shadows, and souls. The underworld (hell, netherworld, or Hades) is located either in remote astronomic regions or somewhere hidden in the underground. Those who had received proper burial rites were allowed to enter. This means that for those that passed away, there must have been a proper way to get there. In Greek mythology, the dead had to cross a certain river(s)—Acheron, Cocytus, Lethe, Styx, or Phlegeton.

Within the everlasting darkness were many chthonian divinities, gods and goddesses considered as the rulers of the underworld in classical mythology: The king of the netherworld was Pluto (and its various ways of spelling, e.g., Plouto), or Hades (δης, Haides, Aides, ιδης, Aidoneus, Klymenos, Clymenus, Dis Pater, Orcus), and his wife, the goddess Persephone (Kore or Proserpina). Hades is the Lord of the Shadows, the King of the Netherworld, the god of the lower world, and the sender of death to the mortals (Albinus, 2000). Hades was also the god of funeral rites. He is mostly seen with his three-headed hellhound Kerberos (latin. Cerberus). In the little book on images of the gods it is said

> ...homo terribilis in solio sulphureo sedens, sceptrum regni in manu tenens dextra: sinistra, animam constringes, cui tricipitem Cerberum sub pedibus collocabant, et iuxta se tres Harpyias habebat. De throno aure eius sulphureo quatuor flumina manabunt, quae scilicet Lethum, Cocytu, Phlegethontem, et Acherontem appellabant, et Stygem paludem iuxta flumina assignabant. . . .

"... an intimidating person sitting on a throne of sulfur, holding the scepter of his realm in his right hand, and with his left strangling a soul. Under his feet, three-headed Cerberus held a position, and beside him, he had three Harpies. From his golden throne of sulfur flowed four rivers, which were called, as is known, Lethe, Cocytus, Phlegethon, and Acheron, tributaries of the Stygian swamp" (as cited in Pluto Mythology, but see Albricus, 1742; Raschke, 1912; Wolf FA, 1839; Wolf RE, 2008; Pepin, 1961).

There are at least two names for the entrances to hell: Charonion (Strabon XIV, 1, 11; Albricus, 1742; Soentgen, 2010) for the cave of Thymbria, between Magnesia on Maeander and Myus; and for Acharaka near Nysa on Maeander (Strabon XIV, 1, 45−47) and Plutonium for Hierapolis (Strabon, XIII, 4, 14; Pepin, 1961; Vitaliano, 1973). Plutonium clearly means a sanctuary for the chthonic god Pluto, whereas Charonion refers to the ferryman Charon.

Extraordinary and fearsome places were often thought to be the entrance to the Netherworld (Zwingmann, 2012; Vorgrimler, 1994; Minois, 2000). Caves, fissures, and fractures, mostly with smelling exhalations, and lakes or streams that were steaming hot or changing color, were such sites. The entrance to the underworld was sometimes also the end of the world. Odysseus had to go there to find Teiresias (Homer, 1900; cf. Cantus XI; Odyssey), where Charon brought him over the river Styx. Quite a few ancient gates to hell are caves (see Zwingmann, 2012 and references therein), some others are lakes or depressions. Caves are uncertain places, entrances to a dark and unknown world. Even more fearful were these places when in addition to the darkness toxic vapors were emitted, able to kill all life. The vapors were called mephitic vapors,[a] which resembled the deadly Hadean breath (or the breath of the hellhound Kerberos). In most cases, these vapors consisted of highly concentrated carbon dioxide (CO_2), sometimes with some sulfur gas (H_2S) impurities (Pfanz et al., 2018). In geothermal fields or when hot water streams were present, large amounts of water vapor were emitted, which led to the fearsome appearance of toxic mist and fog in front of the cave.

The gate to the underworld was always thought to be related to supernatural forces. Miraculous places were often symbolic for these sites. In some cases such entrances to the Hadean underground had probably been found by shepherds or herdsmen who witnessed a strange behavior of their cattle. Herdsmen were astute in nature; even a change or variation in vegetation could hint to them of such extraordinary places (Pfanz et al., 2004; Pfanz, 2008; Soentgen, 2010). Quite often also dead corpses of small animals hinted to the presence of strange and supernatural forces. If priests came to know about such places, they declared those sites as sacred places. Sometimes temples and sanctuaries were built in the surroundings (or even on top of geogenic gas emissions; see Plutonium in Hierapolis/Phrygia).

Alighieri (2001) was most probably the first to relate mythic places and geology (cf. Cantus Inferno XII; XIV and XV). He describes the circum-volcanic conditions quite precisely and not knowing the proper geologic interrelations, he ascribes them to hell. In modern times, most probablyVitaliano (1973) was the first to use the term *geo-mythology*. She was followed by Piccardi and Masse (2007). In the following remarks, we try an amalgamation of mythological and archaeological data with geological, biological, and medical evidence to describe the coherence of the *geo-bio-mythology* of the ancient gates of hell.

13.2 WHY ENTER THE REALM OF THE SHADOWS?

13.2.1 The Souls of the Mortals

For humans that had passed away, the passage to the kingdom of the dead was probably not difficult, if their relatives took enough care during the funeral. The correct funeral procedure was an important help and some sort of guarantee for a safe journey into the underworld. In Greek mythology, the souls of the dead, "the shadows," had to cross the river (swamp) Styx (Acheron). The transfer across the Acheron via ferry was done by Charon, the ferryman of the shadows, who was paid

[a]From Mefitis or Mephitis, a goddess of the evil geogenic smells, who was worshipped in Pompeii, Rocca San Felice, or Rome

by the "danakes (oboli)," mostly golden coins that were put into the mouths of the dead before the funeral.

13.2.2 Incubation and Cure

The cult of worshipping Pluto was most probably imitated from the cults of Serapis in Egypt (Hamilton, 1906). Strabon (XIV) mentions three sites in Asia Minor where Pluto was worshipped. In the region of Nysa and Acharaka he describes a temple for Pluto and Kore and a grotto (the Charonion). Somewhere in the vicinity is Limon (λειμών, the meadow), where similar rites were practiced (Hamilton, 1906). As a site for a third Plutonium, Strabon mentions the famous town Hierapolis with its hot water ponds, its calcareous hot waterfalls (nowadays Pamukkale), and its Charonion (Strabon XIV, 1, 11). Pilgrims prayed for health and cures and/or sometimes asked for prophecies. Priests were dealing with the gods of the underworld. They incubated instead of the patients, and the patients then followed the recipes and cures, which were prescribed according to the dreams of the priests during incubation. The prerequisites of prophecies and dreams were among other parameters (drugs, hallucinogenic plants, and mushrooms) attributed to the presence of geogenic gases. Lack of oxygen and an increase in carbon dioxide mimics hallucinogenic effects (see also De Boer et al., 2001). In this context, a theory has been proposed linking the prophetic power of the Pythia in Delphi's Sanctuary with the occurrence of gases released from a fault ("chasma") (De Boer et al., 2001; Etiope et al., 2006; Piccardi et al., 2008). The Delphi Sanctuary is considered the most important religious location of the ancient Greek world ("omphalos" of the world). The sacredness of the oracle mostly lay on the mantic ability of the Pythia, whose prophecies had a vital role in the political, social, and military scene of the ancient world. De Boer et al. (2001) proposed that the inhalation of the sweet-smelling ethylene, which can cause mild narcotic effects, could be the reason for the inspiration of Pythia. Etiope et al. (2006) propose that if any gas-linked neurotoxic effect of the Pythia needs to be invoked, as suggested by historical tradition, it could be searched for in the possibility of oxygen depletion due to CO_2-CH_4 exhalation in the nonaerated inner sanctum ("adyton").

13.2.3 Necromancy

Only a few extremely brave heroes, like Herakles and Orpheus, dared to make their way down to the netherworld. Orpheus did so to free his wife Eurydike who had died from a snakebite and for Herakles it was one of his twelve tasks (subduing the hellhound Kerberos). Necromancy is a technique to negotiate with the gods of the netherworld or to communicate with the shadows of the deceased. Odysseus wanted to enter the realm of the shadows to meet the ghost of the seer Teiresias. Kirke helped him by instructing Odysseus in the art of necromancy. According to Homer, these rites were performed on the borders of the Underworld, near Tainaron (nowadays Cape Matapan). Other authors claim that Odysseus visited the Nekromanteion at the Acheron River (north-western Greece) or even in Cumae (Campi Flegrei; southern Italy).

13.2.4 The Gate for Chthonic Gods and Ghosts of the Darkness

Yet, the gates to the netherworld were not only used by the dead and shadows or those seeking cure, but by the gods of the netherworld themselves to step out onto the earthly upper world (and down again). The story of the abduction of Persephone is very common in Graeco-Roman mythology. Hades came from the subterranean darkness to abduct and rape Persephone and to take her as his wife and as the Queen of the Netherworld. Homer describes the descent of Hades with Persephone as taking place in Nysa-Acharaka (Phrygia). Yet, some authors still think that Hierapolis may also have played a role. In a Roman version, known as the rape of Proserpina, Pluto abducts Proserpina at the Lago di Pergusa, located in the center of Sicily close to the town of Enna. In this case, the lake was not selected because of its magmatic gas emissions but because of its biogenic sulfur emissions and its color change to purple-red. The red coloration of this circular pond was a hint to chthonic forces (Kondratieva et al., 1976; Rigoglioso, 2005).

13.3 THE GEOLOGIC BACKGROUND

13.3.1 Geogenic Gas Emission: Volcanoes, Faults, and Seismicity

Albertus Magnus writes in this third book (liber III, Meteororum, Tract. II, Caput XII) about toxic gas emanations and strange smells after earthquakes, close to volcanoes or in areas with specific wells and springs:

> Scias etiam, quod frequenter pestilentia praecipue omnem sequitur terraemotum: vapor enim inclusus et privatus sic luce et aere libero grosso est habens quasi veneni naturam, et ideo animalia interficit, praecipue quae terrae quasi semper proximum os tenent, sicut oves: quia antequam totus erumpat vapor, per plures dies semper aliquid eius paulatim per poros terrae evadit, et laedit animalia pastum in loco terraemotus accipientia, et continue os juxta terram habentia: quia ex hoc quasi continue hauriunt vaporem venenosum. Et ideo narrat

Seneca, quod in terraemotu Campano, qui fuit tempore Neronis imperatoris in Pompeiana regione, grex sexcentarum ovium mortuus fuit: et homines et animalia semper pereunt, si prope fuissent et extraxissent per respirationem vaporem illum pestiferum. Ego autem vidi in Paduana civitate Lombardiae, quod puteus ab antiquo tempore clausus inventus fuit, qui cum aperiretur, et quidam intraret ad purgandum puteum, mortuus fuit ex vaope cavernae illius, et similiter mortuus est secundus: et tertius voluit scire quare duo moras agerent, inclinatus ad puteum adeo debilitatus est, quod spatio duorum dierum vix rediit ad seipsum: cum autem exspirasset vapor putrefactus in puteo, factus est bonus et potabilis.

"You have to know in fact that often pestilentia in particular follows the earthquake: in fact, the vapor closed and deprived of light and free air, is rough, and has almost the nature of a poison, and thereafter it kills the animals, mainly those that almost always keep the mouth near the ground, such as sheep: because, before all the vapor bursts, for many days some of it gradually transpires throughout the earth's pores, and damages the animals getting their food in earthquake places, and having their mouth continuously near the soil: such as if, from it, they were continuously gobbling a poisonous vapor. And therefore Seneca tells that in the Campania earthquake, that was upon the time of Emperor Nero in the region of Pompei, an herd of 600 sheep was killed: and men and animals always die, if they were close and they would extract, through respiration, that pestiferous vapor. Moreover, I saw in the city of Padua, Lombardia, that a well, closed since a long time, was found, and when it was opened, somebody entered into it, to clean it, and he died, for the vapor contained in it, and similarly a second one died: and a third one intended to know the reason, and he did lean in the well until he did lose his strength, but in two days he recovered. And then, when putrefied vapor was dissolved, the well became good and drinkable" (translation by A. Raschi, unpublished).

In geodynamic active areas, which are characterized by intense volcanic activity and/or high seismicity and tectonics, many geogenic gases can be released to the atmosphere depending on the local geologic regime (Hansell and Oppenheimer, 2004). Among the major gas species, CO_2 is ubiquitously found in almost all of these areas. Water vapor is found in volcanic and geothermal areas where temperatures at or close to the surface reach the boiling temperature of water. Among sulfur gases, hydrogen sulfide (H_2S) is typical of geothermal areas, while sulfur dioxide (SO_2) is released with high-temperature volcanic gases. The latter can also contain significant quantities of halogens mainly in the form of HCl, HF, and HBr (Aiuppa et al., 2009). All remaining gas species, although relevant for geochemical studies (CO, noble gases, hydrocarbon gases) and/or for their possible long-term effects on human health (Hg, Rn), are found only at trace levels. The only exception is geogenic CH_4 of which emissions may be relevant in sedimentary basins with petroleum production (Etiope and Klusman, 2010; Etiope, 2015).

Concerning the gases measured around or within the gates to hell, CO_2 is, aside from water vapor, the most frequently found gas. As mentioned above, geogenic gas emissions can lead to enhanced carbon dioxide concentrations that influence and threaten life in the neighborhood. Carbon dioxide of geologic origin is formed within the Earth's mantle or crust and is released to the atmosphere not only by volcanoes, but also diffusely from the soil in geothermal areas and seismically active areas. Gases reach the Earth's surface through zones of enhanced vertical permeability within the crust such as volcanic conduits, volcano-tectonic structures, fissures, and faults. In particular, faults generally act as important pathways for CO_2 migration to the surface of carbon dioxide originating by deep thermometamorphism of carbonate rocks or mantle degassing (Chiodini et al., 1999; Mörner and Etiope, 2002). When the gas leaves the lithosphere—pedosphere system it reaches the adjacent atmosphere, where it is normally diluted due to convection triggered by sun or wind. Heavier than air, carbon dioxide can accumulate and reach high levels in valleys, depressions, and poorly ventilated zones, becoming dangerous for organisms. The reason is the specific density of CO_2 gas is 1.5 times heavier than air. This phenomenon is also known from wine cellars, where fermentative CO_2 gas accumulates at the cellar bottom creating a potentially lethal trap for wine makers. Some valleys or dolines (sinkholes) are well known for their dangerous gas atmosphere (Raschi et al., 1997; Kies et al., 2015; Pfanz, 2008). Such gas-filled depressions are known from all over the world. In Rwanda they are called "mazuku" (evil wind). Carcasses of reptiles and birds, as wells as dead bodies of gorillas, and even elephants, are sometimes found in their close vicinity (Vaselli et al., 2002/2003). Soentgen (2010) also describes the possible relation between the gates to hell and places of oracles and geogenic gases.

13.3.2 Reduced Carbon Gases

Aside from carbon dioxide emanations, also the reduced form of carbon (methane CH_4) is degassing around and below archaeological buildings. In the adyton below the temple of Apollo in Delphi, CO_2 was found only in traces but methane and other hydrocarbons were detected that were able to influence the Pythian seer (De Boer et al., 2001; Etiope et al., 2006; Piccardi et al., 2008). Similar findings are reported from Chimaira (Etiope et al., 2011). A comprehensive and detailed description on gas seeps in the ancient world is given in Etiope (2015) and references therein.

FIGURE 13.1   Iron ochre formation within a creek.

13.3.3 Hot Water, Steam, and Geysers

Aside from toxic gas emissions, there is also another geologic reason for the selection of sacred sites. Geothermal energy in hot water creeks, hot water steam, hot water eruptions (geysers), and the reddening of water bodies drew the attention of people, including ancient priests. Red water bodies can have an abiotic or a biotic cause. The red color of a creek is mainly due to the oxidation of reduced iron from great depths; iron oxides and hydroxides are formed (iron ochre; Fig. 13.1). Such places were called "Bullicame" (bubbling water) in the Italian language of the Middle Ages. The term was used either for all stained hot water emanations or specifically for a little geothermal lake and creek close to Viterbo (Dante Alighieri, L'Inferno; Canto XII, 127–129). The physico-chemical action of carbon dioxide to displace and withdraw oxygen is here precisely described. With Dante, CO_2 purely extinguishes flames; with Strabon it extinguishes human lives (see later).

13.3.4 Caves

As already stated, caves and grottos are suitable gates or entrances to the netherworld (Rosenberger, 2001; Ustinova, 2009; Zwingmann, 2012). Yet, it seems that caves are mostly used as oracular sites and only rarely as entrances to hell. Quite interestingly, many ancient gates to hell along the Meander valley in Asia Minor are rated among caves although they are nowadays temples and sanctuaries (cf. Strabon

12.8.17; 14.1.11; Oppe, 1904; Parke, 1985; Ustinova, 2002). The reason for this is that originally temples, theaters, and sacred edifices have overbuilt cave locations. Many prophetic activities and also the necessary prerequisites were performed in chambers and caves (De Boer et al., 2001; Etiope et al., 2006; Piccardi et al., 2008). Incubation and divination, as induced and triggered by sensory deprivation, were often performed in closed rooms as euphoriants and psychotropic gases could exert their influence quite directly on the seer and soothsayer (Clottes, 2004).

13.4 THE PHYSICOCHEMICAL PROPERTIES OF CO_2

13.4.1 Carbon Dioxide is Difficult to Recognize

In contrast to heavily smelling or strongly irritating gases like HCl, HBr, SO_2, or H_2S, carbon dioxide is ascribed as colorless, odorless, and tasteless in its gaseous form (Sax and Lewis, 1989).[b]

This means that if an animal walks into a CO_2 gas cloud, which is gradually increasing in concentration, it will not recognize the danger and become asphyxiated (Krohn et al., 2003; Niel et al., 2007). On the other hand, if one enters a region with a distinct and immediate transition of normal air to a concentrated CO_2 gas lake, the sudden change will clearly be witnessed (see below).

13.4.2 Carbon Dioxide Forms Gas Lakes

The molecular weight of CO_2 is 44 g/mol. Its density of 1.98 kg/m^3 CO_2 is 1.5 times heavier than air (Lide, 2003). Lower-lying areas, holes, valleys, or depressions, but also caves and cellars, will thus be filled with the gas and stable gas lakes can be formed (Pfanz, 2008; Raschi et al., 1997; Bettarini et al., 1999). Under calm weather conditions, such gas lakes form a very distinct boundary between the highly concentrated CO_2 gas lake and the above-lying atmospheric air (see Fig. 13.2). CO_2 concentrations can be as high as 100% within the gas lake and 0.5% in the direct vicinity of the boundary. Some centimeters distant from the interface, CO_2 concentrations normally drop to normal values of 0.04%. At the same time, oxygen concentrations may increase from 2% to 3% within the deadly gas lake to 20.9% in the adjacent atmosphere layering

[b]This is only partly true. If you try to inhale CO_2 concentrations of 35% and higher (don't do this because it is dangerous!), there is the sensation (smell and taste) of champagne or mineral water. The champagne-feeling (or sparkling mineral water-feeling) is easily explained by the extremely high CO_2 concentrations that builds up during opening of the bottle.

FIGURE 13.2   Small CO$_2$ gas-lake resting at the bottom of a gas cave. The dry gas-lake was stained.

on top of the lake. According to what was said above, stable boundaries could easily be recognized by animals and can therefore be avoided. Gas lakes, which would continuously increase in concentration, would be a much larger threat to life. If organisms are trapped within such gas lakes they will lose consciousness within seconds to minutes and then die.

The density of CO$_2$ also leads to the formation of distinct gas creeks. In the geothermal area of the Sousaki volcano (Greece) two caves emit high amounts of CO$_2$ gas ($>90\%$ CO$_2$ within the cave) and a continuous gas creek flows out of the cave's mouth and downhill. This gas creek can be analyzed and stained (Fig. 13.3). On its way downhill, it kills smaller organisms like mice, birds, and insects in its small riverbed.

It has to be added, that CO$_2$ gas lakes are stable only as long as no wind or sun action occurs. This calm and wind-still situation is true for the Hadean grottos below the Apollo temple and the Sanctuary of Pluto in Hierapolis. However, this is also true for shady valley structures, CO$_2$ caves (aragonite cave at Zbrasov, Czech Republic), or for cellars and other basements of buildings. In all other cases, wind turbulence or convection created by solar irradiation will destroy the lake and form normal atmospheric conditions during sunny hours (Pfanz, 2008; Raschi et al., 1997; Bettarini et al., 1999; Kies et al., 2015).

FIGURE 13.3   Dry CO_2 creek creeping out of a CO_2 cave at the crater rim of Sousaki volcano in Greece. The gas is heavier than air and flows downhill. The gas was stained with an orange smoke capsule.

13.4.3 Carbon Dioxide Displaces Atmospheric Oxygen

As CO_2 displaces oxygen, just because of its pure presence in high concentrations, aerobically breathing organisms will run into respiratory problems (Niel et al., 2007; Pilz et al., 2017 and literature therein). As CO_2 is already the highest possible oxidized state of carbon, it is nonflammable and can thus be used to quench flames. This physical property of CO_2 is well known to firefighters; they extinguish fires by displacing oxygen-using nonflammable CO_2. Dante Alighieri described the extinguishing flames on the hot Bullicame creek (Dante Alighieri, L'Inferno, Canto XII, XIV and XV).

13.4.4 Carbon Dioxide Forms an Acid

Gaseous CO_2 is the anhydride of carbonic acid. Furthermore, it is highly soluble in aqueous solutions. Once dissolved, carbonic acid dissociates according to Eq. (13.1) liberating one or two protons.

$$CO_2 + H_2O \Leftrightarrow H_2CO_3 \Leftrightarrow H^+ + HCO_3^- \Leftrightarrow 2H^+ + CO_3^{2-}; \qquad (13.1)$$

Liberation of protons leads to an acidification of the aqueous phase. Depending on the actual buffering capacity, acidification may be larger or smaller (Pfanz und Heber, 1986; Pfanz, 1994). Aqueous phases exist on surfaces and within organisms. The liquid film on eyes, the mucosae in nose and lung, as well as all intercellular compartments are prone to the acidifying effect of CO_2. Itching of organs or the inhibition of enzymatic action may be the consequence (Pfanz und Heber, 1989).

13.5 THE BIOLOGICAL, MEDICAL, AND PHYSIOLOGICAL BACKGROUND

High concentrations of CO_2 can act in two ways. (1) On the one side, high carbon dioxide concentrations displace atmospheric oxygen and (2) on the other side, CO_2 is potentially acidic and deprotonates in solution. In the first case, the physiological effects are hypercapnia, or oxygen deficiencies like hypoxia or even anoxia (Fig. 13.4). In the latter case, cellular compartments get acidified (Pfanz und Heber, 1989) and enzyme action is reduced or blocked (Pfanz, 1994).

Therefore, extreme CO_2 concentrations cannot be tolerated by aerobically breathing creatures; anaerobic organisms (some bacteria and archaea) on the other hand would survive. Mammals already react to CO_2 concentrations as low as 3%−5% (Krohn et al., 2003; Niel et al., 2007; Hill and Flack, 1908, for some exceptions see Russell et al., 2011). Even these rather low concentrations may increase cardiac frequency and respiratory rate if the incubation time is longer than several

FIGURE 13.4 An asphyxiated red-backed shrike (*Lanius collurio*) in a small depression in a mofette meadow in the Czech Republic.

minutes (Pfanz, 2008; Ikeda et al., 1989; D'Alessandro, 2006; D'Alessandro and Kyriakopoulos, 2013). At CO_2 concentrations around 8%–10%, humans are asphyxiated and a longer incubation at 15%–20% inevitably leads to death. A detailed list of human reactions to different CO_2 concentrations is found in the IVHHN Gas Guidelines (2018).

Divers have to cope with oxygen deficiencies and an overdose of CO_2 in their blood. Pregnant women can use the CO_2 trick during the birth process to manipulate their blood pH. Re-inhaling expired CO_2, (holding the hands in front of the mouth) leads to a decrease in blood pH, whereas panting (getting rid of CO_2) alkalizes the blood. In this way, strong pain can be reduced and the oxygen level necessary for mother and baby is conserved. On the other side, sudden infant death syndrome is sometimes related to hypercapnia.

Too much CO_2 leads to dizziness. There are reports of miners digging for coal and lignite who were confronted with sudden leaks of CO_2. Similar fatalities have been described for brewers and fire-fighters (Baxter, 2000).

Also, the Nyos (Cameroon) catastrophic events lead to an input in the knowledge of CO_2 extremes (Sigurdsson et al., 1987). In 1986 more than 1700 people died in a volcanic CO_2 gas accident. A volcanic crater lake was supersaturated with CO_2 being strongly emitted from its bottom into the water (Rice, 2000). Cold temperature and a high water pressure (due to a water depth of 253 m) led to strong accumulation of carbon dioxide in the lake water until super-saturation was reached. Due to an earthquake with a concomitant landslide, the surplus of gas was spontaneously set free and a 30 m high gas cloud was formed upon the lake surface. Due to its special density this deadly gas cloud flowed downhill through a narrow valley reaching several villages, thus killing 1700 people and 5000 cattle. Many of the corpses looked like they were sleeping. Only a few had blisters and bigger skin rashes. The latter were thought to be due to *rigor mortis* and concomitant changes, because the victims were found several days after the accident (Baxter and Kapila, 1989). Similar CO_2 accidents due to volcanic CO_2 exhalations have been reported from Lake Monoun, the Djeng Plateau, and many other places (Vaselli et al., 2002/2003; D'Alessandro, 2006; Le Guern et al., 1982).

The positive CO_2 effects are less but nevertheless worth mentioning. The correct concentration of CO_2 in the breathed air may enhance the blood circulation. Patients with low blood pressure and concomitant cold limbs therefore take "dry CO_2 baths" to speed up the circulation (Hadnagy and Benedek, 1968). The hands and feet of elderly people immersed in gas mixtures with CO_2 concentrations between 25% and 35% (w/v) warm up within a few minutes. Care has to be taken that the deadly gas atmosphere is not directly inhaled. Patients just stand in a 1–1.2 m high, dry CO_2 gas lake. In several countries (Romania,

Bulgaria, Hungary, Slovakia) dry CO_2 gas cures are well-known and some health resorts are famous for their geogenic gas (e.g., Harghita Mountain in Romania; Neda et al., 2008).

However, not only the blood circulation is affected. There are also verbal records on positive effects of high CO_2 on skin diseases caused by bacteria and fungi. The potentially acidic action of CO_2 in aqueous phases (Pfanz, 1994) and within cells may reduce the viability of pathogens, leading to a relief of the pain. The effects of warm mineral and CO_2-containing cure waters are well documented.

Whether these positive effects of geogenic CO_2 gas have really been used in front of the Plutonia has not yet been studied in detail. Yet, the rites of incubation and the rituals around prophecies and seeing strongly hint in this direction (De Boer et al., 2001; Etiope et al., 2005).

13.6 THE KNOWN SITES OF THE ANCIENT GATES TO HELL

In principle, the occurrence of ancient gates to hell reflects the ancient realm of Greece and its colonies (Magna Graecia, Proper Hellas, and Asia Minor). Entrances to the netherworld are therefore known for southern Italy, for the Greek mainland, and for the Asia Minor part of modern Turkey.

13.6.1 Italy (Magna Graecia)

Within the modern borders of Italy, gates to hell may be found in the larger region around Mount Vesuvius. Here many craters of the Phlegraean fields with mofettes, solfatares, and fumaroles, the volcanic lakes of Averno and Agnano but also the Grotta del Cane and Mefite D'Ansanto can be found. For the Grotta del Cane and other gas caves in the vicinity the so-called *spiritus letalis* (deadly haze) is mentioned (Soentgen, 2010).

13.6.1.1 The Phlegrean Fields (Campi Flegrei)

Bordering the gulf of Naples and in sight of Mount Vesuvius the caldera of a super-volcano is located. The region of the Campi Flegrei (burning fields) is punctuated by several volcanoes, older calderas, volcano lakes, and gas-emitting vents (Granieri et al., 2010). The whole area is additionally prone to bradyseismism, meaning that the whole region has an uplift of several meters and then slowly collapses. Within this area are the towns of Pozzuoli, Cumae, and a large part of Naples.

13.6.1.1.1 Lago Averno

In the southern part of the Campi Flegrei close to the coast is a saline lake called "Lago Averno." Its water composition is the result of the mixing of seawater and saline hydrothermal fluids (Caliro et al., 2008). At present, the input of gases is very low and the overturn of its waters depends only on temperature-driven density stratifications. Such overturns happen only in correspondence to particularly cold winters and have as a consequence fish-killing due to the H_2S-rich and O_2-poor composition of the uplifted deepest layers. The H_2S is not only of hydrothermal origin but derives also from sulfate reduction in the deep anoxic part of the lake and the mass of the upwelled gas and the extension of the consequent fish-killing depends mainly on the length of the previous accumulation period (Caliro et al., 2008).

In Roman times, such an overturn mechanism was less probable because the lake had been opened to the sea and transformed into the main military port hosting the imperial fleet. In such conditions, it was less probable that the surface layers would become denser, triggering an overturn of its waters. However, it cannot be excluded that at that time hydrothermal gas inflow or CH_4 production in the deeper anoxic parts was higher, leading to bubble formation or even lake overturn.

Several ancient writers expected the Lago Averno to be a gate to hell. In addition, gas emissions are mentioned: "No bird is able to fly over its surface but will soon die because of toxic vapors" (Vergilius Aeneis, cf. Cantus VI). The name Averno is derived from *aornos*, meaning "birdless" in Greek. In the vicinity, the Roman hero Aeneas met Sybilla to ask her for the way to meet his father (Vergilius, Aeneis, cf. Cantus VI). She explains that it is easy to find the way to Pluto right through Lake Averno. Several times Vergilius mentions toxic exhalations from the lake that killed all life.

13.6.1.1.2 Solfatara and Pisciarelli

More or less in the center of the Phlegrean Fields, directly belonging to the settlement of Pozzuoli, the large crater of Solfatara resembles a hostile, nonhabitable place on Earth. Many mofettes and solfatares emit toxic gases in addition to fumarolic hot water emissions (Granieri et al., 2010). Due to the wide and open morphology of the area, CO_2 generally does not reach dangerous levels but concentrations of a few hundred ppm in excess of normal atmospheric air at 1.5 m height, even during the daytime, testify to strong CO_2 emission (Carapezza et al., 1984). The contemporaneous release of other typical hydrothermal gases (water vapor and hydrogen sulfide) contributes to the image of an entrance to hell's. According to Capaccio (1604) or Galanti and Jagemann (1793) the caldera of Pozzuoli was thought to be the market square in front of hell and the abysm to hell (Fig. 13.5).

FIGURE 13.5   Hot water steam and carbon dioxide emanating from a bubbling pool at Pisciarelli (close to Solfatara) in the Phlegreian Fields, Pozzuoli, Italy.

13.6.1.2 Bullicame

Dante Alighieri described extinguishing the flames on the hot creek Bullicame close to Viterbo (Dante Alighieri, L'Inferno, Canti XII, 127–129, XIV, 88–90 and XV, 1–3). He describes this geothermal phenomenon with additional degassing of sulfurous and carbonic gases in his song of hell as "...a hot stream of blood and sulfur..." Although not directly knowing the gas, Dante clearly describes the physical action of highly concentrated carbon dioxide clouds:

> "Cosa non fu dagli occhi tuoi scorta
> Notabil come lo presente rio,
> Che sopra sè tutte fiammelle ammorta."

> "Nothing has been discovered by thine eyes
> So notable as is the present river
> Which all the little flames above it quenches." *Dante Alighieri, L'Inferno, Canto*
> *XIV, 88–90*

13.6.1.3 Mefite D'Ansanto

In Vergilius (book VII) it is written:

> [565]"est locus Italiae medio sub montibus altis,
> nobilis et fama multis memoratus in oris,
> Amsancti valles; densis hunc frondibus atrum
> urget utrimque latus nemoris, medioque fragosus
> dat sonitum saxis et torto vertice torrens.

hic specus horrendum et saevi spiracula Ditis
monstrantur, ruptoque ingens Acheronte vorago
pestiferas aperit fauces, quis condita Erinys,
invisum numen, terras caelumque levabat. *Vergilius; Aeneis, book VII*

About 80 km east of Mount Vesuvius a place now called *Mefite D'Ansanto* is the eponym of all CO_2 degassing sites. In this area, the goddess Mefitis (Mephitis) was worshipped. For Vergilius this place is the entrance to the underworld (Aeneis book VII). CO_2 degassing is so strong that the water pond at the base is heavily bubbling although it is cold and within 20,000 m^2 no vegetation can be found (Fig. 13.6). Naked soil, void of any life, covers the ground around the venting area. The degassing area is surrounded and framed by reed grass (*Phragmites australis*), several *Juncus*, *Agrostis*, and one endemic *Genista* species, able to tolerate even peak concentrations of CO_2. The gas flow is so high that a dry CO_2 gas creek runs downhill using the riverbed of an ephemeral creek. The area downhill is extremely dangerous to spectators of the mephitic Sanctuary as wind gusts may blow deadly CO_2 gas parcels unexpectedly in any direction (see also Chiodini et al., 2010). In Mefite D'Ansanto, Chiodini et al. (2010) measured a total output of about 1000 t of CO_2 per day over an area of 4000 m^2. Such a huge CO_2 output is not connected to ongoing volcanic activity but the carbon and helium isotopic composition of its gases points to the same mantle origin of the nearby volcanic systems of Vesuvius and the Phlegrean Fields. The massive output, together with the right topographic situation, allows the build-up of sometimes-lethal conditions at the main venting area and along the narrow valley immediately downstream.

FIGURE 13.6   The eponymous CO_2 degassing mofette Mefite D'Ansanto in southern Italy. The area is void of vegetation, indicating extreme CO_2 soil fluxes and concentrations.

Chiodini et al. (2010) found that in the absence of wind, the lethal concentration of 15% of CO_2 at 1.5 m height extends for about 200 m along the valley, while the dangerous threshold of 5% at the same height extends for more than 1 km, reaching far beyond the area devoid of vegetation.

For centuries, Mefite was known as an interesting site for the possibility of finding ancient remnants, in particular coins. The need to get the eyes close to soil led many visitors to breathe the poisonous atmosphere, even on apparently "safe" days. It is therefore not hard to believe that at this site at least 12 persons have died from the gas in the period from the 17th century until present (Chiodini et al., 2010).

13.6.1.4 *Naftia*

About 50 km south-west of Mt. Etna in Sicily is another strong CO_2 degassing site called Naftia, which has been compared by many authors to the Mefite D'Ansanto. This site has been known since ancient times, being described by geographers like Strabon and Diodorus Siculus and in modern times by Ferrara (1805). At this site, abundant gas, mainly composed of CO_2, was bubbling within a shallow lake. In the past, the lake generally dried out during summer leaving two craters from which gas was continuously emitted. Nowadays, the lake has been reclaimed and transformed into arable land, while gas emission still exists (De Gregorio et al., 2002), but has been strongly reduced by industrial exploitation of the natural gas emission.

This site has to be considered a gate to the netherworld being related to the chthonian deities of the Palikoi. Their mythological history can be found in the scripts of many ancient writers (Aeskylus, Vergilius, and Ovid) although in different versions.

The origin of the myth surely predates the Greek colonization of the inner part of Sicily and the Palikoi were previously adored by the indigenous Sikel population. Recent archaeological investigations showed that the site was inhabited since the Neolithic and that the building of a sanctuary started in the 7th−6th centuries BC (Maniscalco and McConnell BE, 2003; Maniscalco, 2008). The Palikoi were the personification of the natural gas emission phenomenon. They were twins representing the two main gas emission points in the lake which created two geyser-like columns and their "come back" from the netherworld was certainly connected to the annually disappearance of the lake that dried out leaving two deep craters in the summer and to their return in the rainy autumn−winter period typical of the Mediterranean climate.

The cult of the Palikoi lasted at least until the Roman imperial period and their sanctuary had great importance, being the site of an oracle, an asylum for run-away slaves, as well as a place where oaths could be tested by different rituals. These rituals were tightly connected to the degassing activity of the lake. In one version of the rituals, oaths were written on tablets and thrown into one of the bubbling craters. If the tablet swam on the water, the oath was considered to be true, but if it sank down, the oath was regarded as perjury. The destiny of the tablet strongly depended on the part of the water surface that it reached. In those parts where no gas was coming up or where the gas pressure was very high the tablet was sustained above the water but where degassing activity was higher, the presence of the bubbles would lower the water density, and consequently its buoyancy, causing the sinking of the tablet.

In another ritual, the person to be tested had to go close to the gas emission and, after making his oath, he had to bow down and touch the crater. Depending on the degassing activity and on the meteorological conditions he could be overwhelmed by the CO_2 and suffocate. This would have been interpreted as a sign of his perjury. In this case, the punishment came directly from the gods, while in the previous case the perjurer was sentenced to blinding or death.

13.6.1.5 Lago di Pergusa

Not directly related to geogenic gas, but probably due to biogenic sulfur gases, is the gate to the netherworld at Lago di Pergusa. Located in the center of Sicily in the neighborhood of the town Enna, this circular brackish lake regularly blooms with algae and bacteria. Semisaline algae from the family Chromatiaceae (*Thiocapsa roseopersicina* and *Thiodictyon elegans*) are the cause of the red coloration (Kondratieva et al., 1976). According to Publius Ovidius Naso (Metamorphosis 5.385–391) it was later thought that the lake was one of the sites where Hades abducted Kore (Prosperina). Close to the lake, there are remnants of a temple devoted to Ceres (the mother of Proserpina).

13.6.1.6 Mount Etna and Stromboli

In medieval times even Mount Etna was regarded as a gate to hell (Classen, 2015). This perception is quite understandable as the frequent eruptive phases of Etna, its spitting of lava and the possibility of having a look into the infinite depth of its inactive craters during times of no activity may have led people to think of the gateway to hell.

In Jules Vernes' book "Journey to the Center of the Earth," a group of people guided by the German professor Otto Lidenbrock descends into the Icelandic volcano Snæfellsjökull. The journey ends when the group is spat out from the netherworld through the open conduits of Stromboli (Verne, 1864).

13.6.2 Greece

13.6.2.1 Eleusis: The Elysian Grotto

The probably most important center to worship Demeter and her daughter Persephone was the Plutonium of Eleusis in Attica (Elefsina in modern Greece). According to several authors, this site is not only the gate to hell but also the most important place where Pluto was thought to abduct and rape Persephone (Von Uxkull, 1957). The proper Plutonium consists of two half-caves and a forecourt (Fig. 13.7). On the right side of the half-cave there is a smaller cave and a hole through which Persephone annually escapes from Pluto's realm to re-enter the desiccated earth at the end of the Mediterranean summer. The appearance of Persephone makes her mother Demeter very happy and therefore "Mediterranean spring" with concomitant sprouting and blooming occurs.

FIGURE 13.7   The Plutonium of Eleusis in Attica where Pluto was thought to have abducted and raped Persephone. The proper Plutonium consists of two half-caves and a forecourt.

Yet, although this Eleusian site is the well-documented *locus typicus* of a Hadean gateway, no degassing of CO_2 could be found within the cave and its surroundings.

13.6.2.2 The Nekromanteion of Acheron-Ephyra

The death oracle (Nekromanteion or also Nekyomanteion) of Acheron is located in the ancient region of Thesprotia (now Prevesa). The region was formerly a swampy area of the Acheron River with its tributaries and lake Acherusia (which doesn't exist nowadays as it was drained in the 1960s). Like Cape Tainaron it is thought that Kirke led Odysseus to the Hadean underworld to meet the seer Teiresias (Homer, 1900; Odysee; Herodot 5.92). In addition, Theseus was thought to visit the underworld in Acheron.

The Acheron River discharges into the Ionian Sea and its most important tributaries are Pyriphlegethon (Fire river, probably due to the occasional fluorescence of its waters), Kokytos (lament, river of mourning), and Vouvo Rema. Close to the confluence of the Acheron, Pyriphlegeton, and Kokytos rivers, a cave is located that is thought to be the abyss to Hades (Dakaris, 2001). The waters of the rivers derive from the underworld river Styx. It is said that at a certain ford of the Acheron River poplars and willows would grow but their fruits would die off (gas?). Moreover, based on Greek Mythology, the Acheron River had bitter waters (specific chemical composition?). During the battle of Titans (Titanomachy), the Titans drank water from the Acheron River to quench, an event that caused the wrath of Zeus, who blew and made the river waters bitter. The ruins of the Nekromanteion are located west of the ancient lake Acherusia close to the village of Mesopotamos. A small statue of Persephone was found and evidently proved the correctness of this as the proper site (Dakaris, 2001). Yet, several authors deny the ruins to be related to the famous Nekromanteion but rather believe it to be an 18th century monastic church (Wiseman, 1998) (Fig. 13.8).

Gas measurements performed in 2016 within the alleged building of the Nekromanteion showed no enhanced soil CO_2 concentrations. When measurements were performed close to the banks of the Kokytos, a tributary of the Acheron River, extremely high amounts of methane but no enhanced CO_2 concentrations were found (Pfanz, Geraga, Dimas, Papatheodorou, unpublished)

FIGURE 13.8   The Nekromanteion at the Acheron valley.

13.6.2.3 Cape Tanairon

At the tip of the peninsula Mani, southern Peloponnese, there seems to be an alternative entrance to hell at which the shadows can avoid crossing the rivers Acheron or Styx. Ovidius (book X, Metamorphosis) describes the death oracle as an entrance to hell. Orpheus was visiting the underworld (Tartarus) via this gate. In addition, Herakles acted at that entrance and forced the three-headed Kerberos out of hell. Although many measurements were carried out at various different loci within this area, no trace of a geogenic CO_2 emission was detected (Pfanz, D'Alessandro, Papatheodorou, Calabrese, Daskalopoulou, 2016, unpublished) (Fig. 13.9).

FIGURE 13.9 A probable gate to hell at Tanairon (Cape Matapan) at Greek Peloponnese.

13.6.2.4 Ermioni

The cave of the Katafyki gorge close to Ermioni is also mentioned as a site for the hell's gate. It is located on a steep slope of the gorge (Fig. 13.10). As it is not easy to reach and as it extends deep into the rock, it may be regarded as an entrance to the netherworld according to Ustinova's definitions (Ustinova, 2009). A preliminary search for geogenic gas emissions was unsuccessful (Pfanz, D'Alessandro, Papatheodorou, Calabrese, Daskalopoulou, 2016 unpublished).

13.6.2.5 Lake Lerna or the Alcyonian Lake

The perilous lake Lerna (lake, river, or spring) and the Alcyonian lake are located close to Dimini on the gulf of Argos in the western

FIGURE 13.10 A cave as a probable gate to hell. It is located in the Katafyki gorge close to Ermioni.

Peloponnese (Zangger, 1991). According to Pausanias Geographica (26.8–36.6 and 37.1–37.5) the Alcyonian lake had an infinite depth and its ground directly led to hell (see also Strabon VIII, 6-8, 8.6.8; Harrison, 1922; Leekley and Noyes, 1976).

Pausanias writes in his description of Greece wrote:

> There is no limit to the depth of the Alcyonian Lake, and I know of nobody who by any contrivance has been able to reach the bottom of it since not even Nero, who had ropes made several stades long and fastened them together, tying lead to them, and omitting nothing that might help his experiment, was able to discover any limit to its depth. This, too, I heard. The water of the lake is, to all appearance, calm and quiet but, although it is such to look at, every swimmer who ventures to cross it is dragged down, sucked into the depths, and swept away.

Several myths exist around the lake, among them the awkward Lernean Hydra being the underwater guard of the proper gate to the subaquatic entrance to the netherworld. According to the *fabulae* of Gaius Julius Hyginus, Hydra was a poly-headed water-monster with a toxic and perilous breath, thus executing a similar task to Kerberos on land (Pausanias Geographica, 26.8–36.6 and 37.1–37.5; Dodwell, 1819; Rose, 1958 and 1963; Marshall, 2002). Another myth around lake Lerna is concerned with the ancient Lernean mysteries devoted to the goddess Demeter (Pausanias, 2.37.1).

As with most entrances to hell, some heroes made their way down and back. In this case, Prosymnus and Dionysos tried to find and free his mother Semele in the underworld (Leekley and Noyes, 1976).

13.6.3 Turkey: Asia Minor

13.6.3.1 Hierapolis

Extreme CO_2 concentrations were found within and around the sanctuary of Hierapolis/Phrygia (Pamukkale, Turkey). Hierapolis is located in the Denizli Graben, which is a geological disturbance zone extending between the Pamukkale and Babadag faults. The first recordings of the town were made by Strabon (XII. 8.17) and also Plinius the Elder (Nat. Hist. V, 105) mentions it. The town, probably a Seleucid foundation in the 2nd century BC (Porter, 2016), developed during the Roman Empire and was famous in the Byzantine times. Hierapolis is cut longitudinally by several parallel fractures of the Pamukkale fault by intraplate tectonics and was destroyed by many earthquake events (Hancock and Altunel, 1997; D'Andria and Silvestrelli, 2000; D'Andria, 2003). Directly built upon such a fault are two buildings: The famous Apollo temple (Negri and Leucci, 2006) and the newly discovered Plutonium, the sanctuary of the Gods of the Underworld, Hades and Kore with the theater above a grotto (D'Andria, 2013).

The existence of deadly vapors (*spiritus letalis, mantikon pneuma, anathymiasis*) around the gates of hell is described by several authors. "Foramina pestilens exhalatur vapor ..." can be read in Seneca (Naturales Quaestiones, libro VL, Chapter 28). Also Plinius (Historia naturalis), Strabon and Vergilius (cf. Cantus VI, Aeneis) mention a deadly haze around Plutonia and Charonia (see also Zwingmann, 2012; Soentgen, 2010).

It has been mentioned by several sources (Zwingmann, 2012; D'Andria, 2003) that strange things were happening at the outlet of the Plutonium. Priests were demonstrating their supernatural power and their equality to the gods by ushering animals like goats and bulls into the antechamber where after a short time the animals showed signs of suffocation, finally dying after several minutes (Strabon XIII, 4, 14, Plinius, Nat. Hist. II, 207−208, and Zwingmann, 2012).

13.6.3.1.1 The Plutonium in Hierapolis

After the excavation held by D'Andria in 2011−13, a subterranean grotto was found below the stone-seats of the Theatron. The grotto belongs to the Sanctuary of Pluto and Kore. The words "Ploutoni kai Kore" in Greek letters engraved into the stone row are still readable. In front of the small hole of the grotto, dead insects and birds can be found. The extremely high CO_2 concentrations and the concomitant low oxygen concentrations within the Hadean antechambers (grottos) below the seat rows in the Plutonium and Apollo temple in Hierapolis/Phrygia are highly toxic. Outside the subterranean grotto, in front of the stone-made sitting rows, solar irradiation and wind do not allow a persistent gas lake. Yet, the permanent flow of CO_2 out of the chamber onto the floor of the theatron forms a transient gas lake. As in many other CO_2 gas lakes, the toxic atmosphere forms in the evening hours, persists throughout the night and is reduced in the morning due to the absorption of the infrared portion of the sunlight (Pfanz, 2008; Bettarini et al., 1999; Etiope et al., 2005; Kies et al., 2015; Pfanz et al., 2018). The great number of corpses of insects and birds prove its existence. Dead birds and more than 70 dead beetles (Tenebrionidae, Carabaeidae, and Scarabaeidae) were found asphyxiated on the floor. Locals report dead mice, cats, weasels, and even asphyxiated foxes. Most animals weren't killed during sunny days, but during the dark hours. Only insects were seen to be asphyxiated even during midday hours. Even then, a thin 5 cm "high-CO_2-gas-layer" covered the ground in front of the grotto of the Plutonium at Hierapolis, killing smaller beasts (Pfanz et al., 2018).

Gas measurements inside the closed subterranean chamber (grotto) revealed CO_2 concentrations of up to 91%. The complete basement of the grotto was completely dark but seemed to be highly humid, due to a warm carbonate-rich creek flowing below it. Also, in front of the actual grotto, deadly CO_2 gas exists. Flooding out of the grotto's mouth, the escaping CO_2 forms a gas lake at the floor. The corpses of animals hint to the absence of oxygen and the presence of high CO_2 that builds up during hours of darkness.

13.6.3.1.2 The Temple of Apollo in Hierapolis

Similar findings were obtained when the grotto under the Temple of Apollo was studied. The sanctuary of Apollo is situated 200 m north of the Plutonium within the same seismic fracture zone and it has been long known for its subterranean pit structure. This pit was interpreted as Plutonium in previous excavations and it is marked as "Plutonium" in tourist guidebooks for Hierapolis. Recent excavation (2011−13) has demonstrated that the proper Plutonium is located in the area south of the Apollo Sanctuary (D'Andria, 2016). The toxicity of its gas atmosphere is known and it was therefore firmly walled in for safety reasons. Within the mouth-hole of the Apollo-Plutonium between 60% and 65% CO_2 was measured; at the same time, oxygen was only 8%−8.5%.

13.6.3.2 Nysa and Acharaka

Close to Acharaka and Nysa ad Maeandrum there was another sanctuary of Pluto and Persephone (Fig. 13.11) and an additional Charonion. It was a well-known site for incubations and health cures. In addition, the abduction of Persephone is described for this place. Strabon (Geographica 14.144) also describes the sacrifice of bulls and other sacred animals. The detailed description of the death of these animals is

FIGURE 13.11   The remnants of the Plutonium at Nysa Acharaka.

consistent with those reported for the Plutonium at Hierapolis. An ancient coin found nearby proves the ritual of sacrificing bulls.

Gas measurements performed in 2015 proved the existence of CO_2 emissions close to the sanctuary (Pfanz and Yüce, unpublished). A stone wall erected in the valley of the nearby sulfur creek emits CO_2 in high amounts. Between the staggered joints and chinks, CO_2 gas is seeping out of the slope behind. Up to 84.1% CO_2 was measured. It would be of great value to thoroughly investigate the area for additional geogenic gas seeps.

13.6.4 Some Modern Gates to Hell

The Darvaza gas crater, located in Turkmenistan near Derweze, originated in 1971, when Soviet geologists drilled a well for methane extraction. As a consequence of vibrations, the rocks collapsed in an underground cavern, thus creating a sort of crater, about 30 m deep. The engineers decided to set the gas on fire, and the flames have been burning since then. The location became an attraction for tourists; locals apparently gave the name "door to hell" to the site.

Similarly, in northern Japan, the Osore Mountain (Osore-Zan), is traditionally considered as an entrance to hell, as a consequence of the many fumaroles and sulfur gas emissions. The local Buddhist temple (Bodai-ji) is characterized by the presence of mediums (itako) that are believed to summon the souls of the dead and to deliver their messages. Intriguingly, the name Osore may come from the Ainu language, and according to some authors, Ainu identify hell as the land of volcanoes (Fackler, 2009).

There is also a modern gate to hell in Czech Republic. Houska Castle (in the city of Blatce, north of Prague), a Gothic castle originally built in 1253–78 during the reign of Ottokar II of Bohemia has a very special myth. Rumors spread of human-animal chimaeras creeping out of a gigantic hole directly leading to hell (Padevet, 2010). The hole is topped by a chapel of the castle. It is also said that the castle is void of a kitchen, of water, and also of inhabitants (David et al., 2004). Its existence is purely aimed at blocking the devil from escaping through the hole. It is unclear whether there are gas emanations within the hole.

Several hell's gates exist on New Zealand or in Africa. The geothermal spa Tikitere close to Rotorua (NZ) promotes "hell's gate sulfur spa" and "hell's gate mud bath". The total enumeration of all modern gates to hell could be endless.

13.7 GATES OF HELL—MAGNA GRAECIA—ASIA MINOR—GREECE (HELLAS)—A SYNOPSIS

In Magna Graecia (southern-most Italy) there are several sites called the gate to hell, door to hell, or entrance to hell that correlate with the existence of CO_2 gas seeps and vents. Among them are the marketplace of hell (the Solfatara—the caldera of the Pozzuoli volcano). Phlegrean Field gas emissions also contain traces of H_2S (hydrogen sulfide), which represents the foul smell of rotten eggs of the Solfatara area. Yet, the main gas emission is due to nonsmelling, invisible CO_2, a gas that killed a couple with their child in 2017 after they had fallen into a quicksand hole. Further sites with a proven correlation of geogenic CO_2 and sanctuaries are Lago Averno, Grotta del Cane at the former Lago Agnano, Naftia, Bullicame, and Mefite D'Ansanto, the sanctuary of the goddess Mephitis.

Also in Asia Minor (Western Turkey), there are at least three well-known sites of ancient gates to hell where geogenic CO_2 emissions have been verified. The sanctuary of Apollo at Hierapolis even has tourist signs that hint to the Plutonium. The gas-emitting grotto below the temple has been sealed off, but the existence of more than 60% deadly CO_2 vapors has been measured (Pfanz et al., 2018). Within the same geologic graben system, Prof. D'Andria and his group (D'Andria, 2013; Simsek and D'Andria, 2017) found a sanctuary of Pluto which shows a perfect correlation between its probable function in antiquity and the measurable geogenic CO_2 emission (Pfanz et al., 2018). At the sanctuary of Pluto geogenic CO_2 escapes in the close neighborhood of the temple (see above).

Quite in contrast and very astonishingly, in Greece proper (Hellas), no correlation has been found so far between geogenic gas exhalation and the known gates to hell. Neither in Cape Tainaron, nor at the Nekyomanteion in the Acheron delta, nor in the famous Eleusis were any hints of gas emissions found (Pfanz, D'Alessandro, Papatheodorou, unpublished). The reason why no geogenic gas has been found at these places remains unclear. It might be that there was never any degassing in the vicinity of these sites. Another theory is that there has been degassing in ancient times but gas seepage has stopped for seismic or tectonic reasons.

Recent studies about gas manifestations in Greece (Daskalopoulou et al., 2018) evidenced that most of the mentioned hell gates are in areas where CO_2 degassing is unlikely for geological reasons. In these areas there is conversely a greater probability of methane emissions. The only exception is the site of Eleusis which is not far from the volcanoes of Sousaki and Methana, which both have active CO_2 degassing areas (D'Alessandro et al., 2006, 2008). Further investigations may help in elucidating any causal relationships.

13.8 THE HISTORICAL RELEVANCE

Several gates to hell are described in the ancient literature. Not many of them have been seriously checked for geogenic gas emissions. In some places, it was clearly demonstrated that deadly high CO_2 emissions occur exactly as described by the ancient writers (Pfanz et al., 2018). A *mantikon pneuma*, a prophetic mist or smoke from the soil or an *anathymiasis* (an exhalation) is ascribed to Plutarch who was an eyewitness of the functionality of Delphi (Cross and Aaronson, 1988; Soentgen, 2010). In Hierapolis/Phrygia the source of this deadly gas is clearly the subterranean grotto which is considered to be the entrance to the Hadean underworld (Strabon XIII, 4.14, Plinius, Nat. Hist. II, 207—208; Zwingmann, 2012; D'Andria, 2013; Pfanz et al., 2018). All these authors and many others describe the deaths of small birds; local people sold them to spectators and ancient pilgrims of the sanctuary. The animals were thrown into the gaseous atmosphere just below their seats of the theatron; they watched how they died (Zwingmann, 2012). Strabon and Plinius also describe the sacrifice of bulls or goats within the grotto. Priests were demonstrating their supernatural power and their equality with the gods by ushering animals like goats and bulls into the Plutonium. They exactly describe how the bulls showed signs of suffocation, how their body vibrated, and how these robust herbivores died within minutes. Yet, the Galloi, the castrated priests of the Mother Goddess (Cybele) survived (Strabon XIII, 4.14; Plinius, Nat. Hist. II, 207—208; Zwingmann, 2012). The priests often stood on slightly elevated places. They knew about the deadly vapors. They probably also knew the exact height of the deadly gas lake. During the sacrifice of animals, they tried not to breathe deeply.

The toxic atmosphere in front of the gates to hell was thought to resemble the deadly smell of hell's unpleasant atmosphere but it also may be attributed to the voracious breath of the ferocious hellhound Kerberos. He was posted to prevent ghosts of the dead from leaving the netherworld. Several sources exist which describe the deadly smell from the mouth of the terrible, three-headed hell guard (Pausanias, Description of Greece 2. 35.10). Even the saliva of this wild beast was poisonous. When Herakles drove Kerberos out of Hades (which was one of his twelve tasks), the dog was spitting slime on the ground. From his toxic saliva the deadly toxic plant aconite or wolfsbane (*Aconitum napellus*) was born (Hesiod, Theogony 769 and the internet source: Greek gods). There is even a relation between the breath and slime of Kerberos and the hallucinogenic effects described in the neighborhood of the caves to hell: "... *(the Erinys) Tisiphone brought with her poisons too of magic power (to invoke madness): lip-froth of Cerberus, the Echidna's venom, wild deliriums, blindness of the brain, crime and tears, and maddened lust for murder; all ground up, mixed with fresh blood, boiled in a pan of bronze, and stirred with a green hemlock stick.*" (Ovid, Metamorphoses 4.500 ff. and the internet source: Greek gods).

It seems that in ancient times, natural phenomena could not easily and satisfactorily be explained (Masse et al., 2007). If they were additionally associated with pain and death, they were attributed to the deadly necrotic and chthonic forces living in unpleasant places of no return. The invisible haze resembled ghosts, gods, or the breath of the ferocious hellhound. It was something unexplainable, something supernatural, and something fearsome. It can therefore be assumed that the ancient word for "gas" is god, death, or mephitis. Nowadays these phenomena can easily be described and analyzed by physico-chemical means and quantitatively explained with a knowledge of geology, chemistry, and biology. Yet, it is still surprising how precise and exact, and without great exaggeration, that ancient writers were able to correctly describe these exciting natural (or supernatural?) phenomena.

Acknowledgments

We want to thank the Governorship of Denizli city, the Director of Hierapolis/Pamukkale and the provincial culture and tourism directorate of Ankara and Denizli. We would like to acknowledge the kind help of Dr. Kalliope Papaggeli (Director of Elefsina archaeological site), Dr. Pio Panarelli, Dr. Kadir Özel, Dr. Marco Esposito (Hierapolis excavation site), Dr. Giovanni Chiodini and Dr. Stefano Caliro (Osservatorio Vesuviano, Napoli).

References

Aiuppa, A., Baker, D.R., Webster, J.D., 2009. Halogens in volcanic systems. Chem. Geol. 263, 1–18.

Albertus Magnus, liber III Meteororum, Tract. II, Caput XII.

Albinus, L., 2000. The House of Hades. Studies in Ancient Greek Eschatology. Aarhus University Press, Aarhus.

Alighieri, Dante, 2001. Die göttliche Komödie. Hölle. Faber&Faber, Leipzig.

Baxter, P.J., 2000. Hunter's diseases of occupations. In: Baxter, P.J., Adams, P.H., Aw, T.C., Cockcroft, A., Harrington, J.M. (Eds.), Gases. Arnold, London, pp. 123–178.

Baxter, P.J., Kapila, M., 1989. Acute health impact of the gas release at Lake Nyos, Cameroon, 1986. J. Volcanol. Geotherm. Res. 39, 265–275.

Bettarini, B., Grifoni, D., Miglietta, F., Raschi, A., 1999. Local greenhouse effect in a CO_2 spring in Central Italy. Ecosystems Response to CO_2. The Maple Project Results. European Commission, Luxembourg, pp. 13–23.

Caliro, S., Chiodini, G., Izzo, G., Minopoli, C., Signorini, A., Avino, R., et al., 2008. Geochemical and biochemical evidence of lake overturn and fish kill at Lake Averno, Italy. J. Volcanol. Geotherm. Res. 178, 305–316.

Capaccio G,C., 1604. Puteolana Historia. C. Vitalis, Napoli

Capacio, J.C., 1604. Puteolana Historia. C Vitalis, Napoli.

Carapezza, M., Gurrieri, S., Nuccio, P.M., Valenza, M., 1984. CO_2 and H_2S concentrations in the atmosphere at Solfatara di Pozzuoli. Bull. Volcanol. 47, 287–293.

Chiodini, G., Frondini, F., Kerrick, D.M., Rogie, J., Parello, F., Peruzzi, L.F., et al., 1999. Quantification of deep CO_2 fluxes from Central Italy. Examples of carbon balance for regional aquifers and of soil diffuse degassing. Chem. Geol. 159, 205–222.

Chiodini, G., Granieri, D., Avino, R., Caliro, S., Costa, A., Minopoli, C., et al., 2010. Non-volcanic CO_2 earth degassing: Case of Mefite D'Ansanto (southern Appennines), Italy. Geophys. Res. Lett. 37, L11303.

Classen, A., 2015. Handbook of Medieval Culture. Walter de Gruyter GmbH & Co KG, Berlin, Boston, 978-3-11-026730-3p. 664.

Clottes, J., 2004. Hallucinations in caves. Cambridge Archaeol. J. 14, 81–82.

Cross, T.M., Aaronson, S., 1988. The vapours of on entrance to Hades. Antiquity 62, 88–89.

Dakaris, S.I., 2001. Das Nekromanteion am Acheron. Zaravinos, Athens.

D'Alessandro, W., 2006. Gas hazard: an often neglected natural risk in volcanic areas. In: Martin-Duque, J.F., Brebbia, C.A., Emmanouloudis, D.E., Mander, U. (Eds.), Geo-Environment & Landscape Evolution II. WIT Press Southampton, Ashurst Lodge, Ashurst, Southampton, pp. 369–378.

D'Alessandro, W., Kyriakopoulos, K., 2013. Preliminary gas hazard evaluation in Greece. Nat. Hazard. 69, 1987–2004.

D'Alessandro, W., Brusca, L., Kyriakopoulos, K., Rotolo, S., Michas, G., Minio, M., et al., 2006. Diffuse and focussed carbon dioxide and methane emissions from the Sousaki geothermal system, Greece. Geophys. Res. Lett. 33, L05307.

D'Alessandro, W., Brusca, L., Kyriakopoulos, K., Michas, G., Papadakis, G., 2008. Methana, the westernmost active volcanic system of the south Aegean arc (Greece): insight from fluids geochemistry. J. Volcanol. Geotherm. Res. 178, 818–828.

D'Andria, F., 2003. Hierapolis of Phrygia (Pamukkale). An Archaelogical Guide, Pamukkale.

D'Andria, F., 2013. Il Ploutonion a Hierapolis di Frigia. Istanbuler Mitt. 63, 157–217.

D'Andria, F., 2016. Nature and cult on the Ploutonium of Hierapolis. Before and after the colony. In: Simsek, C., D'Andria, F. (Eds.), Landscape and History in the Lykos Valley: Laodikea and Hierapolis in Phrygia. Cambridge Scholars Publ., Cambridge.

D'Andria, F., Silvestrelli, F., 2000. Ricerche Archeologiche Turche Nella Valle del Lykos, Galatina (Lecce), Congedo.

Daskalopoulou, K., Calabrese, S., Grassa, F., Kyriakopoulos, K., Parello, F., Tassi, F., et al., 2018. Origin of methane and light hydrocarbons in natural fluid emissions: A key study from Greece. Chem. Geol. 479, 286–301.

David, P., Soukup, V., Čech, L., 2004. Wonders of Bohemia, Moravia and Silesia. Euromedia Group, Knizni Klub, Praha, 978-80-242-2455-8p. 80.

De Boer, J.Z., Hale, J.R., Chanton, J., 2001. New evidence for the geological origins of the ancient Delphi oracle (Greece). Geology 29, 707–710.

De Gregorio, S., Diliberto, I.S., Giammanco, S., Gurrieri, S., Valenza, M., 2002. Tectonic control over large-scale diffuse degassing in eastern Sicily (Italy). Geofluids 2, 273–284.

Dodwell, E., 1819. A Classical and Topographica Tour Through Greece. Rodwell and Martin, London.

Etiope, G., 2015. Seeps in the ancient world: myths, religions, and social development. In: Etiope, G. (Ed.), Natural Gas Seeps. Springer International Publ., New York, Heidelberg, pp. 183–193.

Etiope, G., Klusman, R.W., 2010. Microseepage in drylands: flux and implications in the global atmospheric source/sink budget of methane. Global Planet. Change 72, 265–274.

Etiope, G., Guerra, M., Raschi, A., 2005. Carbon dioxide and radon geohazards over a gas-bearing fault in the Siena Graben (Central Italy). Terr. Atmos. Ocean Sci. 16, 885–896.

Etiope, G., Papatheodorou, G., Christodoulou, D., Geraga, M., Favali, P., 2006. The geological links of the ancient Delphic Oracle (Greece): a reappraisal of natural gas occurrence and origin. Geology 34, 821–824.

Etiope, G., Schoell, M., Hosgormez, H., 2011. Abiotic methane flux from the Chimaera seep and Tekirova ophiolites (Turkey): understanding gas exhalation from low temperature serpentinization and implications for Mars. Earth Planet. Sci. Lett. 310, 96–104.

Fackler, M., 2009. As Japan's Mediums Die, Ancient Tradition Fades. The New York Times ISSN 0362-4331.

Ferrara, F., 1805. Memorie sopra il Lago Naftia nella Sicilia meridionale: sopra l'ambra siciliana sopra il mele ibleo e la città d'Ibla Megara sopra Nasso e Callipoli, Palermo, Dalla reale stamperia. pp. 216.

Galanti, G.M., Jagemann, C.J., 1793. Neue Historische und Geographische Beschreibung Beider Sicilien, Vol 4. Sigfried Lebrecht Crucius, Leipzig.

Geographica 14.144.

Granieri, D., Avino, R., Chiodini, G., 2010. Carbon dioxide diffuse emission from the soil: ten years of observations at Vesuvio and Campi Flegrei (Pozzuoli), and linkages with volcanic activity. Bull. Volcanol. 72, 103–118.

Hadnagy, C., Benedek, G., 1968. Information of action mechanism of mofettes in Covasna. Arch. Phys. Ther., Baleol. Klimatol. 20, 229–233.

Hamilton, M., 1906. Incubation – or the Cure of Disease in Pagan Temples and Christian Churches. Henderson WC & Son, St. Andrews.

Hancock, P.L., Altunel, E., 1997. Faulted archaeological relics at Hierapolis (Pamukkale), Turkey. J. Geodyn. 24, 21–36.

Hansell, A., Oppenheimer, C., 2004. Health hazards from volcanic gases: a systematic literature review. Arch. Environ. Health 59, 628–639.

Harrison, J.E., 1922. Prolegomena to the Study of Greek Religion, third ed. Cambridge Uni. Press, Cambridge University Press, Cambridge.

Herodot liber5, lines 92ff.

Herodot 5,92.

Hill, L., Flack, M., 1908. The effect of excess carbon dioxide and of want of oxygen upon the respiration and the circulation. J. Physiol. 37, 77–111.

Homer, 1900. [c. 700 BC]. The Odyssey (S. Butler, Trans.). Longmans, Green & Co., London, UK.

Ikeda, N., Takahashi, H., Umetsu, K., Suzuki, T., 1989. The course of respiration and circulation in death by carbon dioxide poisoning. Forensic Sci. Int. 41, 93−99.

IVHHN Gas Guidelines, ⟨http://www.ivhhn.org/images/pdf/gas_guidelines.pdf⟩ (accessed 15.02.2018)

Kies, A., Hengesch, O., Tosheva, Z., Raschi, A., Pfanz, H., 2015. Diurnal CO_2-cycles and temperature regimes in a natural CO_2 gas lake. Int. J. Greenhouse Gas Control 37, 142−145.

Krohn, T.C., Kornerup Hansen, A., Dragsted, N., 2003. The impact of low levels of carbon dioxide on rats. Lab. Anim. 37, 94−99.

Kondratieva, E.N., Zhukov, V.G., Ivanovsky, R.N., Petushkova, Y.P., Monosov, E.Z., 1976. The capacity of phototrophic sulfur bacterium *Thiocapsa roseopersicina* for chemosynthesis. Arch. Microbiol. 108, 287−292.

Leekley, D., Noyes, R., 1976. Archaeological Excavations in Southern Greece. Lincoln, Park Ridge, New Jersey.

Le Guern, F., Tazieff, H., Faivre-Pierett, R., 1982. An example of health hazard: people killed by gas during a phreatic eruption: Dieng Plateau (Java, Indonesia), February 20th 1979. Bull. Volcanol. 45, 153−156.

Lide, D.R., 2003. CRC Handbook of Chemistry and Physics, 84th ed. CRC Press, Boca Raton, FL.

Maniscalco, L. (Ed.), 2008. Il santuario dei Palici: un centro di culto nella Valle del Margi. Collana d'Area. Quaderno n. 11. Regione Siciliana, Palermo.

Maniscalco, L., McConnell, B.E., 2003. The Sanctuary of the Divine Palikoi (Rocchicella di Mineo, Sicily): fieldwork from 1995−2001. Am. J. Archaeol. 107, 145−180.

Marshall, P.K. (Ed.), 2002. Hyginus: Fabulae. Saur Verlag, Munich.

Masse, W.B., Barber, E.W., Piccardi, L., Barber, P.T., 2007. Exploring the nature of myth and its role in science. In: Piccardi, L., Masse, W.B. (Eds.), Myth and Geology, 273. Geolog. Soc., London, pp. 9−28. , Special Publ.

Minois, G., 2000. Die Hölle. Kleine Kulturgeschichte der Unterwelt. Herder, Freiburg im Breisgau.

Mörner, N.A., Etiope, G., 2002. Carbon degassing from the lithosphere. Global Planet. Change 33, 185−203.

Neda, T., Szakacs, A., Cosma, C., Mocsy, I., 2008. Radon concentration measurements in mofettes from Harghita and Covasna Counties, Romania. J. Environ. Radioact. 99, 1819−1824.

Negri, S., Leucci, G., 2006. Geophysical investigation of the temple of Apollo (Hierapolis, Turkey). J. Archaeol. Sci. 33, 1505−1513.

Niel, L., Stewart, S.A., Weary, D.M., 2007. Effect of flow rate on aversion to gradual-fill carbon dioxide exposure to rats. Appl. Anim. Behav. Sci. 109, 77−84.

Nietzsche, F., 1883−1891. Band 1−4 Also Sprach Zarathustra - Ein Buch für Alle und Keinen. Verlag Ernst Schmeitzer, Chemnitz.

Ovid Metamorphosis Liber IV, 6 Athamas and Ino, pp. 500ff.

Oppe, A.P., 1904. The chasm of Delphi. J. Hellenistic Stud. 24, 214−240.

Padevet, J., 2010. Cesty s Karlem Hynkem Máchou. Academia, Praha, pp. 126−128.

Parke, H.W., 1985. The Oracles of Apollo in Asia Minor. Croom Helm, London.

Pausanias, Description of Greece - Greek Geography C 2nd A.D.

Pausanias Geographica. Liber 2, Chapter 26.8−36.6 and 37.1−37.5

Pepin, J., 1961. Mythe et allegorie: les origines grecques et les contestations judeo-chretiennes. In: Roques, R. (Ed.), Revue de L'histoire des religions, vol. 159, Persee, Lyon, pp. 81−92.

Pfanz, H., 1994. Apoplastic and symplastic proton concentrations and their significance for metabolism. In: Schulze, E.-D., Caldwell, M.M. (Eds.), Ecophysiology of Photosynthesis. Ecological Studies 100. Springer Verlag, Berlin, Heidelberg, pp. 103–122.

Pfanz, H., 2008. Mofetten – kalter Atem Schlafender Vulkane. RVDL-Verlag, Köln.

Pfanz, H., Heber, U., 1986. Buffer capacities of leaves, leaf cells, and leaf cell organelles in relation to fluxes of potentially acidic gases. Plant Physiol. 81, 597–602.

Pfanz, H., Heber, U., 1989. Determination of extra- and intracellular pH values in relation to the action of acidic gases on cells. In: Linskens, H.F., Jackson, J.F. (Eds.), Modern Methods of Plant Analysis NS; Vol 9. Gases in Plant and Microbial Cells. Springer Verlag, Berlin, Heidelberg, pp. 322–343.

Pfanz, H., Vodnik, D., Wittmann, C., Aschan, G., Raschi, A., 2004. Plants and geothermal CO_2 exhalations – survival in and adaptation to a high CO_2 environment. In: Esser, K., Lüttge, U., Kadereit, J.W., Beyschlag, W. (Eds.), Progress in Botany, 65. Springer Verlag, Berlin, Heidelberg, pp. 499–538.

Pfanz, H., Yüce, G., Gulbay, A.H., Gokgoz, A., 2018. Deadly CO_2 gases in the Plutonium of Hierapolis (Denizli, Turkey). Archaeol. Anthropol. Sci. Available from: https://doi.org/10.1007/s12520-018-0599-5.

Philosophus A. 1742. De deorum imaginibus libellus. Chapter VI: De Plutone.

Piccardi, L., Masse, W.B. (Eds.), 2007. Myth and Geology. The Geological Society, London.

Piccardi, L., Monti, C., Vaselli, O., Tassi, F., Gaki-Papanastassiou, K., Papanastassiou, D., 2008. Scent of a myth: tectonics, geochemistry, and geomythology at Delphi (Greece). J. Geol. Soc. 165, 5–18.

Pilz, M., Hohberg, K., Pfanz, H., Wittmann, C., Xylander, W.E.R., 2017. Respiratory adaptations to a combination of oxygen deprivation and extreme carbon dioxide concentration in nematodes. Respi. Physiol. Neurobiol. 239, 34–40.

Plinius.Historia Naturalis, Liber II, 207-208, liber V, 105, liber VII, 73.

Pluto Mythology, 2018. ⟨http://www.en.wikipedia.org/wiki/Pluto⟩. (mythology).

Porter, S.E., 2016. In: William, B. (Ed.), The Apostle Paul: His Life, Thought, and Letters. Eerdmans Publishing Company, Grand Rapids.

Publius Ovidius Naso. Metamorphosis Liber 5, lines 385–391

Raschi, A., Miglietta, F., Tognetti, R., van Gardingen, P.R., 1997. Plant responses to elevated CO_2. Evidence from Natural CO_2 Springs. Cambridge University Press, Cambridge.

Raschke, R., 1912. De Alberico Mythologo. Diss. Univ., Breslau.

Rice, A., 2000. Rollover in volcanic crater lakes: a possible cause for Lake Nyos type disasters. J. Volcanol. Geotherm. Res. 97, 233–239.

Rigoglioso, M., 2005. Persephone's sacred lake and the ancient female mystery religion in the womb of sicily. J. Feminist Stud. Religion 21, 5–29.

Rose, H.J., 1958. Fabulae (Genealogiae) of Gaius Julius Hyginus. Mnemosyne 42–48. Fourth Series 11.1.

Rose, H.J. (Ed.), 1963. Hygini Fabulae. second ed. A.W. Sijthoff, Leiden.

Rosenberger, V., 2001. Griechische Orakel – Eine Kulturgeschichte. Theiß, Darmstadt.

Russell, D., Schulz, H.-J., Hohberg, K., Pfanz, H., 2011. The collembolan fauna of mofette fields (natural carbon-dioxide springs). Soil Organisms 83, 489–505.

Sax, N.I., Lewis, R.J., 1989. Dangerous Properties of Industrial Materials, seventh ed. Van Nostrand Reinhold, NY.

Seneca. Naturales Quaestiones, libro VL, Chapter 28.

Sigurdsson, H., Devine, J.D., Tchoua, F.M., Presser, T.S., Pringle, M.K.W., Evans, W.C., 1987. Origin of the lethal gas burst from Lake Monoun, Cameroun. J. Volcanol. Geotherm. Res. 31, 1–16.

Simsek, C., D'Andria, F., 2017. Landscape and History in the Ykos Valley: Laodikea and Hierapolis in Phyrygia. Cambridge Scholar Publishing, Newcastle upon Tyne, ISBN (13): 978-1-4438-9859-1.

Soentgen, J., 2010. On the history and prehistory of CO_2. Found. Chem. 12, 137–148.

Strabon Geography Liber XIII, Chapter 4, 14ff, liber XIV Chapter 1, 11ff and 45-47

Strabon Geography Liber 8, Chapter 6, 8ff

Ustinova, Y., 2002. Either a demon or a hero, or perhaps a god. Mythological residents of subterranean chambers. Kernos 15, 267–288.

Ustinova, Y., 2009. Cave experiences and ancient Greek oracles. Time Mind 2, 265–286.

Vaselli, O., Capaccioni, B., Tedesco, D., Tassi, F., Yalire, M.M., Kasarerka, M.C., 2002/2003. The evil winds (mazukus) at Nyiragongo volcano (Democratic Republic of Congo). Acta Vulcanol. 14/15, 123–128.

Verne, J., 1864. Voyage Au Centre De La Terre. Pierre Jules Hetzel, Paris.

Vitaliano, D.B., 1973. Legends of the Earth: Their Geologic Origins. Indiana University Press, Bloomington.

Von Uxkull, W., 1957. Die Eleusinischen Mysterien. Versuch Einer Rekonstruktion. Avalun Verlag, Büdingen-Gettenbach.

Vorgrimler, H., 1994. Geschichte der Hölle. Fink, München.

Wiseman, J., 1998. Rethinking the "Halls of Hades". Archaeology 51, 12–17.

Wolf, F.A., 1839. In: Gürthler, J.D., Hoffmann, S.F.W. (Eds.), Vorlesungen über die Alterthumswissenschaft, Vol. 3. Lehndholdsche Buchhandlung, Leipzig.

Wolf, R.E., 2008. The Vatikan Mythographers. Fordham. Univ. Press, New York.

Zangger, E., 1991. Prehistoric coastal environments in Greece: the vanished landscapes of Dimini Bay and Lake Lerna. J. Field Archaeol. 18, 1–15.

Zwingmann, N., 2012. Antiker Tourismus in Kleinasien und auf den vorgelagerten Inseln. Habelt Verlag, Bonn.

Further Reading

Greek gods, 2018. ⟨http://www.theoi.com/Khthonios/Haides.html⟩.

Pfanz, H., Yüce, G., D'Andria, F., D'Alessandro, W., Pfanz, B., Manetas, Y., et al., 2014. The ancient gates to hell and their relevance to geogenic CO_2. In: Wexler, P. (Ed.), History of Toxicology and Environmental Health. Toxicology in Antiquity, Vol. I. Academic Press, Amsterdam, pp. 92–117.

Smets, B., Tedesco, D., Kervyn, F., Kies, A., Vaselli, O., Yalire, M.M., 2010. Dry gas vents ("mazuku") in Goma region (North-Kivu, Democratic Republic of Congo): formation and risk assessment. J. Afr. Earth. Sci. 58, 787–798.

Spiller, H.A., Hale, J.R., de Boer, J.Z., 2002. The Delphic oracle: a multidisciplinary defense of the gaseous vent theory. Clin. Toxicol. 40, 189–196.

Stewart, I.S., Piccardi, L., 2017. Seismic faults and sacred sanctuaries in Aegean antiquity. Proc. Geol. Assoc. 128, 711–721.

Vergilius Mar, P., 2005. In: Fink, G. (Ed.), Aeneis. Artemis and Winkler, Düsseldorf.

von Diest, W., 1913. Nysa ad Maeandrum. Jahrbuch des Kaiserlich Deutschen Archaeologischen Instituts. Ergänzungsheft X. Georg Riemer Verlag, Berlin.

Lead Poisoning and the Downfall of Rome: Reality or Myth?

Louise Cilliers and Francois Retief

Honorary Research Associate, Department of Greek, Latin and Classical Studies, University of the Free State, Bloemfontein, South Africa

Toxicology in Antiquity
DOI: https://doi.org/10.1016/B978-0-12-815339-0.00014-7 **221**

Lead, one of the seven metals of antiquity, has been mined since the fourth millennium BCE. In the Graeco–Roman era, the use of lead increased remarkably and the metal probably became a health hazard to the population between 500 BCE and 300 CE (Waldron and Wells, 1979). Pliny the Elder recognized the toxicity of lead, but the clinical picture of chronic lead poisoning was not clearly described. Despite this, there have been hypotheses that lead toxicity contributed significantly to the decline and fall of the Roman Empire (Retief and Cilliers, 2000).

In this study, the nature and extent of lead poisoning in ancient Rome is reviewed with emphasis on its possible influence on the long-term development of the region.

14.1 THE LEAD INDUSTRY IN ANCIENT ROME

14.1.1 Lead Production

Lead was originally mined from ores like *cerusite* (lead carbonate) and *galenite* (lead sulfide), which also contained silver, copper, gold, and even arsenic. Lead was produced in foundries near mines, and it is very probable that workers were exposed to extensive contamination from lead powder and fumes (Nriagu, 1983).

During the Roman era, lead production increased significantly. At its peak, the Roman Empire probably produced 80,000 t of lead annually, comparable to two-thirds of the lead production in the United States in the 1970s. This amounts to approximately 4 kg of lead per capita annually (Woolley, 1984; Needleman and Needleman, 1985).

Besides *cerusite* and *galena*, the Romans were aware of other lead products: *litharge* (lead oxide, yellow lead) was used in the building trade; *cerusa* (lead carbonate, white lead) was an industrial product, occasionally used as a medicine; sugar of lead (lead acetate) was commonly used as a sweetener, and in the preservation of wine and food (Waldron, 1973).

14.1.2 Uses of Lead

The word "plumbing" is derived from the Latin *plumbum* (lead). In water conduits, lead was extensively used to line or seal earthenware pipes, aqueducts, tanks, and reservoirs. Domestic containers and cooking utensils commonly contained lead. Bronze and copper pots were frequently lined with lead (or a lead–silver alloy) to counter the unpleasant copper taste in the food. Lead provided a sweetish flavor. Pewter (lead and tin alloy) utensils were in common use. Containers for wine and olive oil were frequently made of lead or had a lead lining

FIGURE 14.1   A corner shop where bypassers could get a mug of wine from the lead containers in the counter.

(Fig. 14.1). Wine was commonly fermented in vats with a lead lining (Needleman and Needleman, 1985; Waldron, 1973).

Sugar was unknown and honey was expensive. The Romans, thus, made common use of grape must concentrate as a sweetener. This was prepared by way of a cooking process, preferably in lead-lined copper pots, resulting in *sapa* or *defrutum*. This lead-contaminated product was used as a sweetener in food and wine, as a preservative in canned fruit, and as an inhibitor of fermentation in the wine industry. Lead acetate (sugar lead), prepared by adding acetic acid to lead, was also used as a sweetener.

Lead preparations (*cerusa*, lead carbonate, in particular) were used as facial powders, ointments, eye medicine, and white paint. It is also suggested that lead preparations were used for contraception, and the Roman botanist Dioscorides prescribed *litharge* for skin ailments; the medical use of lead has, however, remained limited through the ages. The Roman population also came in contact with lead in various other ways, for example, handling of lead ornaments; lead in roofing materials and gutters; coins; projectiles used in times of war; and in the shipping and building industry in general [Gilfillan, 1965; Beck, 2005 (Pedanius Dioscorides of Anazarbus, Bk. V. 87)].

14.2 THE EFFECT OF LEAD ON HUMANS

14.2.1 Metabolic Effect

Lead has extensive toxic effects on human tissue. It blocks enzymatic action (sulfhydryl radicals in particular) and has a direct toxic effect on organs such as the liver and kidney, as well as the nervous system, bone, hair, and hemopoietic organs. Lead is absorbed slowly through the skin but rapidly through the lungs and intestinal tract. In the body, redistribution of lead occurs and, eventually, 90% of that not excreted by the kidney accumulates in bones. Skeletal lead gives an indication of total lead absorption, and is not systemically toxic (Drasch, 1982; Retief and Cilliers, 2000).

14.3 CLINICAL PICTURE OF LEAD TOXICITY (RETIEF AND CILLIERS, 2000)

Acute poisoning presents as abdominal pain, nausea and vomiting, painful mouth and throat with a metallic taste, thirst, and eventually diarrhea. Large doses may cause shock, peripheral paresthesia, pain, and muscular weakness. In severe cases, hemolytic anemia and renal failure may lead to death after a day or two.

Chronic poisoning presents with various symptom complexes: an abdominal syndrome with anorexia, a metal taste, gray "lead line" on the gums, severe constipation, and acute colic; a neuromuscular syndrome characterized by peripheral paralysis (for example, "wrist drop"), asthenia, and encephalopathy (especially in children); a hematological syndrome with anemia due to bone marrow depression, or hemolytic anemia due to production of weak red blood cells. Gout may occur ("saturnine gout") due to renal damage. A characteristic pallor, unassociated with anemia, is also recognized.

14.4 ARCHEOLOGICAL DETERMINATION OF LEAD TOXICITY

In children, lead toxicity manifests as diagnostic "lead bone lines" in X-rays, but this investigation is rarely used by archeologists (Jarcho, 1964).

The lead content of bone can be determined by atomic absorption spectrophotometric analysis, and gives an indication of the individual's total lead exposure during life. Possible contamination from soil surrounding the skeleton must be excluded. It is important to remember that bone lead will not increase during acute poisoning, and lead distribution is not uniform through the skeleton—it is higher in the skull, ribs, and vertebrae than in the pelvis and arm bones. For a comparable lead intake, bone lead is higher in children than adults, and in older specimens of bones compared to younger ones. Lead content of hair also reflects systemic lead load but is of limited value because of the brief lifespan of hair (Grandjean, 1978).

14.5 OCCURRENCE OF LEAD TOXICITY

14.5.1 Sources of Toxicity

We have convincing evidence that lead production and utilization increased significantly during Graeco–Roman times. According to some scholars, lead production in Europe and the Mediterranean area rose 10 times during the second millennium BCE, to reach a peak by 500 BCE. Subsequently, it decreased progressively up to 1000 CE, to reach levels similar to those of the third millennium BCE (Nriagu, 1983).

Sources of lead contamination in Roman times may be summarized as follows:

- Domestic use of water from lead and lead-lined water conduits. It has, however, been postulated that this was a limited source of contamination, because flowing domestic water absorbs little lead from pipes; calcium precipitates from hard water would also have decreased contamination. The Roman architect, Vitruvius, however, claimed that water from earthenware pipes was healthier than that from leaden pipes. He also noted lead contamination of water originating from areas with mines and foundries.
- Lead from medicines and cosmetics would have caused toxicity only if consumed orally by intent or accidentally. Lead absorption through the skin is minimal [Nriagu, 1983; Retief and Cilliers, 2000; Vitruvius, 1931 (Bk. 3 cc. 6, 10, 11)].
- Contamination of food, wine, and olive oil due to preparation in pewter or lead containers, and addition of sugar lead (*sapa*) were important causes of lead poisoning. It has been calculated that 50%–60% of a free adult Roman's lead intake probably came from wine. The nobility and the rich, who drank up to 2 L wine per day, would thus have been predisposed to lead poisoning. Pewter ware was extensively used by the middle class. It is calculated that the Roman

aristocrat probably took in 250 mg of lead per day, the plebeian 35 mg, and the slave 15 mg. This compares with 30–50 mg of lead per day for the average contemporary adult in the United States. WHO considers 45 mg per day as the maximal lead intake for a healthy individual (Nriagu, 1983; Needleman and Needleman, 1985).

Workers in lead mines, foundries, and lead production works were prominently exposed to lead contamination. Approximately 80,000 laborers were annually exposed to contamination at lead works, and a further 60,000 at production units. Hygiene was primitive, although protective clothing for the head and face was in use, and probably protected against toxic inhalations. Pliny the Elder described the toxic effects of foundry gases on humans and animals [Pliny, 1952 (Bk. XXXIII.31); Nriagu, 1983].

14.5.2 Proof of Lead Poisoning

14.5.2.1 Clinical Picture

Ancient writers knew that lead products were poisonous. Dioscorides writes that cerusa taken orally was potentially fatal, and that certain sweet wines affected the abdomen and nerves negatively. Pliny the Elder commented on the poisonous nature of lead fumes, the toxicity of certain lead preparations, and the negative effect of sapa-containing wines on certain persons; the encyclopedist Celsus refers to the toxicity of white lead; Vitruvius wrote that lead pipes caused illness (but went on to design water conduits with prominent lead components) [Vitruvius, 1931 ([Granger F. Transl.] Bk. VIII.3.6); Pliny, 1952 (XXXIV.50); Beck, 2005 (Pedanius Dioscorides of Anazarbus, V.9.103)].

Nicander, a Greek poet of the 2nd century BCE, is generally credited with the first description of acute lead poisoning. This episode refers to an illness caused by cerusa (lead carbonate): acute abdominal upset, vomiting, painful muscles with progressive paralysis, involuntary eye movements, disturbed balance, hallucinations, and death.

However, chronic lead poisoning as we understand it was not clearly described in this era. Gout, which may be associated with chronic lead poisoning, was, of course, well described. The first full classical description of chronic lead poisoning was by Paul of Aegina in the 7th century CE (Major, 1959).

14.5.2.2 Archeological Findings (Grandjean, 1978; Retief and Cilliers, 2000)

Analysis of bone lead values relevant to the present study were reported by P. Grandjean. Additional analyses include those of Sudanese skeletons from as early as 3300 BCE; prehistoric Danish skeletons; British skeletons from the Roman era; skeletons from the North American colonial era; Peruvian skeletons from 500 to 1000 CE; and European skeletons from the 18th century BCE to the 20th century CE.

Findings may be summarized as follows:

- With Sudanese and Peruvian data taken as "zero values" (from before the era of lead contamination), modern bone lead values represent a 20- to 100-fold increase.
- In spite of a wide intragroup variation, affluent communities showed higher lead concentrations than the poor, and urban communities higher concentrations than rural communities.
- During the late Roman period, lead concentrations in Roman occupied areas were 41%–47% of contemporary levels in Europe. After 500 CE these values fell to 13% of modern levels, but during the Middle Ages they rose again to values comparable to the levels of ancient Rome.
- Lead levels in the capital, Rome, were not significantly higher than in legionary cities like Augsburg.

14.6 DISCUSSION

It is clear that the production and utilization of lead increased significantly from the second millennium BCE to approximately 500 CE. This resulted in significant environmental lead contamination, and the question arises as to what extent this negatively affected the population of ancient Rome.

During the past century it has been suggested that lead contamination possibly hastened the decline and fall of the Roman Empire. It was speculated that lead poisoning specifically affected the aristocracy, diminishing their fertility and reproduction, and so decreasing Roman leadership. J.D. Nriagu and others argued that use of lead-contaminated water supplies, cooking procedures, and wine production specifically exposed the leaders to lead contamination.

Scarborough and Needleman and Needleman, however, warn that the extent of lead contamination should not be overestimated. While *sapa* prepared in lead containers undoubtedly contained much lead, lead containers were not necessarily used by the majority of the population. The amount of *sapa* added to wine was also not standardized. Pliny the Elder claimed that there were 185 types of wine in his day, each with its own composition. He preferred unsweetened wine. If sweetened wines were the most serious cause of lead poisoning, the extent of poisoning would thus depend on the distribution and use of types of wine (Scarborough, 1984).

Gilfillan believes that lead poisoning decreased fertility and caused miscarriages and abortion, thus precipitating the decrease in aristocracy evident from the 1st century BCE(Gilfillan, 1965). Aristocratic names like Julius and Cornelius, for instance, disappeared systematically. However, lead's actual influence on pregnancy is quite uncertain—only exposure to high doses of lead has been shown to affect pregnancy. Needleman and Needleman argue strongly that the actual cause of the gradual diminution of aristocratic family names and the decrease in family sizes is complex in nature, involving many factors besides possible lead contamination (Needleman and Needleman, 1985).

If lead contamination did in fact comprise a serious community problem in ancient Rome, the question arises as to why the typical clinical picture of chronic lead poisoning was not described, while the toxicity of lead was indeed appreciated. Nicander's poem describing acute lead poisoning appeared in the 2nd century BCE but Paul of Aegina's description of recognizable chronic poisoning did not appear until the 7th century CE. Vague descriptions of abdominal pain and muscular weakness by Pliny the Elder and Vitruvius do not constitute convincing evidence of the community's appreciation of lead poisoning. It is possible that gout, common in Roman times, could, however, have been a manifestation of subclinical lead poisoning.

Archeological evidence based on skeletal bone lead confirms that the mean lead content of the Roman population was less than half that of the modern European (Drasch, 1982). This does of course not mean that episodes of lead poisoning did not occur or that lead poisoning could not have occurred endemically in parts of the Empire. The affluent Roman took in more lead than the less privileged. Studies from Augsburg and Britain showed skeletal lead contents comparable with that of Rome. While lead contamination in antiquity was predominantly the result of oral intake, it was probably supplemented by pulmonary contamination from lead fumes in foundries. The high skeletal lead content of the modern European is the result of inhalation of lead-containing petrol fumes.

In summary, we accept that increased lead production occurred during the second millennium BCE, reaching a peak during the Roman Empire. This caused significant contamination of the population, and the affluent citizenry in particular. Although clinical lead poisoning probably occurred sporadically, archeological evidence indicates that the mean skeletal lead content of the population was less than half that of the modern European in the same countries. The typical clinical picture of chronic lead poisoning was not described before the 7th century CE. It is thus unlikely that lead poisoning played a significant role in the decline and fall of the Roman Empire toward the 5th century CE.

References

Beck L.Y., (Trans.) Pedanius Dioscorides of Anazarbus. De materia medica. Olms-Weidmann, Hildesheim, Zurich, New York, 2005. Bk. V. 87.

Drasch, G.A., 1982. Lead burden in pre-historical, historical and modern bodies. Sci. Total Environ. 24, 199–231.

Gilfillan, S.C., 1965. Lead poisoning and the fall of Rome. J. Occup. Med. 7, 53–60.

Grandjean, P., 1978. Widening perspectives of lead toxicity. Environ. Res. 17, 303–321.

F. Granger (Trans.) Vitruvius. On architecture. Loeb Classical Library, Harvard University Press, Cambridge, MA, 1931. Bk. 3 cc. 6, 10, 11.

Jarcho, S., 1964. Lead in the bones of pre-historic lead-glaze potters. Am. Antiq. 30 (1), 94–96.

Major, R.H. (Ed.), 1959. Classic Descriptions of Disease. CC Thomas, Springfield, IL.

Needleman, L., Needleman, D., 1985. Lead poisoning and the decline of the Roman aristocracy. Classical Views 29, 64–94.

Nriagu, J.O., 1983. Occupational exposure to lead in ancient times. Sci. Total Environ. 32, 105–116.

Pliny, 1952. Natural History (H. Rackham, Trans.). Loeb Classical Library, Harvard University Press, Cambridge, MA. Bk. XXXIII.31.

Retief, F.P., Cilliers, L., 2000. Loodvergiftiging in antieke Rome (Lead poisoning in ancient Rome). Acta Academica 32 (2), 167–184.

Scarborough, J., 1984. The myth of lead poisoning among Romans: an essay reviewed. J. Hist. Med. 39, 469–475.

Waldron, H.A., 1973. Lead poisoning in the ancient world. Med. Hist. 17, 391–399.

Waldron, T., Wells, C., 1979. Exposure to lead in the ancient populations. Trans. Stud. College Physicians Phila. 1, 102–115.

Woolley, D.E., 1984. A perspective of lead poisoning in antiquity and the present. Neurotoxicology 5 (3), 353–361.

Further Reading

Drasch, G.A., 1982. Lead burden in prehistorical, historical and modern human bodies. Sci. Total Environ. 24, 199–231.

Emsley, J., 1994. Ancient world was poisoned by lead. New Sci. 142, 14.

Gilfillan, S.C., 1965. Lead poisoning and the fall of Rome. J. Occup. Med. 7, 53–60.

Hodge, A.T., 1981. Vitruvius, lead pipes and lead poisoning. Am. J. Archaeol. 85, 486–491.

Needleman, L., Needleman, D., 1985. Lead poisoning and the decline of the Roman aristocracy. Classical Views 29, 64–94.

Nriagu, J.O., 1983. Saturnine gout among Roman aristocrats. N. Engl. J. Med. 308 (11), 660–663.

Retief, F.P., Cilliers, L., 2000. Loodvergiftiging in antieke Rome (Lead poisoning in ancient Rome). Acta Academica 32 (2), 167–184.

Scarborough, J., 1984. The myth of lead poisoning among Romans: an essay reviewed. J. Hist. Med. 39, 469–475.

Waldron, H.A., 1973. Lead poisoning in the ancient world. Med. Hist. 17, 391–399.

Woolley, D.E., 1984. A perspective of lead poisoning in antiquity and the present. Neurotoxicology 5 (3), 353–361.

Poisons, Poisoners, and Poisoning in Ancient Rome

Louise Cilliers and Francois Retief

Honorary Research Associate, Department of Greek, Latin and Classical Studies, University of the Free State, Bloemfontein, South Africa

OUTLINE

Toxicology in Antiquity
DOI: https://doi.org/10.1016/B978-0-12-815339-0.00015-9

231

15.1 SOURCES

Our knowledge of poisonous substances known to the ancient Romans is derived from the records of various contemporary writers. The Greek scholar, Theophrastus, associate and successor of Aristotle as head of the Lyceum (4th century BCE) led the way in identifying plants with medicinal (and poisonous) properties. In the 1st century CE, Dioscorides of Anazarbus wrote his famous herbal, *De materia medica*, which superseded all existing literature in classifying remedies and drugs from the vegetable, animal, and mineral kingdoms. This work, which dealt with close to 1000 drugs, became the standard text for centuries to come. Information on poisons can also be gleaned from the writings of the poet Nicander (2nd century BCE), the army physician Scribonius Largus (1−50 CE), the encyclopedist Pliny the Elder (23−79 CE), another encyclopedist, Cornelius Celsus (1st century CE), and the famous physician and philosopher Galen (2nd century CE) (Hornblower and Spawforth, 1996).

The Latin word *venenum* is ambiguous and can mean remedy or poison. In fact, in a Roman court jurists demanded that the user of the word *venenum* must add whether it was beneficial or harmful (Kaufman, 1932). The ambiguity is due to the fact that the difference between remedy and poison did not necessarily lie in the substance itself but in the dosage. This is clearly illustrated by Dioscorides when he describes the properties and uses of the opium poppy (*Papaver somniferum*), widely employed as a soporific and analgesic; he cautions that if administered in a more concentrated form and greater dosage, it plunges the patient into lethargy and stupor and can even kill. Thus, although in the 16th century Paracelsus more formally articulated the principle that the dose makes the poison, it was understood in practice well before his time.

15.2 POISONS

Poisons were of vegetable, animal, or mineral origin (Horstmanshoff, 1999; Cilliers and Retief, 2000). Vegetable poisons were best known and most frequently used. They included plants with belladonna alkaloids such as henbane, thorn apple, deadly nightshade, mandrake, aconite, hemlock, hellebore, yew extract, and opium. It has been suggested that cyanide was in Roman times extracted from the kernels of certain fruits like almonds, but strychnine was unknown to the ancients. The use of figs and mushrooms to dispose of a person was almost certainly based on applying poisons to these foods. If inherently toxic mushrooms had been used, the rapid death of the victim would implicate the *Amanita*

FIGURE 15.1 Colored miniature of the human-like roots of mandragora in the 14th-century manuscript *Medical Housebook of the Centuri family*. Austrian National Library, Vienna.

muscaria or *Amanita pantherina* kind. Mandragora (mandrake is the common name for this plant genus), with its human-like root, was steeped in superstition: there was a widely shared belief that gathering the root was dangerous, as the plant, when uprooted, uttered a shriek, which caused the death or insanity of those who heard it. However, it was one of the first drugs used effectively as anesthetic and analgesic, rather than as a fatal poison (Cilliers, 2012, pp. 31–44) (Fig. 15.1).

Hemlock was well known in antiquity, and used as early as the 5th century BCE by the law courts of Athens as a legal mode of execution. Socrates died this way in 399 BCE. It causes a gradually ascending paralysis and in the end asphyxia when the respiratory organs become paralyzed, and is said to result in an easy and painless death (Cilliers, 2012, pp. 40–42) (Fig. 15.2).

Aconite was referred to as "the stepmother's poison" or the "mother-in-law's" poison. It was probably used extensively. As little as 3–6 mg is lethal for an adult. It causes rapid onset of symptoms of death due to cardiovascular collapse and respiratory paralysis.

Mineral poisons, for example, salts of lead, mercury, copper, arsenic, and antimony, were known but virtually never used. Fumes from the

FIGURE 15.2   Hemlock.

lead smelting process and from silver and gold mines were recognized as toxic. Poisonous animals were studied by Nicander in his two books on antidotes. The poisons included the venoms of snakes, scorpions, spiders, and insects such as the Spanish fly, as well as such unlikely "poisons" as bull's blood, toads, and salamanders.

The ancients differentiated between three kinds of poisons, namely acute poisons killing rapidly, chronic poisons causing physical deterioration, and chronic poisons causing mental deterioration. Professional poisoners like Locusta, Martina, and Canidia, the infamous trio of women poisoners in Roman times, were often requested by their clients to prepare poisons to suit their specific needs (Cilliers and Retief, 2000, p. 90). Poisons were usually administered with food or drink—and for this reason official tasters, *praegustatores* (slaves or freedmen), were employed by the nobility and the wealthy. They became so common that they formed a collegium with a *procurator praegustatorum*. Poisons were also administered by way of enemas, poisoned needles, or a poisoned feather to induce vomiting, as in the case of the emperor Claudius.

15.3 POISONS USED

Pinpointing drugs used in murders (or suicides) recorded in history is difficult, especially when done retrospectively. Descriptions of the mode of death and *post mortem* changes considered typical of poisoning, for example, darkening of the skin and early bloating due to delayed putrefaction, are hardly reliable when related by historians whose sources were at best secondhand. Furthermore, it is likely that combinations of substances were used—it was believed that more substances would be more effective.

In the vast majority of poisonings recorded in ancient Rome, we do not know the nature of the poison or poisons used. However, in the following instances, specific poisons were mentioned (for more details see Cilliers and Retief, 2000, pp. 96−7):

1. Catuvolcus, the British king mentioned in Caesar's writings, committed suicide by drinking the *sap of the Yew tree*.
2. The emperor Claudius was killed by poisonous mushrooms or, more probably, *poisoned mushrooms*.
3. The philosopher Seneca, forced by Nero to commit suicide, took *hemlock* (after unsuccessfully cutting his wrists).
4. In his novel, *Metamorphoses* (also known as *The Golden Ass*), Apuleius tells of a young man rendered unconscious by *mandragora*.
5. Rumors claim that the emperor Titus was poisoned with *sea-hare* (a marine gastropod) by his younger brother Domitian.
6. Commodus, son of the emperor Marcus Aurelius, killed the Praetorian Prefect, Motilenus, with poisonous figs or *poisoned figs*.
7. In his satires, Juvenal tells of the woman, Pontia, who boasted among her friends of her ability to kill with *aconite*. Plutarch, on the other hand, tells of one Orodes who was cured of dropsy by taking *aconite* (Fig. 15.3).

FIGURE 15.3   Aconite/monkshood.

15.4 INCIDENTS OF POISONING DURING THE ROMAN REPUBLIC

The following survey of recorded crimes of poisoning reveals the pervasive influence of contemporary sociopolitical circumstances, as well as superstition.

- The first recorded incident of poisoning in Rome can be traced back to 331 BCE. The city was stricken by a serious pestilence, and rumor had it that the high mortality rate was partly due to poisoning. Though not convinced of the truth of this story, the historian Livy states that a servant girl informed the *curule aedile* that a number of married women conspired to kill a great number of men by means of poison. Twenty prominent women were arrested, but during the trial they stated that they were preparing medications, not poison. When the women were asked to drink these medications themselves, they asked for time to deliberate the request, but soon returned and drank the potion; all died. Eventually, 170 other women were found guilty and executed (Livy. History of Rome, Foster, 1926, Bk. 8 c. 18).
- During the Second Punic War, the city of Capua revolted against Rome, hoping that Hannibal would come to their aid. It did not happen, and in 211, Capua was taken after a siege. Knowing that they could not expect any mercy, the leader, Virrius, decided to commit suicide, and tried to persuade the senate to do the same. Twenty-seven of the 80 senators assented and followed Virrius to his home. A meal was prepared during which much wine was imbibed, followed by the poison. They then movingly bade each other farewell and departed to their respective homes. When the Romans entered the city the next morning, the 27 senators had all died. The remaining 53 senators were executed by the Romans (Livy. History of Rome, Foster, 1926, Bk. 26 cc. 13–14).
- The historian Livy relates the following moving incident: near the end of the Second Punic War the Roman general, Scipio, defeated the Numidian king, Syphax. Syphax's wife, Sophonisba, had in the meantime become involved with Masinissa, one of Scipio's generals. In a desperate attempt to prevent his beloved from being made a Roman captive, Masinissa married Sophonisba. However, just after the ceremony, he was informed that the marriage would not save her. He thus went to his tent and prepared a poisonous potion which he sent to Sophonisba with the message that his commander had robbed him of the opportunity of fulfilling the first promise that he as bridegroom owed his bride, but that he would stand by his second promise, namely to provide her with the opportunity of evading Roman captivity. She drank the potion and, in a last message to

Masinissa, commented on the irony that the same day would commemorate her unconsummated marriage and burial (Livy. History of Rome, Foster, 1926, Bk. 30 cc. 12–15).

- In 154, two ex-consuls were poisoned by their wives. The senators were becoming uneasy, and assumed jurisdiction over all cases requiring public investigation, such as treason, conspiracy, assassination, and poisoning. At the end of the 2nd century, a court dealing with cases involving poisoning was instituted, and in 80 BCE the dictator Sulla promulgated various laws against poisoning. A person would henceforth be guilty if he prepared, sold, or had in his possession dangerous substances (Scullard, 1982).
- Cicero's speeches provide further evidence of the high incidence of poisoning in his time. In one of the court cases he refers to Oppianicus, who poisoned his wife, brother, and pregnant sister-in-law in order to inherit money; other cases involve Domitius, who murdered his cousin, Fabricius because the latter wanted to poison a friend, and Catiline, who was guilty of poisoning various people (Cicero. Orations. Pro Cluentio et al., vols. IX, X, and XV).
- Pliny tells us that before the battle of Actium (31 BCE) Marcus Antonius was so suspicious that he refused any food from Cleopatra unless it was tasted by his personal bodyguard. The queen, however, managed to smuggle a poisoned jug past the bodyguard, but at the last moment, because of her love for Antony, prevented him from drinking it. The cup was then given to a prisoner, who immediately died after drinking it. Plutarch is responsible for the story that Cleopatra committed suicide by the bite of an adder, but it is more likely that she took poison—a snake large enough to kill the queen and two ladies in waiting would have been too large to smuggle past the guards posted there by Octavian to prevent her from committing suicide (Plutarch, 1920; Retief and Cilliers, 1999).

This brings us to the end of the Republican era. Four years after the battle of Actium, Octavian became Augustus, the first emperor of the Roman Empire.

15.5 POISONERS AND INCIDENTS OF POISONING DURING THE EMPIRE

15.5.1 Augustus (27 BCE–14 CE)

The Julio-Claudian dynasty which started with Augustus, was infamous for numerous cases of poisoning occurring in the imperial family. Livia, the formidable wife of the emperor, was said by the 2nd century CE historian, Dio Cassius (1924), to have poisoned Augustus's grandchildren, Gaius and Lucius Caesar, in order that her own son, Tiberius, would succeed the emperor. There was also a rumor that she had poisoned her aged husband with poisoned figs, but there is no evidence for this (Tacitus, 1931, Bk. 1, cc. 5.1−3). A scandal was exposed when a friend of Augustus, Nonus Asprenas, was arraigned in the senate for the poisoning of 130 guests at a dinner party.

15.5.2 Tiberius (14−37 CE)

In 19 BCE, Tiberius's nephew, the popular general Germanicus, died in Asia Minor amid suspicious circumstances after a quarrel with the governor, Piso, and his wife, Plancina. Germanicus's wife, Agrippina, and his friends stated that he was poisoned by Piso, helped by the infamous poisoner, Martina. The case was referred to the senate in Rome, but on her way to Rome, Martina suddenly died; her body showed no signs of suicide, but during the autopsy, poison was discovered, concealed in a knot in her hair. Poisoning could not be proved during the trial, but Piso was heavily censured and then committed suicide (Tacitus, 1931, Bk. III. cc. 7 and 12−15).

In 23 CE, Tiberius's son, Drusus, died after a long illness. It was later revealed that he was poisoned by the Praetorian Prefect, Sejanus, who was the lover of the victim's wife, Livilla. Sejanus's wife affirmed under torture that Livilla had prepared a poison which had the effect of physical deterioration over a long period (Tacitus, 1931, Bk. IV. c. 8). It is ironic that Germanicus, who probably died a natural death, was believed to have been poisoned, while Drusus, who was poisoned, was believed to have died a natural death.

Sejanus then overreached himself by trying to take over the government while Tiberius was in Capri, but the plot was exposed, and Sejanus and his supporters were all arraigned and executed. Tiberius now became increasingly paranoid and Tacitus relates that many people were accused of treason during mass trials, and committed suicide by poisoning to prevent their families from losing their possessions to the accusers. When the emperor died in Capri, there were rumors that he was poisoned by Gaius (Germanicus's son, also called Caligula), but there is little evidence for this (Suetonius Tiberius, 1998).

15.5.3 Gaius (Caligula) (37−41 CE)

Dio Cassius tells that this mentally disturbed emperor collected poisons and killed gladiators, jockeys, and horses in his attempt to manipulate these sports in order that his favorites could win (Dio Cassius, 1924, Bk LIX c. 14). The satirist, Juvenal, claims that his insanity was the result of a love potion administered by his wife, Caesonia (Satires, 2004).

15.5.4 Claudius (41–54 CE)

Claudius probably suffered from congenital cerebral palsy but proved an unexpectedly efficient emperor. Dio Cassius relates that he dumped Gaius's poisons into the sea, which caused the death of scores of fish (Dio Cassius, 1924, Bk LIX c.14). It is generally believed that Claudius's fourth wife, Agrippina, poisoned him with poisonous or poisoned mushrooms obtained from the poisoner Locusta, to make room for her son, Nero. The 65-year-old ailing emperor did not die immediately, and the physician Xenophon had to use a poisoned feather, ostensibly to make him vomit to get rid of the poison, to finally dispatch him (Tacitus, 1931, Bk. XII. cc. 66–67).

15.5.5 Nero (54–68 CE)

After the accession of Nero, poisoning among the aristocracy showed a marked increase. The first years of this emotionally unstable young emperor's reign, when the philosopher Seneca was his counselor, were irreproachable, but Seneca's influence started waning as Nero grew up. Seneca was eventually unfairly accused of conspiracy and forced to commit suicide by poison; after opening his arteries he took hemlock which also did not have the desired immediate effect and he was thus smothered in a steam bath (Tacitus, 1931, Bk. XV cc. 60–62). The later years of Nero's principate, when he was under the influence of the evil Tigellinus, Prefect of the Praetorian Guard, became an orgy of murders. The historians Suetonius (Suetonius (Nero) Tiberius, 1998, c. 35.5), Tacitus (1931, Bk. XIII, c.1), and Dio Cassius (1924, Bk. LXI c.7.4) tell of the deaths (by poison) of Nero's stepbrother, Britannicus; Silanus, governor of Asia; Domitia, his aunt whose riches he desired; Burrus, Prefect of the Praetorian Guard; and Pallas and other freedmen. Nero also got rid of his meddlesome mother, Agrippina, but could not use poison since she had for some time been using antidotes to make her immune; he thus had her brutally beaten to death after she had swum ashore when the planned shipwreck which was to have disposed of her went awry (Tacitus, 1931, Bk. XIV, cc. 7–8).

Nero's poisons were usually prepared by Locusta. Her death penalty was suspended after the murder of Britannicus and various others, and she was appointed the emperor's adviser on poisons, and allowed to establish a school of poisoning where she could train others in her art. Locusta tested her poisons on animals and convicted criminals (Suetonius, 1998).

15.5.6 The Flavian Dynasty (69–96 CE)

According to historians the number of poisonings decreased after Nero's death, but there were still some incidents during the next two centuries which will be dealt with. Nero's death was followed by the notorious Year of the Four Emperors (69 CE) when armies from all over the empire tried to raise their general to the throne. Then followed the Flavian dynasty, of which the cruel, unstable younger son *Domitian*, became emperor in 81 CE. He was accused of poisoning his elder brother, the emperor *Titus* (who died unexpectedly at the age of 42), with sea-hare (a marine gastropod). It is, however, more likely that Titus died of malaria while besieging Jerusalem (Thompson, 1931). Another suspicious death was that of Agricola, the popular Roman governor in Britain, who died in 93 CE amid suspicious circumstances. It was well known that Domitian was jealous of him, and when physicians from Rome visited Agricola during his last illness, the rumor originated that the emperor had arranged that he be poisoned. However, Tacitus could find no conclusive evidence (Tacitus Agricola, 1970).

15.5.7 Hadrian (117–138 CE)

This capable emperor, the second of the so-called Five Good Emperors (98–180 CE), had an unhappy married life, probably because of his homosexual inclination. When his wife, Sabina, died in 136 CE, it was widely believed that he had poisoned her. At a later stage when the emperor himself suffered from serious edema, he tried to commit suicide but the weapon was taken from him; he then asked his physician to provide poison, but rather than accede to the request, the physician killed himself (Spartianus, 1976).

15.5.8 Commodus (180–192 CE)

Unlike his father, Marcus Aurelius, Commodus was a failure as emperor. It is said that he poisoned the Prefect of the Praetorian Guard, Motilenus, with poisoned figs. Eventually, during a conspiracy, Commodus's favorite concubine, Marcia, tried to poison him, but the poison only made him drowsy, and thereafter he began vomiting. When it became evident that Commodus was not going to die, he was strangled by the athlete Narcissus (Lampridius, 1995).

15.5.9 Caracalla (211–217 CE)

The reign of Caracalla began with the murder of his brother, Geta, who was designated co-emperor in the will of their father, Septimius Severus. Geta was popular and had many supporters, who were mercilessly dispatched after his death, mostly by the sword, but also by poisoning. It was said that 20,000 people died during Caracalla's reign (Lampridius, 1995, c.3).

15.6 CONCLUSION

The survey of recorded crimes by poison underlines the pervasive influence of contemporary sociopolitical circumstances (Cilliers and Retief, 2000, pp. 98—9). Incidents of poisoning in early Rome usually coincided with crises, such as wars and epidemics; superstition and the belief that scapegoats would solve the problem would have been determining factors. This phase was gradually superseded by a very well-documented era in which countless individual cases of poisoning are reported, and which reached a peak during the reign of the Julio-Claudian emperors in the 1st century CE. Apart from the influence of personal factors such as paranoid and mentally unstable emperors, the transition from Republic to Empire was a politically unstable period which created much tension, and thwarted ambition led to numerous political intrigues and murders. During the reign of the so-called Five Good Emperors in the 2nd century CE, commonly regarded as the most prosperous period in Roman history, there was a significant decline in poisoning and suicides by poison. Granted that this period is less well documented, the political stability of this era, which would have brought a greater degree of peace of mind, could have been a determining factor in the decline of violent deaths.

References

Cicero. Orations. In: Pro Cluentio, et al. (Eds.), In Catilinam I—IV, Philippics [H.G. Hodge, W.C.A. Kerr, Trans.]. Loeb Classical Library, vols. IX, X, and XV. Harvard University Press, Cambridge, MA.

Cilliers,, L., 2012. Anaesthesia and Analgesia in Ancient Greece and Rome (c. 400 BCE—300 CE). In: Askitopoulou,, H. (Ed.), Proceedings of the Seventh International Symposium on the History of Anaesthesia. Crete University Press, Heraklion, pp. 31—44. , here p. 36.

Cilliers, L., Retief, F.P.R., 2000. Poisons, poisoning and the drug trade in ancient Rome. Akroterion 45, 91—95.

Dio Cassius, 1924. Roman History [E. Cary, Trans.]. Loeb Classical Library, vol. VII. Harvard University Press, Cambridge, MA (Bk. 56 c. 30.2.).

Hornblower, S., Spawforth, A., 1996. The Oxford Classical Dictionary, third ed Oxford University Press, Oxford.

Horstmanshoff, H.F.J., 1999. Ancient medicine between hope and fear: medicament, magic and poison in the Roman Empire. Eur. Rev. 7 (1), 37—51. here p. 43.

Kaufman, D.B., 1932. Poisons and poisoning among the Romans. Classical Philol. 27, 156—167. here p. 156.

Lampridius, 1995 Commodus. In: Chronicle of the Roman Empire [C. Scarre, Trans.]. Thames & Hudson, London (c.9).

Livy, 1926. History of Rome. [B.O. Foster, Trans.] Loeb Classical Library. Harvard University Press, Cambridge, MA (Bk.8 c.18).

Plutarch, 1920 Antony [B. Perrin, Trans.]. Loeb Classical Library, vol. IX. Harvard University Press, Cambridge, MA (c.86).

Retief, F.P., Cilliers, L., 1999. Dood van Kleopatra [Death of Cleopatra]. Geneeskunde 4 (1), 8–11.

Satires, 2004. [Braund S.M., Trans.]. Loeb Classical Library. Harvard University Press, Cambridge, MA, pp. 610–626, Sat. VI, lines.

Scullard, H.H., 1982. From the Gracchi to Nero: A History of Rome from 133 BCE to CE 68. Methuen, London, Here p. 86 re the Quaestio de sicariis et veneficiis.

Spartianus, 1976. Augustan History. Lives of the Later Caesars [Birley A., Trans.]. Penguin Classics, cc. 23–26.

Suetonius Tiberius, 1998. [Bradley K.R., Trans.]. Loeb Classical Library, vol. I. Harvard University Press, Cambridge, MA, c.73.

Suetonius, Claudius, 1998. [Bradley KR, Trans.]. Loeb Classical Library, vol. II. Harvard University Press, Cambridge, MA, c.44; Ibid. Nero c.33.3.

Tacitus Agricola, 1970. [Ogilvie R.M., et al., Trans.]. Loeb Classical Library, vol. I. Harvard University Press, Cambridge, MA, c.43.

Tacitus, 1931. Annals, [Moore Ch., Trans.]. Loeb Classical Library, vol. III. Harvard University Press, Cambridge, MA, pp. 1–3, Bk. 1, cc. 5.

Thompson, C.J.S., 1931. Poisons and Poisoning. Macmillan & Co., New York, NY.

16

Chemical and Biological Warfare in Antiquity

Adrienne Mayor

Classics Department and Program in History and Philosophy of Science,
Stanford University, Stanford, California, United States

Biological weapons employ viable, living organisms, such as pathogens, venoms, and toxic plants, insects, or animals, while poisonous gases, dust, or smoke, and incendiary materials to burn, blind, choke, or asphyxiate foes constitute chemical weapons. In antiquity, a wide range of substances, from toxic plants and venomous insects and reptiles to infectious agents and noxious chemicals, were weaponized in Europe, the Mediterranean, North Africa, the Middle East, Central Asia, India, China, and in the Americas. Natural toxins were exploited to wage the earliest forms of biological and chemical warfare. Literary and archeological evidence for the concept and practice of toxic warfare can be traced back thousands of years. For example, in Asia Minor cuneiform tablets of about 1200 BC record that that the Hittites deliberately drove victims of plague into enemy territory. True scientific understanding of toxicology, epidemiology, and chemistry were not required to carry out such practices, nor did the perpetrators need to possess advanced technology. Instead, the use of toxic arms and tactics was based on centuries of observation and experimentation with easily available toxic materials. A strategy based on attacking an opponent's biological vulnerabilities with poison could give a commander an advantage when his men faced troops that were superior in numbers, courage, military prowess, or technology. Nevertheless, the use of toxic weapons also entailed practical and ethical dilemmas even in antiquity.

The concept of poisoned projectiles is embedded in the ancient Greek language. *Toxicon*, the word for "poison," derives from *toxon*, the word for "arrow." It is likely that the first projectiles treated with poisonous substances were devised for hunting animals and then later turned toward warfare. The bow and arrow was a highly effective delivery system for toxins at an early date. Even a scratch from a poisoned arrowhead or spear could be fatal.

16.1 THE CONCEPT OF TOXIC WEAPONRY IN GRECO-ROMAN AND INDIAN MYTHOLOGY

Greek mythology offers further evidence of the antiquity of the concept of toxic weapons. The great mythic hero Hercules, for example, created the first biological weapon in Western literature. After destroying the terrible many-headed serpent, the Hydra, Hercules dipped his arrows in the monster's venom; thereafter his quiver was filled with a never-ending supply of Hydra-envenomed arrows. Homer's *Iliad*, an oral epic first written down in the 8th century BC, contains indirect allusions to the use of toxic projectiles in the legendary Trojan War. Homer described black (rather than red) blood oozing from wounds, doctors sucking out poisons from arrow wounds on the battlefield, and never-healing wounds: these details are hallmarks of poisoning by snake venom. In the *Odyssey* (2.225—30) Homer describes the Greek hero Odysseus seeking lethal plant juices to treat his arrows intended for enemies. According to ancient legend, Odysseus himself was killed by a poisoned weapon—a spear tipped with the toxic spine of the marbled stingray, a common species in the Mediterranean (Mayor, 2009).

In the *Aeneid* (9.770—74), an epic poem by Virgil recounting the legendary history of Rome, poisoned spears were employed by the early

Romans. Poisoned weapons also appear in the great mythological epic of India, the *Rigveda*. These Greek and Indian myths and legends are thought to reflect the early invention of biological weaponry in ancient cultures. The mythic examples also offered models for and may have reflected the actual practice of chemical and biological warfare (Mayor, 2009).

16.2 POISONS FROM PLANTS IN HISTORICAL WARFARE

In antiquity, dozens of toxic Eurasian plant species were known and used as medicines in very tiny dosages. These powerful plants were also gathered to make arrow poisons or other biological weaponry used in historical battles. "Hellebore" was one of the most popular plant drugs, classified by the ancient Greeks as black hellebore (the Christmas rose of the buttercup Ranunculaceae family, *Helleborus niger*) and white hellebore (the lily family, Liliaceae). Each of these two unrelated plants is laden with powerful chemicals that cause severe vomiting and diarrhea, muscle cramps, delirium, convulsions, asphyxia, and heart attack. Hellebore was the poison sought by Odysseus in Homer's *Odyssey*, and it was one of the arrow drugs used by the Gauls, among other groups. Hellebore was also used to poison the wells of besieged cities (see below) (Fig. 16.1).

Aconite or monkshood (wolfsbane) was another favorite plant toxin. Aconitum (buttercup family) contains the alkaloid aconitine, a dangerous poison. In high doses it causes vomiting and paralyzes the nervous system, resulting in death. Aconite was employed by archers of both ancient Greece and India, and its use in warfare continued into modernity. In 1483, for example, in the war between the Spanish and the Moors, the Arab archers wrapped their arrow points with aconite-soaked cotton. During World War II, Nazi scientists experimented with aconitine-treated bullets (Mayor, 2009).

Another arrow poison in antiquity was henbane (*Hyoscyamus niger*), a sticky, nasty-smelling weed containing the powerful narcotics hyoscyamine and scopolamine. Henbane causes violent seizures, psychosis, and death. Other plant juices used on projectiles in antiquity included yew (*Taxus*), hemlock (*Conium maculatum*), rhododendron, azalea, and deadly nightshade or belladonna, which causes vertigo, extreme agitation, coma, and death. According to Pliny the Elder, a natural historian of Rome in the 1st century AD, the archaic Latin word for deadly nightshade was *dorycnion*, "spear drug." Pliny (*Natural History* 21.177–79) noted that this fact indicated that the plant was used to treat weapons at a very early date in Italy.

FIGURE 16.1   Toxic plant hellebore.

16.3 SNAKE VENOM ARROWS

Snake venom was another dreaded arrow poison used by several groups in antiquity. Because snake venom is digestible it could be safely used for hunting game. But if venom could be introduced into the bloodstream of a human enemy, it guaranteed a painful death or a never-healing wound. Numerous poisonous snakes exist around the Mediterranean and in Africa and Asia. According to ancient Greek and Roman sources, archers who steeped their arrows in viper or snake venom included the Gauls, Dalmatians and Dacians of the Balkans, the Sarmatians of Persia (Iran), Getae of Thrace, Slavs,

Armenians, Parthians between the Indus and Euphrates, Indians, North Africans, and the Scythian nomads of the Black Sea region and the Central Asian steppes. According to Strabo, a Greek geographer, the people of Ethiopia dipped their arrows "in the gall of serpents." Strabo also claimed that the arrow poison concocted by the Soanes of the Caucasus was so noxious that the odor alone was injurious. The Roman historian Silius Italicus (Punica 1.320–415) described envenomed arrows used by the archers of Morocco, Libya, Egypt, and Sudan. Ancient Chinese texts tell us that arrow poisons were also in use in China, and in the New World Native Americans used snake, frog, and plant poisons on projectiles for both hunting and warfare (Mayor, 2009; Sawyer, 2007; Jones, 2007).

Simple and complex ways of making envenomed arrows are recorded in Greek and Latin texts. Snake venom crystallizes and so can cling and remain viable on wooden, bone, and metal points for a considerable time. One of the most feared arrow drugs in antiquity was complicated *scythicon* created by the Scythians. They mixed venom with bacteriological pathogens from animal dung, human blood, and putrefying viper carcasses (Aelian *On Animals* 9.15, Ovid *Tristia; Letter from Pontus*). With even a superficial wound from a *scythicon*-treated arrow, the toxins would begin taking effect within an hour. Envenomation would be compounded by shock, and necrosis and suppuration would be followed by gangrene and tetanus. *Scythicon* ensured an agonizing death.

Several snake species were available to the Scythians: the steppe viper *Vipera ursinii renardi*, the Caucasus viper *Vipera kasnakovi*, the European adder *Vipera berus*, and the long-nosed or sand viper *Vipera ammodytes transcaucasiana*. The natural historian Aelian (3rd century AD Rome) described one of the most fearsome poisons of India, derived from the venom and rotting carcasses of the so-called white-headed purple snake. From Aelian's detailed description, herpetologists identify the purple snake as the rare, white-headed viper discovered in South Asia in the late 1880s, *Azemiops feae*.

The army of Alexander the Great in his conquest of India in 327–325 BC encountered a different snake-venom weapon. According to the historians of his campaigns, Quintus Curtius, Diodorus of Sicily, and others, the defenders of Harmatelia (Mansura, Pakistan) had dipped their arrows and swords in an unknown snake poison. Any man who suffered even a slight wound felt immediately numb and experienced stabbing pains and convulsions. The victim's skin became pale and cold and he vomited bile. Soon, a black froth exuded from the wound. Purplish-green gangrene spread rapidly, followed by death. Modern historians have assumed that cobra venom was used at Harmatelia, but the very detailed descriptions of the ghastly symptoms and deaths suffered by Alexander's soldiers points to another snake. Cobra venom would bring a relatively painless death, from respiratory paralysis. But the common Russell's viper of Pakistan and India causes the very same symptoms suffered by Alexander's men: numbness, vomiting, severe pain, black blood, gangrene, and death (Mayor, 2009).

16.4 PLAGUE AND CONTAGION

Modern historians have considered the Mongols' ploy of catapulting of bubonic plague victims over the walls at Kaffa on the Black Sea in 1346 as the first recorded act of biological warfare. But empirical under-standings of contagion developed much earlier in history. The Hittites use of plague was mentioned above. In another early example, in Mesopotamia in about 1770 BC cuneiform tablets warned that disease could be spread by fomites, infectious pathogens on clothing, bedding, other items. Jewish legends about King Solomon suggested that he hid plague in sealed jars in the Temple of Jerusalem to infect Babylonian and Roman invaders. During the Peloponnesian War (4th century BC), the Athenians suspected that the Spartans had spread the great plague (apparently smallpox) by poisoning their wells. The accusation shows that people feared such tactics, whether they really occurred or not.

The Latin phrase *pestilentia manu facta* ("man-made pestilence") was coined by Seneca (*On Anger* 2.9.3). Two incidents of deliberately trans-mitted contagion were reported during the reigns of Domitian (AD 90) and Commodus (AD 189). According to Dio Cassius (*Epitome* 67.11 and 73.14), saboteurs pricked people with tiny needles dipped in deadly sub-stances, causing epidemics and panic. Whether true or not, such rumors play a role in bioterrorism. The Great Plague of AD 165−180 (again, probably smallpox), was spread from Babylon (modern Iraq) to Syria, Italy, and as far as Germany by Roman soldiers returning from the war undertaken to control Mesopotamia. According to historians of that era, the epidemic began when some Roman soldiers looted a treasure chest in an enemy temple in Babylonia. These historical accounts imply that the chest had been deliberately booby trapped with plague-laden items, a plausible notion. The local population would have had some immunity to the epidemic, while the invading Roman army would have been immunologically naive and vulnerable. At the very least, the ancient reports demonstrate that the idea of deliberately spreading epidemics among the enemy was widely contemplated by that time (Mayor, 2009).

In ancient India, Kautilya's *Arthashastra*, a warfare manual dating to the 4th century BC, suggested ways of infecting enemies with illnesses such as fevers, wasting lung disease, and rabies. The *Arthashastra* also gives numerous recipes for poisoning the food and water of the enemy (Kautilya, 1992).

16.5 POISONING WATER SOURCES AND FOOD SUPPLIES

Tainting water and food supplies was another ancient poison tactic. A legendary Greek account set in about 1000 BC tells how King Cnopus conquered Erythrae (western Turkey) by consulting a "witch" from Thessaly, Greece. Her strategy was to poison a sacrificial bull. She then tricked the Erythraeans into eating the poisoned meat. The enemy was easily routed by King Cnopus's army.

The earliest historically documented case of poisoning drinking water in Greece occurred in about 590 BC during the First Sacred War. Athens and allied city-states were besieging Kirrha, the strongly fortified town that controlled the road to Delphi, the site of the sacred Oracle of Apollo. Athens and her allies argued that Kirrha had offended the god and was therefore to be totally destroyed. To break the siege deadlock, the commander of the league of allies ordered his men to gather and crush a great quantity of hellebore plants which grew in profusion in the area. They placed the hellebore in the water pipes supplying Kirrha. The soldiers manning the walls—and the entire population of Kirrha— fell violently ill after drinking the poisoned water. Athens and the league overran the city and slaughtered the combatants and the civilians alike. After the Sacred War, Athens and her allies had second thoughts. They agreed among themselves not to interfere with water supplies should they ever find themselves at war with each other (Mayor, 2009).

At least one Roman general was known to have poisoned wells. Manius Aquillius had ended a long-drawn-out war to quell insurrections in the Roman province of Asia Minor in 129 BC, by pouring an unknown poison into the springs supplying the rebelling cities. According to the Roman historian Florus (1.35.5–7), however, his victory brought dishonor to Rome because of the resort to underhand biological tactics that also affected non-combatants. Two Carthaginian generals, Himilco and Maharbal, reportedly overcame enemies in North Africa by tainting wine with mandrake, a heavily narcotic root of the deadly nightshade (Mayor, 2009).

In the 1st century BC, naturally occurring toxic honey was deployed against the army of the Roman general Pompey during the war against King Mithradates VI of Pontus on the southern coast of the Black Sea (northeastern Turkey). In 65 BC, Mithradates' allies gathered wild honeycombs and placed them along the Romans' route and waited in ambush. In that region, wild bees gather nectar from profuse rhododendron blossoms; the honey does not harm the bees but contain devastating neurotoxins dangerous to animals and humans. The Roman legionnaires eagerly ate sweet honey, and then they began collapsing with vertigo, vomiting, paralysis, and diarrhea. Mithradates' allies lying in ambush came and slaughtered 1000 of Pompey's men (Mayor, 2009).

16.6 VENOMOUS INSECTS, SNAKES, AND SCORPIONS

Stinging insects, such as wasps, deadly vipers, and scorpions could be enlisted as allies in war. Some scholars suggest that in Neolithic times, people threw hives filled with furious bees at enemies, who were thrown into chaos by the painful stings. Much later, in the Roman period, catapults were used to hurl beehives at foes. In Central Mexico, the ancient Mayans devised ingenious booby traps to repel besiegers on their fortress walls, consisting of dummy warriors with heads made of gourds filled with angry hornets (Mayor, 2009; Lockwood, 2009).

In the Aegean off the coast of Turkey in the 2nd century BC, the famous Carthaginian general Hannibal found himself outnumbered during a naval battle against Pergamon, ruled by Eumenes II. Hannibal ordered his men ashore to collect live vipers and pack them into clay pots. Then as the enemy ships approached the Carthaginians vessels, Hannibal's men hurled the pots. As the catapulted pots smashed on the ships' decks, releasing masses of snakes, the Pergamene sailors panicked and were easily defeated (Mayor, 2009).

A similar tactic using insects and arthropods saved the fortified city of Hatra (Iraq), in AD 198–199. Besieging Roman legions led by Emperor Septimius Severus were forced to retreat after the Hatreni defended their walls with terracotta pots filled with live scorpions, assassin bugs, and other poisonous insects from the surrounding desert (Fig. 16.2). According to the Roman historian Herodian, as the insects rained down on the Romans scaling the walls, they "fell into the men's

FIGURE 16.2   Re-creation of a clay pot filled with live scorpions, like that used at Hatra against the Romans.

eyes and exposed parts of their bodies, digging in, biting, and stinging the soldiers, causing severe injuries." The terror effect would be highly effective, no matter how many men were actually stung. Scorpion stings inject a complex combination of toxins, causing intense pain, thirst, great agitation, muscle spasms, convulsions, slow pulse, irregular breathing, and torturous death. Assassin bugs, large bloodsucking insects with sharp beaks, inflict an extremely painful bite and inject a lethal nerve poison that liquefies tissues. It is possible that *Paederus* rove beetles (Staphylinidae) were also collected in the desert by the defenders of Hatra. Pederin, the virulent poison secreted by predatory rove beetles was well known in ancient India and China. Pederin is one of the most powerful animal toxins in the world, causing severe blistering of the skin and eyes. In the bloodstream, its toxicity is more potent than cobra venom (Mayor, 2009).

16.7 AEROSOL AND INCENDIARY WEAPONS

Choking fogs and asphyxiating clouds of smoke, dust, and gases were effective chemical weapons in antiquity. One of the earliest documented examples of toxic aerosols occurred during the Peloponnesian War in 429 BC. That year Sparta was besieging the city of Plataia. The Greek historian Thucydides reported that the Spartans created a massive fire just outside Plataia's walls. The Spartans fueled the fire with great quantities of resinous pinetree sap (crude turpentine pitch) and sulfur. The combination of pitch and sulfur created great clouds of toxic sulfur dioxide gas, fumes that can be fatal when inhaled in large amounts. A few years later, in 424 BC, the Boeotians, allies of the Spartans, invented a "flamethrowing" machine. They used the contraption to propel billows of noxious smoke from charcoal, resin, and sulfur against the walled city of Delium.

Archeologists have discovered physical evidence to show that in AD 256, the Sassanians attacking Roman-held Dura-Europos (on the modern border of Iraq and Syria) created a similar incendiary mixture of sulfur and pitch that resulted in a deadly gas enveloping a siege tunnel, suffocating nineteen Romans and one Sassanian. The skeletons of 20 victims and residue of sulfur crystals and pitch burned in braziers to create poisonous sulfur dioxide gas were found in the tunnel (Mayor, 2009; James, 2011).

Aeneas the Tactician, a Greek writing in 360 BC on how to defend against sieges, suggested the use of incendiaries made with pitch, hemp, and sulfur. Roman historians tell how burning chicken feathers created irritating, choking fumes that could be propelled by bellows into enemy siege tunnels.

In 80 BC in Spain, the Roman general Sertorius deployed choking clouds of dust to defeat the Characitani tribe, who had taken refuge in inaccessible limestone caves. The fine white soil around the cave formations consists of limestone and gypsum. Sertorius ordered his masked soldiers to sweep up great heaps of the powder in front of the caves. Then when the wind was just right, the Romans stirred up the dust with poles to raise great clouds of caustic lime powder, a severe irritant to the eyes and lungs. The Characitani surrendered (Plutarch *Sertorius*). A similar choking method was used in China to quell an armed peasant revolt in AD 178, when "lime chariots" equipped with bellows blew limestone powder into the crowds. The powdered lime interacts with the moist membranes of the eyes, nose, and throat with a corrosive, burning effect, blinding and suffocating the victims (Mayor, 2009).

In the Middle East where crude petroleum is abundant, naphtha (the volatile and toxic light fraction of oil) was used as a weapon by itself and later, in the 7th century AD, perfected in Greek fire, whose exact recipe and delivery system have been lost (Forbes, 1964; Partington, 1999). Naphtha was ignited and poured on attackers with devastating effects (Pliny *Natural History* 2.235). The ancient Indians and Chinese added combustible chemicals to their incendiaries, usually explosive saltpeter or nitrite salts, a key ingredient of gunpowder. The Chinese also used a variety of plant, animal, and mineral poisons, such as arsenic and lead, in making smoke and fire bombs even more harmful. In the New World and in India, the seeds of toxic plants and hot peppers were burned to rout attackers (Mayor, 2009; Sawyer, 2007).

16.8 PRACTICAL ISSUES AND ETHICAL QUALMS

The toxicity of plants, venoms, and other poisons used in armaments often imperiled those who wielded them. Toxic weapons are notoriously difficult to control and often resulted in the destruction of noncombatants as well as soldiers, especially in siege situations. Both the ancient mythology and the ancient history of poison weapons abound with incidents of accidental self-injuries, friendly fire problems, and unintended collateral damage. Windborne toxins also incurred obvious "blowback" problems. Kautilya addressed this issue in his *Arthashastra*, and recommended that protective salves and other remedies be applied before deploying poisonous smokes (Kautilya, 1992).

The use of poisons in warfare led people to seek antidotes. Ancient Greek and Latin sources list hundreds of substances—from rust filings to poultices made from medicinal plants—that were thought to counteract specific weaponized poisons. Some believed that one could become invulnerable to toxins by ingesting minute amounts of various poisons over time. Mithradates VI of Pontus (d. 63 BC) was an early experimenter in creating a "universal antidote" of more than 50 poisons and antidotes in tiny amounts. It later became known as the Mithridatium. The recipe is lost but pills purporting to be the Mithradatium were ingested by every Roman emperor from Nero onward and then by European royalty including Charlemagne and Henry VIII, in the hope of achieving immunity to poison (Mayor, 2010).

Toxic weapons were surrounded by ambivalence in antiquity, even though there were few rules of war governing their use. In many cultures, the idea of weapons that delivered hidden poisons to make an enemy defenseless or experience excessive suffering aroused moral criticism. Nevertheless, their use was rationalized in many recorded instances. The ancient Greeks considered poisoned projectiles a cowardly weapon, for example, but their greatest mythic heroes, Hercules and Odysseus, resorted to such arms, and well poisonings and toxic aerosols were documented in historical Greek conflicts. Many Romans deplored poisoned arrows and poisoning water and food supplies (e.g., Tacitus *Annals* 3.1; 5.2). Yet Roman generals occasionally turned to such strategies. In India, the Hindu *Laws of Manu* (dating to 500 BC) forbade the use of poisoned arrows (although it recommended spoiling the enemy's food and water). In the same era, Kautilya's *Arthashastra* extolled the advantages of poisoning projectiles, food, and water and asphyxiating foes with chemical- and disease-laden clouds of smoke. Notably, Kautilya stressed the deterrent effect of publicizing the horrid ingredients of one's toxic arsenal. That psychological strategy was apparently taken up by the Scythians and others who broadcasted their recipes for poison arrows. In China, Sun Tzu's *Art of War* (500 BC) praised deceptive terror strategies based on fire and other Chinese war treatises give recipes for making toxic aerosols and incendiaries. On the other hand, however, humanitarian codes of war in China (450–200 BC) forbade ruses of war and harming noncombatants (Mayor, 2009; Sawyer, 2007; Kautilya, 1992).

A common rationale for the use of toxic weapons was self-defense. Besieged cities and desperate populations turned to biological weapons as a last resort, while some commanders resorted to poisons out of frustration, hoping to break stalemates or long sieges. Some situations—such as holy wars, quelling rebellions, and fighting people considered "barbarians" or somehow less than human—encouraged the indiscriminate use of unfair toxic weapons against entire populations. The threat of horrifying chemical or biological weapons could discourage potential invaders or bring quick capitulation. Some generals had no compunctions about using any weapons at hand. Some cultures considered poison arrows to be acceptable weapons in both hunting and warfare.

Human ingenuity in weaponizing natural forces in antiquity is impressive. Many ancient toxic weapons and tactics anticipated, in substance or principle, almost every basic form of biological and chemical weapon known today, from spreading plague to poisoning water. Asphyxiating clouds of smoke can be considered the precursors of mustard and other toxic gases first used in World War I. In the 4th century BC, the defenders of Tyre (Lebanon) catapulted red-hot sand onto Alexander the Great's men causing gruesome deep burns and agonizing death. That weapon can be seen as the ancient prototype of the modern thermite bombs of World War II and white phosphorus bombs of the present day. The unquenchable burning, adhering effects of ancient petroleum-naphtha incendiaries anticipated the modern invention of napalm so notorious during the Vietnam War. Today's advanced "calmative mists" or sedative-drugged water supplies, and the top-secret insect- and animal-based weapons being developed by modern military scientists also have antecedents in the ancient world (Mayor, 2009).

Even the dangers of self-injury and the perilous problems of disposing of poison weapons were anticipated in antiquity. The ancient Greek myth of the Hydra with its ever-proliferating heads with venom-dripping fangs is a fitting symbol of the dilemmas of creating toxic armaments. Hercules had to dispose somehow of the immortal central head of the Hydra, which could not be destroyed. His solution was to bury the eternally dangerous Hydra head deep underground and he placed a huge boulder as a marker over the spot to warn people away. An analogous geological solution is used today to dispose of toxic and nuclear weapons. The eternal, indestructible materials are to be buried deep underground in the deserts of New Mexico and other sites in the American West, necessitating warnings to future generations about the danger of the biochemical agents. Notably, a model for halting the proliferation of toxic weaponry is also found in Greek myth. The young archer who inherited Hercules' Hydra-venom arrows had experienced accidental grievous injury from the arrows himself, before he deployed them against the Trojans. After the Trojan War, the old veteran decided to dedicate the terrible poison arrows in a temple of Apollo, the god of healing, rather than passing them on to the next generation of warriors (Mayor, 2009).

References

Forbes, R., 1964. Bitumen and Petroleum in Antiquity, second ed. Brill, Leiden.
James, S., 2011. Stratagems, combat, and "chemical warfare" in the siege mines of Dura-Europos. Am. J. Archaeol. 115, 69–101.
Jones, D., 2007. Poison Arrows: North American Indian Hunting and Warfare. University of Texas Press, Austin, TX.

Kautilya, 1992. "Arthashastra." Trans. L.N. Rangarajan. Penguin Classics, Delhi.

Lockwood, J., 2009. Six-Legged Soldiers: Using Insects as Weapons of War. Oxford University Press, Oxford.

Mayor, A., 2009. Greek Fire, Poison Arrows and Scorpion Bombs: Biological and Chemical Warfare in the Ancient World. Overlook Press, New York, NY.

Mayor, A., 2010. The Poison King: The Life and Legend of Mithradates, Rome's Deadliest Enemy. Princeton University Press, Princeton, NJ.

Partington, J., 1999. A History of Greek Fire and Gunpowder. Johns Hopkins University Press, Baltimore, MD.

Sawyer, R., 2007. The Tao of Deception: Unorthodox Warfare in Historic and Modern China. Basic Books, New York, NY.

Asclepius and the Snake as Toxicological Symbols in Ancient Greece and Rome

Gregory Tsoucalas[1] and George Androutsos[2]

[1]History of Medicine, Department of Anatomy, School of Medicine, Democritus University of Thrace, Alexandroupolis, Greece [2]Emeritus Professor of the History of Medicine, Medical School, National and Kapodistrian University of Athnes, Athens, Greece

The potential for myth-making has always resided in human consciousness as a means of coping with and explaining the unknown. The mind's capacity for scientific reasoning does not necessarily contradict the need for myth. Mythology constructs a world of logic and illogic, enriched with interconnected stories to conjure up viable personal and social environments (Tsatsos et al., 1986). The unlikeliness of a snake with human attributes led to the myth of Asclepius and his companion snake (Fig. 17.1). The snake, by using its venom, could both inflict poisonous wounds and be therapeutic. Asclepius, one of the band of the mythic Greek heroes known as the Argonauts, was expert in all facets of drugs and poisons. Eventually, myth had him evolving into the snake God, and so it was that the snake became the eternal symbol of medicine and toxicology (Tsoucalas, 2004).

The 5th century BC in Ancient Greece marked the reappearance of an earlier custom of praising the gods with votive offerings, while palliative measures comprised of magical and divine forces once again

FIGURE 17.1   Asclepius, a painting by Theodoros Ntolatzas, 2009, Athens, Greece.

became the mainstream approach in medicine. The social changes that ensued at the end of the 5th century BC were favorable for the myth-making evolution of god-therapists who could help restore health. The individual in need of treatment sought a deity who could heal and cure (Kerényi, 1947).

Soon, various political and social systems were established to incorporate the individual into the larger community. The people's frustration with the existing deistic framework was especially evident in rural areas of Greece and had its largest impact on the interface of religion and medicine. The Olympian gods with their heroic, brutal, chaotic, and bombastic characteristics were no longer able to meet the needs of the common people, and, as a result, the gods began to take on a less intimidating, softer character (Bengston, 1991; Boardman, 1967). During this period, medicine was linked with religion and was under the patronage of one god, Asclepius. The main purpose of the Asclepiadai (Greek: Ασκληπιάδαι)—the physician-priests in the temples of Asclepius—was to renew the human being and keep him healthy. The possibility of divine intervention in dealing with humanity's infirmities was fairly well accepted. Said originally to be a mortal, Asclepius was son of Apollo and a mortal woman, and was ultimately transformed into the primary god for healing. Some say this occurred following his death and resurrection when he became, in a sense, an alter ego of the healing god Apollo (Constantelos, 1988). The Ancient Greeks, in worshiping a god who had ascended from their own mortal species, thereby acquired a more personal contact with the divine when it came to dealing with their health problems. This relationship was marked by ceremonies in which votive offerings were combined with a purification bath, herbal fumigation, and sacred temple animals such as the dog, the turtle dove, the rooster, and, above all, the snake (Aravantinos, 1907).

Asclepius is often pictured with a snake-entwined staff (Papahatzis, 1986). He was undeniably the best known practitioner of medicine in mythology and was said to be able to resurrect the dead, which reflected his strong chthonic nature, reinforced by iconography depicting him with his steadfast companion, the snake (Kerényi, 1947; Holtzman, 1984). Today, about 50 statues depicting Asclepius with his companion are on display in museums around the world (Hamarneh, 1986). The sacred snake was not simply an emblem for Asclepius. He himself, when he entered the sacred grounds of his temples, would appear transformed into a holy snake, ready to cure the believers (Graeciae descriptio, 1903). Moreover, the snake was always depicted as the guardian of the temples of the gods, as well as the Oracle of Delphi, where a giant snake, the "Dragon Python," protected the sacred ground (Apollodorus, 1854). Numerous questions persist: Was the snake ascended from a simple guardian to a healing god? Was the snake

Asclepius's alter ego? Was Asclepius himself a famous mythical snake that had once affected a miraculous cure with its venom and then came to be worshiped in human form? Was the snake a poisonous deadly creature, or was the snake a healing demigod? Hypotheses abound, but the answers remain lost forever in the dense fog of Greek mythology. About 200 years later the snake-god, traveling the long mythical route from Epidaurus to Rome, was adopted by the Romans, who called him Aesculapius and made him the symbol of medicine for the entire Roman Empire. Although the Romans worshiped many healing gods, notably Dea Prema, Bona Dea, Carna, and Deus Subigus, Aesculapius was so dominant and revered that he became the first healing god assigned a holy temple on Tiber Island (Pollak, 1969).

According to Greek myth, Asclepius was the son of Apollo and the nymph Koronis (Graeciae descriptio, 1903; Hesiod, 2007). In his youth he became the greatest pupil of the centaur Chiron, the famed teacher of physicians (Tsoucalas, 2004; Pherecydes, 1824; Philostratus Fl, 1844). On one hand, Apollo was a fearsome god of pestilence; capable of annihilating people by the score; on the other hand, he was a healing god, worshiped by the Greeks in temples where believers, pleading for a cure of their physical ailments, would participate in purification and hypnotization ceremonies (Bernheim and Zener, 1978). It was the noble Chiron, though, expert in all aspects of medicine, who was the primary tutor of Asclepius (Tsoucalas, 2004). Meanwhile, the goddess Athena gave Asclepius blood from the mythical creature, Medusa, whose head was covered not with hair but with snakes. Athena taught Asclepius how to use the blood from the right vein (snake) as a healing elixir and the blood from the left vein as a deadly poison to "harm people" (Stagiritis, 1818). It was through this action that the goddess Athena bestowed divine powers on the mortal serpent.

Asclepius thus became the great healer, able to manage the fate of the sick. He could save them or he could condemn them to Hades; he could even bring the dead back to life. Asclepius was almost equal in stature to the Olympian gods and was a famous physician on Mount Pelion, where he had been tutored by Chiron. Soldiers wounded in fierce battles, suffering from serious wounds inflicted by bronze swords or by stones hurled at them, sick people distressed by their suffering, people burned from the summer sun or frozen from the winter cold—all sought relief on Pelion. Asclepius used spells, songs, elixirs, ointments, and drugs, and succeeded in curing them all (Krug, 1993; Doukas, 1842). Knowing that snakes were the guardians of the temples, the gatekeepers that knew everything about the gods, the temples, the priests, and the common people, Asclepius, in his human form, decided to raise a snake of his own. He most probably chose one from Epidaurus, where the snakes were known to be tamer (Tziarou, 2006), and brought it with him to Pelion to serve as his companion (Tsoucalas, 2004). His life forever after would be closely connected with the snake.

Asclepius, with his ability to resurrect the dead, matching the power of Hades, could transform himself into the holy snake, the guardian that soon became a demigod to be worshiped alongside him in his temples, the Asclepeiia (Greek: Ασκληπιεία) (Graeciae descriptio, 1903). This transformation suggests that the first form of the new healing god was probably that of a serpent. When summoned by his priests, the Asclepiadai, he would arrive in the form of a giant snake to practice his healing skills. On one occasion the snake-god entered his temple in Epidaurus, where a poor man suffering from a malignant ulcer in his foot, was ready to die. The snake slowly moved toward him and licked the ulcer, while the fluids (i.e., the venom) cured the malignancy (Epigraph, 115–118). On another occasion, an infertile woman summoned Asclepius for aid. Again in the form of a giant snake he entered the temple, and by cuddling the woman's abdomen he bestowed upon her a fertile future: She would give birth to five strong babies (Epigraph, 117–119).

The fame of the snake-god, Asclepius, reputed to deliver miraculous remedies for almost all ailments, spread throughout Ancient Greece. Votive offerings with spiraling snakes were devoted in large numbers in his name. The poisonous snake-god with his venom could provoke a quick death to express the gods' wrath, or alternatively, he could palliate the sufferer. Asclepiadai reportedly cared for the snakes inside a round peribolos, the abaton (Greek: άβατον: a sanctuary in the center of a temple without a roof) (Campbell, 1955). They also knew how to extract venom from the mouth cavity of the snake and used it in small doses as a therapeutic drug (Stagiritis, 1818; Angeletti, 1991). For centuries, the snake had served Apollo and had learned theurgic medicine. For years it was used in the medical training of Asclepius by the Centaur Chiron, an expert both in botany and in the preparation of drugs and poisons. Finally, as previously noted, Athena taught Asclepius how to use Medusa's blood. In his human form, Asclepius practiced the use of drugs to heal and harm. In the end, the snake-god could cure the diseases of the body, spirit, and soul, as well as control the erotic passions. He could hypnotize and treat the sufferers by appearing in their dreams. Asclepius was the founder of a line of physicians, while his alter ego, the snake, became for all time the symbol of healing and toxicology (Pollak, 1969).

In 293 BC, after 2 years of a devastating epidemic in a Roman province, all hopes were on Asclepius. It was his moment of glory to enter the Roman Pantheon, to make his breakthrough to eternity. For 2 years Lazio (the region of Italy where Rome is located) had suffered horrific deaths among its population, as well as anemia and other diseases. The plethora of the dead, the despair of disease, the suffering of the ill, and the fatigue of the grave diggers drove the Romans to the sanctuary of the Delphic Oracle. The Oracle's prophecy was clear "That which you seek isn't here. You must search for it elsewhere. You don't need Apollo, but in reality you need his son. Go and beg his son." A Roman contingent therefore traveled all the way to Epidaurus where they asked the local lords for aid. While the lords were discussing whether to help them, the snake-god appeared in the dreams of the Romans, advising them: "Don't be afraid, I will abandon my statue and come with you." Subsequently, Asclepius, in the form of a snake, boarded the Roman ship for the long voyage to Rome. Through the River Tiber, he reached the shore near the eternal city, where, as a serpent, he jumped into the water and swam to the island of Tiber. Resuming, at that point, his human form, he cured the sick and saved the people from the pestilence (Pollak, 1969; Ovidius, 1922; Brouwer, 1989; Fairbanks, 1907).

The name Asclepius itself reveals a strong connection to snakes. The first part, "Ascl" (Greek: Ασκλ), derives from the word *Ascalavo* (Greek: Ασκάλαβο), which means snake (Greek: όφις or φίδι), while the second part "epius" (Greek: ήπιος) means meek or gentle. The Greek word for snake, όφις (ofis), derives from the Greek word ώφθην (ofthin), which means he who sees everything, the guardian. The sacred snake was also mentioned as a dragon (Greek: δέρκομε, derkome), which means he who possesses excellent vision and understanding, that is, all seeing and all knowing (Pollak, 1969). In Ancient Greek "giras" (Greek: γήρας) means the skin that the snake periodically sheds, but it also means "old age" (Ramoutsaki et al., 2000). Asclepius was an omniscient snake-god, who could see through every disease and could manipulate the healing properties of all drugs; a god who could rejuvenate himself at will by shedding his old skin, the "giras."

It was important for a pharmacist or physician of the era to have deep knowledge of both remedies and poisons. By understanding the snake's own immunity to its deadly venom, many have tried to devise the ultimate potion to treat all kinds of poisoning. King Mithridates the 6th (132–64 BC), in an attempt to protect himself from poisoning, and wanting to strengthen his immune response concocted an antidote consisting of 54 different substances. In an attempt to make himself invulnerable to poisons, he drank small portions of the antidote. This antidote became known as Mithridatiki. It was patterned after the Theriac, an antidote of 64 different substances, prepared in 63 BC, by the pharmacist Krateus, Mithridates' advisor (Vasileiou and Papavasileiou, 1979; Gottlob, 1827; Karaberopoulos et al., 2012). The epic poem by Nicander (197–130 BC), *Theriaca*, includes a chapter titled "Opfiaca" (Greek: Οφιακά, όφις = snake) and deals with the healing potions against snake venom. In time, antidotes aiming to heal all poisons were named "Theriac," the ultimate cure, and many, as Mithridates did, tried to formulate such a cure-all (Tsoucalas, 2004).

In Ancient Greece, many figures associated with the underworld were also relevant to fertility. The chthonic goddess Demeter, for example, was also the protector of fertility, while Artemis and Hecate (similar or identical goddesses) were protectors of newborns, and chthonic Apollo was also a god of light (poetic paradox), the healing god of the Ancient Greeks. The snake, apart from crawling on the ground, could enter the earth connecting the mortal world with the underground world of Hades. Its reemergence above ground was the sign of a new birth. The Ancient Greeks were innovative sea pioneers, but they were also an agricultural civilization that strongly valued the earth and the work of sowing and harvesting—a cycle of death and rebirth. For them, the snake encapsulated the most significant values of their culture (Rethimiotakis, 1997).

In Minoan Crete, the Mother goddess was a symbol of fertility and palliative medicine. She was depicted holding two snakes in her hands, and in some cases she held poppies. The healing powers of the opium poppy were known in Minoan culture and were connected with the snake. The lekythi, a trefoil flask, which had been used to store what was probably liquid opium, usually depicted a snake under its handles (Ramoutsaki et al., 2000), suggesting a relationship between opium and snake venom. On the island of Kos, the homeland of Hippocrates, every year the inhabitants celebrated "the rod's ascension," indicating the significance of the physician's walking stick. This wooden shaft was once a weapon for warriors, but later it became not only a supportive instrument, but also a tool symbolizing the physician's readiness to heal, a symbol of life. By adding the snake, a second symbol of life, Asclepius was acknowledged as the primary god of healing, the custodian of health (Kerényi, 1947; Pollak, 1969). Coins from the Greek island of Cos, which lies on the Aegean Sea a few miles from Crete, often depicted a bearded man holding a walking stick with a snake coiled around it (Krug, 1993; Ramoutsaki et al., 2000). That may have been the origin of the iconography of the physician and the staff-entwined snake. Later, coins from Pergamum and Rome depicting curved snakes, a rod with a snake, or the figure of Asclepius with both of his symbols, appeared with greater frequency (Hart, 1965).

The snake has been recognized as a medical emblem for more than 2500 years. It is featured entwined around a staff of knowledge and wisdom in most images and statues depicting Asclepius (Antoniou et al., 2011). The snake became the symbol of rebirth, the provider of innovative remedies, the bringer of the theriac, and the healing demigod companion of Asclepius (Aesculapius). It was an early example of a binary god who could use venoms and had knowledge of their circulation inside the veins. The snake represented the flow of energy. Being wrapped around the crook of Asclepius, the snake had supported Asclepius in his long voyages. The serpent demigod, who became a binary chthonic god with both serpent and human form, became the insignia of medicine and toxicology for millennia to come. Indeed, it is a totem of life.

References

Angeletti, L.R., 1991. Views of classical medicine. Theurgical and secular rational medicine in the healing-temples of Ancient Greece. Forum (Genova) 2, 1–11.

Antoniou, S.A., Antoniou, G.A., Learney, R., Granderath, F.A., Antoniou, A.I., 2011. The rod and the serpent: history's ultimate healing symbol. World J. Surg. 35, 217–221.

Apollodorus, 1854. Bibliotheca 'A. Taubner, Leipzig.

Aravantinos, A., 1907. Asclepius and Asclepieia. Drougoulinou, Leipzig.

Bengston, H., 1991. History of Ancient Greece. Melisa-Gavriilis, Athens.

Bernheim, F., Zener, A.A., 1978. The Sminthian Apollo and the epidemic among the Achaeans at Troy. Trans. Am. Philol. Assoc. 108, 11–14.

Boardman, J., 1967. Excavations in Chios 1952–1955. Greek Emporio, Athens.

Brouwer, H.J., 1989. Bona Dea. E. J. Brill, Leiden.

Campbell, J., 1955. Spirit and Nature. Routledge and Kegan Paul, London.

Constantelos,, D., 1988. The Interface of Medicine and Religion in the Greek and the Christian Greek Orthodox Tradition. Greek Orthodox Theological Review, Athens.

Doukas, N., 1842. Comments on Pindar. Adreou Koromila, Athens.

Epigraph IG IV(2),1 [Epidauros] 121 115–118, https://inscriptions.packhum.org/biblio.html.

Epigraph IG IV(2),1 [Epidauros] 122 117–119.

Fairbanks, A., 1907. The Mythology of Greece and Rome. Appleton, New York, NY.

Gottlob, K.K., 1827. Galen, Opera Omnia. Cnoblochii, Leipzig.

Graeciae descriptio, 1903. Teubner, Leipzig, https://www.abebooks.com/book-search/title/pausanias-descriptio-graeciae/.

Hamarneh, S.K., 1986. Background of History of Arabic Medicine and the Allied Health Science. Yarmouk University, Yarmouk (Jordan).

Hart, G.D., 1965. Asclepius, god of medicine. Can. Med. Assoc. J. 92, 232–236.

Hesiod, 2007. Fragmenta. Pirinos Kosmos, Athens.

Holtzman, B., 1984. Asklepios. LIMC II, Munich–Zurich.

Karaberopoulos, D., Karamanou, M., Androutsos, G., 2012. The theriac in antiquity. Lancet 379, 1942–1943.

Kerényi, K., 1947. Der göttliche Artz: Studien über Asklepios und seine Kultstätte. Ciba, Basel.

Krug, A., 1993. Heilkunst und Heilkult: Medizin in der Aentike. Beck, Munich.

Ovidius, P., 1922. Metamorphosis. Cornhill Publishing Co, Boston.

Papahatzis, N., 1986. Asclepius the physician. Greek Mythology, the Gods. Ekdotiki Athinon, Athens.

Pherecydes, 1824. Fragmenta. Altera, Leipzig.

Philostratus Fl, 1844. Olearius. Meyer & Zeller, Zurich.

Pollak, K., 1969. Die Heilkunde der Antike. Griechenland-Rom-Byzanz. Die Medizin in Bibel und Talmud. Econ Verlag, Dusseldorf.

Ramoutsaki, I., Haniotakis, S., Tsatsakis, A., 2000. The snake as the symbol of medicine, toxicology and toxinology. Vet. Hum. Toxicol. 42 (5), 306–308.

Rethimiotakis, G., 1997. Minon shrine. Archaeol. J. 134, 177.

Stagiritis, A., 1818. Ogygia. (On doctors)In: Varth, John (Ed.), Tsvekiou, Vienna.

Tsatsos, K., Kakridis, I., Kiraikidou-Nestoros, A., 1986. Analysis and Interpretation of the Greek Myth. Ekdotiki Athinon, Athens.

Tsoucalas, I., 2004. Pediatrics from Homer Until Today. Science Press, Skopelos–Thessaloniki.

Tziarou, K., 2006. Asclepius, the god of medicine. Med. Podium J. 3, 94–101.

Vasileiou, R.P., Papavasileiou, I.T., 1979. Handbook of the History of Medicine. Athens University Press, Athens.

18

Anthropogenic Air Pollution in Ancient Times

László Makra

Institute of Economics and Rural Development, Faculty of Agriculture,
University of Szeged, Szeged, Hungary

Toxicology in Antiquity
DOI: https://doi.org/10.1016/B978-0-12-815339-0.00018-4

267

18.1 POLLUTION OF THE ENVIRONMENT IN ANCIENT TIMES

Environmental pollution is coeval with the appearance of humans. When *Homo sapiens* first lit fire, its smoke provided the first medium of environmental pollution. The burning of fuels for heating and cooking has contributed to the air pollution of inner spaces. The walls of caves, inhabited several thousands of years ago, are covered with thick layers of soot. The presence of smoke must have made breathing difficult and must have irritated the eyes in the confined space as well. Most of the lungs of mummified bodies from the Paleolithic (however, there are few of them) have a black tone. In the first inhabited areas smoke was not driven away (one of the practical reasons for this might have been protection against mosquitoes) and the people dwelling in these inner areas found shelter in the smoke (McNeill, 2001). [Millions of people live this way today. In 1993, when we were in Nepal trekking in the Langtang National Park, we visited many small villages and found accommodation at local houses on the southern slopes towards the High Himalayas. The smoke of fire is not driven away from the buildings there, even today. The walls of the houses are built of metamorphic slates with no bonding, carrying roofs covered with rush matting. When there is a fire, it produces a thick smoke inside the house, irritating the eyes and making breathing difficult. It is impossible to sleep; one can only stay there for a short time. And from the outside, the house looks as if it is on fire, with smoke streaming out through the slits and gaps of the walls.] Humans seem to have been living together with this unhealthy form of air pollution for many thousands of years.

Indoor air pollution, and especially particulate matter, was a significant problem in ancient times. Animal and vegetable oils were burned to provide light; furthermore, the houses were heated by burning wood and animal dung. All these materials produced high quantities of soot and toxic gases (Colbeck and Nasir, 2010). Capasso (2000) examined skeletons buried by the volcanic eruptions of Vesuvius and found evidence of inflammation of the pulmonary tract. Histological assessment of the lungs of ancient human mummies has shown that anthracosis (accumulation of carbon in the lungs caused by inhaled smoke or coal dust) was a regular disorder in many ancient societies, due to long exposure to smoke from domestic fires. Smoke protected against mosquitoes and other insects, however it greatly increased the risk of chronic respiratory diseases (Colbeck and Nasir, 2010).

Environmental pollution was responsible for several kinds of illnesses in the early times. The very first polluting material might have been human feces. Bowel bacteria living in the human body, such as

Escherichia coli, might have easily gotten into the drinking water, infecting early humans. This type of *"environmental pollution"* causes illnesses affecting several millions of people even today. In China, where a comprehensive system was developed for waste disposal even in ancient times, the use of human feces as a fertilizer was an important element of agriculture even many thousands of years ago. The high productivity of the alluvial plain in the eastern part of the country has been maintained this way for over 4000 years. This tradition is being followed even today in several regions of China. As Han Suyin says, *"In Chengtu (the capital of Sechuan Province) those families, who owned the city sewers and this way could sell the accumulated faeces in the countryside, belonged to the richest ones even in the 20th century (till 1949)"* (Markham, 1994). The fertilization of rice-paddies with feces contributed largely to the pollution of ground water which, in this way, is not suitable for drinking in the whole of tropic Asia. However, if water is boiled, resulting in the precipitation of salts, it becomes totally tasteless. The tradition of flavoring boiled water with tea leaves comes from China, and it started spreading throughout the empire about 2000 BC and then all over Asia (Makra, 2000).

Dust pollution also appeared in ancient times. Part of this was of a natural origin, and Chinese and Korean writers wrote about yellow dust (often referred to as Kosa dust), that was transported by air currents to thousands of kilometers from Inner Asia to East Asia (e.g., Chun, 2000). According to Janssens, in the New Stone Age people mining flint from the embedding limestone in the stone mines day after day might have suffered from silicosis, e.g., in Obourg (Markham, 1994). The day-long inhalation of dust must have been the underlying factor. Sometimes the geographical location of an area was responsible for an outbreak of certain diseases. Investigations have revealed that near Broken, in what is now Zambia, Hominides who lived about 200,000 years ago suffered from lead poisoning. The reason for this illness was the transport of lead into the spring located next to the cave from a neighboring streak of ore (Markham, 1994).

The harmful activities of ancient civilizations caused long-lasting changes to the environment, the effects of which can be experienced even today. However, these effects appeared only at a regional scale, without causing any global changes. Increasing soil alkalinity on the floodplains of the Tigris and Euphrates Rivers between 3500 BC and 1800 resulted in a gradual decrease in the productivity of Sumerian agriculture. Water used for irrigation raises the ground water table, and if the redundant water is not driven away by channels, then soils become saturated with water, resulting in the dissolution of salts and their precipitation on the surface in the form of an impermeable layer. The Sumerian people noted this process as the following: *"the soil surface turned white."* Water used for irrigation gradually makes the region more and more unsuitable for agricultural production by leaching the soils. This phenomenon largely contributed to the decline of Sumerian culture (Markham, 1994; Mészáros, 2002). Babylonian and Assyrian law included clauses that affected neighbors' property. Although the earliest laws, those of Hammurabi (23rd century BC), relate mostly to water (Driver and Miles, 1952), smoke was typically treated in the same way in ancient law (Brimblecombe, 1987b). Around 200 AD the Hebrew *Mishnah* and its interpretation through the Jerusalem and Babylonian *Talmud,* detail pollution issues (Mamane, 1987).

In ancient times, air pollution had substantial consequences only in the cities. The air of these early towns, as in some current settlements as well, was filled with the penetrating smell of decaying organic domestic waste, rotting meat, as well as human feces. During a siege, when there was no opportunity to remove these waste materials, they emitted aggressive smells and unbearable conditions prevailed in these settlements. According to Egyptian historical records, when Nubian troops enclaved the city of Hermopolis, which is situated on the left bank of the Nile half-way between Theba and Memphis, the inhabitants surrendered pleading for mercy, rather than further bear the smell of their own town air (Brimblecombe, 1995). In ancient cities, pollution deriving from unpleasant odors was generally seen as important. Aristotle (384–322 BC) set a rule in his work *Athenaion Politeia*, according to which manure should be placed outside the town, at least 2 km away from the town walls (Mészáros, 2001). Smoke stained marble in antique towns giving it a grayish tone. This annoyed several classic poets as well [e.g., Horace (65 BC–8 AD)] and made, among others, ancient Jews introduce a list of relevant laws (Mamane, 1987). In ancient times, smoke and soot represented the two major media of air pollution.

There are several examples of environmental pollution in China too. Before the Tang era (618–907 AD), the firs on the mountains of Shantung were logged and burned; then in the Tang era the slopes of the Taihang mountains, located at the borderline of the Shansi and Hopei provinces, became barren (Schäfer, 1962). Similarly, during the Tang dynasty, forests were cut around Loyang, the capital, in a circle with a radius of 200 mi. The trunks of the trees were mostly used as firewood and partly burned in order to get ink for government offices (Epstein, 1992).

Urban air pollution depends on the dimensions of the given settlement, on the extension of the built-up territory, as well as on the nature of the industrial activity, especially the use of traditional fuels. As urbanization progressed in China, the Mediterranean basin, and northwestern Africa, from about 1000 AD, more and more people lived in smoky and sooty surroundings. Maimonides, the philosopher and physicist (1135–1204 AD), who had comprehensive experiences on the towns of that era from Cordoba to Cairo, found that urban air is *"stuffy, smoky, polluted, obscure and foggy."* Furthermore, he thought that this condition is caused by *"dullness preventing understanding, lack of intelligence and amnesia"* of the inhabitants (Turco, 1997).

On the other hand, traffic and transportation difficulties restricted the rate of air pollution within the cities. Industrial activities consuming the most energy (e.g., the production of tiles, glass, pottery, bricks, and iron casting) were located near the forests, since the transportation of fuels in large quantities to the cities would have been too expensive. This way, though air pollutants of industrial origin made the air smelly, only a few people inhaled it. Port cities were partial exceptions, as ships could transport wood and charcoal more economically. Hence, Venice could maintain its glass industry, ensuring its energy supply by the transportation of wood from distant places. However, the majority of urban air pollution derived from household fuels, such as manure or wood, but sometimes smokeless charcoal as well (McNeill, 2001). The air of the Chinese cities might have been extremely polluted too, since the developed water transport system (Big Channel) enabled the use of large quantities of fuel, at least in the Sung capital, Kaifeng. This city, 500 km south of Beijing, was probably the first in the world to convert its energy supply from wood to coal. The transition occurred at the end of the 11th century, when the city had about one million inhabitants. However, the coal-heating period was short, because Mongolian troops destroyed Kaifeng in 1126 and those who remained in the city died from plague in the early 13th century (Hartwell, 1967).

Intensive environmental pollution appeared simultaneously with the development of societies. Extensive environmental losses occurred even in the earliest societies. Air and water were polluted, soils were destroyed, and many animal and plant species were exterminated. However, environmental changes caused by the earliest societies were generally minor, with a rapid restoration of the original conditions. Thanks to this, many people are not aware of the environmental losses triggered by the activities of these early societies. Consequently, they tend to be more forgiving towards these early societies than towards modern societies living in an urban environment. At the same time, there are examples of such environmental activities in ancient times, which resulted in long-lasting changes, the signs of which are observable even today. The cutting down of forests in large areas for building ships in ancient times might have contributed to the decrease of woodland coverage in the Balkan Peninsula, and in the territory of Greece. However, the drier summers and droughts in the Mediterranean might also have contributed to this large-scale reduction of woodland areas (Karatzas, 2000). Nevertheless, this latter fact has no relation to human activities. In Greece, due to the small amount of summer precipitation, stunted plants and bushes develop that are suitable only for the grazing of sheep and goats. These animals, by overgrazing the slopes of mountains, increase soil erosion. The thin soil layer, which becomes loose, is transported from the slopes by winter runoff, creating barren limestone surfaces quite rapidly as a final and complete stage of erosion.

There are several examples of deforestation in other regions as well. During the reign of King Solomon, cedar woodlands covered an area of 5000 km². Cedar woodlands were first mentioned in the literature between 2500 and 2300 BC. However, recently very few cedars are found there. In the golden age of the Roman Empire, the main road from Baghdad to Damascus was shadowed throughout by cedars. Today, the road between these cities is surrounded by desert (McNeill, 2001).

Several cultures emphasize that one should live in harmony with the environment. However, even in those societies where this idea has been mentioned since antiquity (e.g., in Asian societies), schemes to improve the environment frequently give way to financial demands.

Ancient Romans called air pollution "gravioris caeli" (heavy heaven) or "infamis aer" (infamous air) (Colbeck and Nasir, 2010). "The smoke, the wealth, the noise of Rome ..." held no charms for the Roman poet Horace (65 BC − 8 AD), who described the blackening of houses and temples by smoke (Costa, 1997). Air pollution problems of ancient times are mentioned even in the poems of classical poets. Horace wrote that Roman buildings turned more and more dark from smoke, and this phenomenon might also have been observed in many ancient cities. Seneca (4 BC−65 AD), the teacher of Emperor Nero (37−68 AD), was in poor health all his life and his physician frequently advised him to leave Rome. In one of the letters he wrote to Lucilius in 61 AD, he mentions that he must escape from the gloomy smoke and kitchen odors of Rome in order to get better (Heidorn, 1978).

Sextus Julius Frontinus (∼30−100 AD), once governor of Britain, later oversaw the water supply to imperial Rome (recorded his book: De Aquaeductu Urbis Romae) and believed his actions also improved Rome's air. Civil claims over smoke pollution were brought before Roman courts almost 2000 years ago (Brimblecombe, 1987b). According to Roman law, cheese-making manufacturers should be established in such way that their smoke would not pollute other houses. Much similar material is available from the very important book Pan's Travail (Hughes, 1993) and Brimblecombe (1987a).

The Roman Senate introduced a law about 2000 years ago, according to which: "Aerem corrumpere non licet," namely "Polluting air is not allowed." The Institutes issued under the Roman emperor Justinian in 535 AD were used as a text in law schools. Under the section Law of Things, our right to the air is clear: "By the law of nature these things are common to mankind − the air, running water, the sea, and consequently the shores of the sea." (Lib. II, Tit. I: Et quidem naturali iure communia sunt omnium haec: aer et aqua profluens et mare et per hoc litora maris).

18.2 LEAD IN ANCIENT TIMES

18.2.1 Lead Mining and Exploitation

In the ancient Mediterranean, mining and metallurgy played a primary role in the economy. According to Xenophon (434–359 BC) and Lucretius (98–55 BC), the smoke of lead mines in Attica was harmful to human health (Weeber, 1990).

Lead is extracted from its most important ore, galena. The lead content of galena is 86.6%, but it also contains arsenic, tin, antimony, and silver. The major part of silver production globally comes from galena, and not from silver ore, since mining and exploitation of galena is much more significant. A long time after the introduction of silver coin as a currency (about 2700 BC), the primary aim of galena mining was to extract silver, and lead was considered to be only a byproduct (Boutron, 1995).

The oldest lead object found by archeologists is a string of beads worn in Anatolia some 8000 years ago. Its use as jewelry suggests that this was a time when lead was still new and rare (Eisinger, 1984). At the same time, lead mining started about 4000 BC. Considerable exploitation began 1000 or so years later; when a new smelting technology was introduced in order to extract lead (and silver) from sulfide ores of lead. The exploitation of lead ores and the use of lead became more and more important during the Copper, Bronze, and Iron Ages (Nriagu, 1983a). This progress was promoted by the introduction of silver coins and the development of Greek civilization (during that time lead production was 300 times greater than that of silver). Lead production reached its maximum of 80,000 MT (metric ton)/year in the golden age of the Roman Empire, which was about the same magnitude as that of the Industrial Revolution some 2000 years later (Hong et al., 1994). The most important lead mines were situated in the Iberian Peninsula, the Balkans, in the territory of ancient Greece, and in Asia Minor (Nriagu, 1983a). Lead production suddenly decreased after the fall of the Roman Empire, and reached its minimum at about 900 AD, with a mass of only some 1000 MT/year. Then, the production began to increase again, thanks to the new lead and silver mines opened in Central Europe after about 1000 AD.

18.2.2 The Utilization of Lead

Lead mines during the Roman period were a plentiful source of the metal, as Pliny describes lead being found *"in the surface stratum of the earth in such abundance that there is a law prohibiting the production of more than a certain amount"* (Waldron, 1973). In Roman times lead was the most popular metal, and was widely used in everyday life. It has a number of useful properties that suit it to a relatively low level of technology. It melts at low temperatures, it is malleable and easy to work, it can be readily cast and joined, and it is resistant to corrosion. Thus it comes as no surprise that it found such widespread application in the ancient world (Waldron, 1985). Its compounds were used as face powders, lipstick, or mask paint, as well as a coloring agent in paints. Furthermore, lead was used for preserving foods; it was even portioned to wine in order to prevent its fermentation. Lead compounds were used as a birth control medicine (for exterminating sperm) and as a kind of spice, too. Cups, jugs, pots, and frying pans were made of lead alloys. Coins were also made of lead, as well as of alloys of lead and other metals such as copper, silver, and gold. Since it resists corrosion and can be processed easily, lead was extensively used in shipbuilding, house building, and in the construction of water pipes. During house building, hot lead was poured between limestone/marble blocks and in this way it served as a binder. In ancient Rome and in other cities of the Roman Empire, the construction of water pipes was the most important field of lead utilization. Also, in Babylon a water pipe made of lead was used for watering the hanging gardens built by king Nabu-kudurri-usur [also spelled Nebuchadnezzar (605−562 BC)]. Because of the abovementioned facts lead is frequently referred to as a Roman metal (Markham, 1994; Eisinger, 1984).

18.2.3 Illnesses Caused by Lead

Both lead and its compounds are poisonous. Thanks to its volatility (lead vapor), as well as the volatility of some of its compounds [e.g., as a petrol additive $(Pb(C_2H_5)_4)$] or solubility [e.g., $Pb(CH_3COO)_2$], it can easily be absorbed into the human body. Symptoms of lead poisoning include headache, nausea, diarrhea, fainting, and cramp.

The Romans knew that lead was a dangerous metal, as they noticed the symptoms people who worked in lead mines suffered from. Pliny wrote: *"red-lead is a deadly poison and should not be used medicinally"* as well as warning that the *"exhalations from silver mines (i.e. galena mines) are dangerous to all animals"* (Waldron, 1973). Furthermore, the geographer (Strabo, 1988) described (about 7 BC) the high chimneys required to disperse the air pollutants during silver production in Spain. However, since lead was used extensively in everyday life, its dangers were not considered. Lead was believed to be less dangerous if it got into the body only in small doses. As carbon dioxide dissolved in water interacts with lead in water pipes resulting in a possible enrichment of these dissolved lead compounds in the body, this process can easily

FIGURE 18.1 Pont du Gard, an ancient Roman aqueduct in France. Piped water in ancient Rome—supplied to the city via lead pipes called fistulae—contained 100 times more lead than water drawn directly from local springs. That amount of lead in the water may have been a major public health issue. Source: *https://pixabay.com/en/pont-du-gard-prov-ence-france-533365/*.

lead to so-called *"lead disease,"* a consequence of which might be paralysis. The ancient writers Xenophon and Lucretius observed the noxious emissions from metal mines in Greece, and Pliny declared that smelter emissions were dangerous to animals, especially dogs (Nriagu, 1990; Morin, 2009). The presence of lead in food and drinking water might have led to infertility or stillbirth (Goldstein, 1988). Nevertheless, mineworkers suffered mostly from the harmful effects of lead. Hence, Romans generally made slaves work in mines. In Greek—Roman times, according to estimates, several hundred thousand people (mainly slaves) died of acute lead poisoning caused by the mining and smelting processes (Nriagu, 1983a,b; Hong et al., 1994). The credit for the first direct clinical account of lead poisoning has in recent times been accorded to Hippocrates (Waldron, 1973).

The use of lead water systems represented a hazard to health, but both the Romans and Greeks exposed themselves to a far greater risk (Figs. 18.1 and 18.2). They found that by coating their bronze or copper cooking pots with lead or lead alloys, not only was the leaching of copper from the pot prevented, thus avoiding spoiling the taste of the food, but also these were of great value in preparing wine and grape syrup (sapa), which was used almost exclusively as a sweetening agent. Pliny also advocated this dangerous practice. He wrote *"Preference should be given to lead vessels ... in boiling defrutum and sapa"* (Waldron, 1973). One property of lead is that it inhibits enzyme activity. Hence, sapa prevented fruit from souring and fermenting and was used extensively as a preservative. In addition, sapa was found to improve the quality of a poor wine and to prolong the length of time for which any wine could

FIGURE 18.2   Ancient Roman lead water pipes (1st century) with inscription "IMP. VESP.VIIII.T.IMP.VII.COS.CN.IULIO.AGRICOLA.LEG.AUG.PR.PR." *"(made)* when the Emperor Vespasian was consul for the ninth time and the Emperor Titus was consul for the seventh time, when Gnaeus Iulius Agricola was imperial governor *(of Britain)."* Chester (England), Grosvenor Museums. Source: *By Wolfgang Sauber (Own work) [GFDL (http://www.gnu.org/copyleft/fdl.html) or CC BY-SA 3.0 (https://creativecommons.org/licenses/by-sa/3.0)], via Wikimedia Commons."*

be kept (Waldron, 1973). The adoption of using lead in sweetening wine in medieval Europe caused widespread illness (Eisinger, 1991).

To some authors not normal behavior of Emperors Caligula (12–41 AD) and Nero might also be the consequences of lead poisoning (Goldstein, 1988). No aspect of the history of lead is likely to provoke as much interest and controversy as the prevalence of chronic lead disease (derived from extensive lead mining and the widespread use of devices made of lead) in antiquity and the suggestion that it played an important part in the decline of the Roman Empire (Nriagu, 1983a,b; Hong et al., 1994). On the other hand, there are some reasons this idea has been largely discredited: (1) the lead-related gluttony and other excesses of the Julio-Claudian and Flavian emperors are difficult to reconcile with the loss of appetite and constipation which are among the prominent symptoms of chronic plumbism; (2) the Empire attained its greatest wealth, power, and extent under Trajan and other effective emperors, who also consumed foods and drinks prepared in leadened devices (Waldron, 1985); and (3) most of the lead in the bones from citizens of Rome is due to postmortem absorption (Fouache, 2013). It is suggested that the gradual depopulation contributed to the failure of the Western Empire and that lead-induced sterility may have played only a part in this (Eisinger, 1984). The rise of Christianity might have also been a major reason (Zeek, 1986). The decline of the Roman Empire is a phenomenon of great complexity and it is too simple to ascribe it to a single cause.

18.2.4 Lead Pollution of Ancient Tooth Samples in the United Kingdom

English researchers, co-fellows of the Natural Environment Research Council and the British Geological Survey, studied the concentrations of lead in tooth enamel of Romano-British and early medieval people from various sites in the United Kingdom. Then, they compared the lead concentrations present in these people to that of both their prehistoric forebears and modern people living in the United Kingdom at the time of the study. According to an extensive study on the tooth enamel, the lead concentrations of adults living in the United Kingdom carried out in the early 1980s displayed spatial variations with an average of 3 ppm. Some more recent analyses on modern children's teeth found lead concentrations with averages around a few tenths of ppm suggesting, as indicated also by the atmospheric data, that modern lead exposure is decreasing. On the other hand, Neolithic people living before the use of metals, had tooth enamel lead concentrations that averaged around 0.3 ppm. These concentrations are only a tenth of the average for modern people and possibly similar to those in modern children.

While analyzing tooth enamels of Roman, Anglo-Saxon, and Viking people living in the United Kingdom, the researchers found individuals with tooth lead concentrations greater than 10 ppm, and occasionally even higher values. Concentrations of this magnitude among modern people can be associated with occupational or acute exposure, and suggest that lead pollution was a significant problem for both Roman and the early medieval ancestors of the British citizens.

The explanation may be the fact that England, Scotland, Wales, and Ireland are all rich in natural lead deposits. Furthermore, each of these countries has abundant ores, which have been mined since antiquity. Probably, it was partly the richness of the country's lead ores, with their associated silver of course, which led to Rome's initial interest in the conquest of this most northerly reach of the Empire. It is also known that the Romano-British, Anglo-Saxon, and Viking people inhabiting the area of the United Kingdom were exposed predominantly to lead from the ore sources, because of the characteristic isotopic composition of the lead remaining in their teeth.

On the other hand, high exposures were detected not only among the people actively involved in lead mining, smelting, or metal working, but in the tooth enamel of children too. Thus a high lead concentration was considered to be an environmental rather than an occupational problem.

18.2.5 Lead Pollution on Regional and Hemispheric Scales

In 1957—58, as part of the International Geophysical Year, the first extensive research programs were launched to analyze information stored in many hundreds of thousands of years old snow and ice layers of Greenland and the Antarctic. The aim of this research was to establish a possible hemispheric scale of air pollution for a time period spanning many thousands of years. Later, the ice cores from this area gave substantial information on the atmospheric effects of human activities (e.g., Boutron et al., 1991, 1993, 1994). In Greenland, the deepest boring corresponds to an interval of 7760 years, which is well before the time when silver was first smelted from galena. We can speak about background levels of the atmospheric lead concentration up to this period (Boutron, 1995).

The chemical analysis of an ice core of 9000-ft deep (1 ft = 30.48 cm) from Greenland enabled the collection of information on the atmospheric pollution for the past 7760 years. According to this, the lead concentration in the atmosphere before the beginning of lead production, when atmospheric lead was derived only from natural sources, was low. At this time the percentage of atmospheric lead as compared to normal soil was near 1 (0.8), which indicated that this lead derived from soils and rocks. Three thousand years ago, the lead concentration of the atmosphere practically corresponded to the levels measured at the beginning of lead production. This means that anthropogenic lead emission was still negligible up to this time, considering the amount of lead getting into the atmosphere naturally. The atmospheric concentration of lead started to increase in the 5th century, and during Greek—Roman times (between 400 BC and 300 AD) the enrichment factor of lead reached a value of 4 and remained at the same high level for seven centuries. Basically, a four times higher lead concentration was detected for this period in the snow and ice layers of Greenland compared to the earlier, natural values. This is the earliest detected hemispheric scale air pollution, almost 2000 years before the Industrial Revolution and well before any other polluting effect (Hong et al., 1994).

In the golden age of the Roman Empire, about 2000 years before, 5% of the total processed lead (being about 80,000 MT) got into the atmosphere, which might have resulted in an atmospheric emission peak of 4000 MT/year (Hong et al., 1994). Regarding the economic development of the Roman Empire, Scheidel (2009) associated the lead pollution level of the atmosphere with the annual number of shipwrecks (trade volume) and meat consumption (animal bones). Lead emission deriving from metal processing caused important local and regional air pollution all over Europe, which can be detected, e.g., in the lacustrine deposits of southern Sweden (Renberg et al., 1994). Furthermore, these emissions significantly polluted the troposphere over the Arctic (Hong et al., 1994).

Rosman examined the possible sources of lead pollution in the ancient atmosphere. According to the analysis of lead isotope ratios in ice cores, the mines in the territory of Spain proved to be the main sources of atmospheric lead. These mines were supervised by Carthago between, the Pun Empire between 535 and 205 BC, and then were followed by the Romans till 410 AD. About 70% of lead in the ice layers of Greenland in the period between 150 BC and 50 AD comes from the mines of Rio Tinto, in the south-eastern part of Spain (Rosman et al., 1993).

During the Greek–Roman age, an important part of the fourfold increase in lead concentration in the troposphere over Greenland came from lead/silver mining and processing. During the Roman Empire, 40% of the lead production in the world was from Spain, Central Europe, Britain, Greece, and Asia Minor (Nriagu, 1983a). Lead was smelted in open furnaces, for which the rate of emission was not checked. The escaping small aerosol particles could easily reach the Arctic region, via the routes which have become known only recently (Hong et al., 1994).

After the fall of the Roman Empire, the atmospheric lead concentration suddenly dropped to the background level, which was characteristic of 7760 years ago. Then, in the Medieval and Renaissance Ages it began to increase again and 471 years before, it reached double the concentration that was detected during the Roman Empire (Boutron, 1995). In the 17th century scientists widely identified such illnesses in mining areas that were thought to arise from dispersion of toxic elements (Brimblecombe, 1987b). Afterwards, this increase was continuous following the Industrial Revolution. From the 1930s till about 1960, snow and ice samples in Greenland indicated a rapid increase. This can be traced back to the antiknock additives of leaded fuels, which were first used in 1923 (Nriagu, 1990). On a global scale, two-thirds of the leaded additives were used by the United States in the 1970s, 70% of which directly entered the atmosphere from the vehicle exhaust gases. Atmospheric lead concentrations measured in the 1960s were about 200 times higher than natural values. This is one of the most serious global pollution problems ever recorded (Boutron, 1995). The sudden decrease observed from 1970 can be traced back to an increasing use of unleaded fuels. Recently, all gasoline sold in the United States, and a gradually increasing ratio sold in Europe, is unleaded (Nriagu, 1990). Recently, Eurasia has been responsible for 75% of the global atmospheric lead concentration (Rosman et al., 1993).

Lead pollution in the atmosphere has been detected over the Antarctic since the beginning of the 20th century. The use of leaded fuels and then their reduction can also be detected. Furthermore, it can be established that an important part of anthropogenic lead comes from

South America (Boutron, 1995). At the same time, natural changes in lead concentration (and other heavy metals) were also considerable over the Antarctic during past ages. Low concentration values were detected in the Holocene period, while the lead concentration was two orders of magnitude higher than this during the last glacial maximum, about 20,000 years ago (Boutron and Patterson, 1986).

18.3 COPPER IN ANCIENT TIMES

18.3.1 Copper Mining and Exploitation

At first (about 7000 years ago), copper was produced from native copper (Fig. 18.3). This was the main technique for about 2000 years. Following this period, the discovery and introduction of a new smelting technique of oxide and carbonate ores as well as the appearance of tin-bronze initiated the real Bronze Age. Then, the production increased continuously. In the period between 2700 and 4000 BP (before the present), total production was about 500,000 MT (Tylecote, 1976; Morin, 2013).

Copper production suddenly increased in Roman times. In this period, copper alloys were used more and more intensively and frequently, both for military and civil purposes (e.g., minting). Production reached its maximum 2,000 years ago with a mass of about 15,000 MT/year. In this period, the main copper mines were situated in the territory of Spain (half of the total production of the world derived from this country, from the regions of Huelva and Rio Tinto), in Cyprus, and Central Europe (Hong et al., 1996a). The total production, in the period

FIGURE 18.3   A glacially rounded copper nugget, as an example of the raw material worked by people of the Old Copper Complex. Source: *By Rob Lavinsky (CC BY-SA 3.0). (https://eos.org/articles/miners-left-pollution-trail-great-lakes-6000-years-ago).*

between 2250 and 1650 BP, was about 5 million metric tons (Healy, 1988).

Generally speaking about any metals, peaks/decreases in production correspond to the booms and busts of the economy. This statement is valid for both the Roman Empire and China as well. A decrease in mining of all metal ores, including copper, started with the weakening of the Roman Empire. After the fall of the Empire, copper production decreased significantly in Europe. World production stagnated with a mass of about 2000 MT/year until the 8th century and then started to increase again. This increase in Europe is especially due to the opening of new mines in the 9th century in Germany, and in the 13th century in Sweden (the latter particularly in the region of Falun) (Pounds, 1990).

Outside the Roman Empire, important copper production occurred in Southwest Asia and in the Far East. When the Han dynasty (206 BC−220 AD) extended its influence over Southwest Asia, copper production of China was about 800 MT/year. In the medieval age, most of global production came from China (during the rule of the northern Sung dynasty). In this period, Chinese production reached its maximum of 13,000 MT/year, and this resulted in the peak world production of 15,000 MT/year in 1080s AD. Most of the copper was used for minting (Archaometallurgy Group and Beijing University of Iron and Steel Technology, 1978). For some hundred centuries after this period, production suddenly dropped (about 2000 MT/year in the 14th century), and then started to increase again from the Industrial Revolution until recently. (As a comparison: total global copper production was 10,000 MT/year at the beginning of the Industrial Revolution.) In Japan, pollution from the extensive production of copper used in the manufacture of giant Buddhist statues gave rise to extensive environmental pollution beginning in the 8th century (Satake, 2001).

18.3.2 Copper Pollution on Regional and Hemispheric Scales

Before the beginnings of anthropogenic use of copper, about 7000 years ago, total atmospheric copper was derived from natural sources and the situation did not change until 2500 years ago. Since 2500 BP, the atmospheric copper concentration has increased as a consequence of large-scale copper pollution in the northern hemisphere (Hong et al., 1996b).

Copper emissions since ancient times until the recent period have been the results of mining and metallurgical activities. Other anthropogenic activities (e.g., the production of iron and nonferrous metals, wood burning) contribute to these emissions only to a lesser extent.

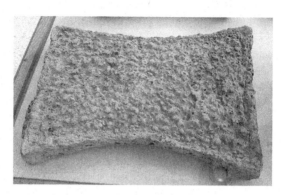

FIGURE 18.4   Copper age. A corroded copper ingot from Zakros, Crete, shaped in the form of an animal skin typical in that era. Source: *By Wikipedia user Chris 73 (https://commons.wikimedia.org/wiki/File:Minoan_copper_ingot_from_Zakros,_Crete.jpg).*

Emissions concerning production, in connection with a significant technological development, have considerably changed during the past 7000 years. In ancient times, due to the primitive smelting procedures, the emission factor was about 15% (Morin, 2013; Hong et al., 1996a,b). At the start, several extraction processing steps of sulfide ores (roasting, smelting, oxidation, and cleaning) were performed in open furnaces (Fig. 18.4). Emissions were ignored until the Industrial Revolution. From this time on, more developed furnaces and more advanced metallurgical procedures started to spread. Since the middle of the 19th century, the processing procedure has been reduced to five steps. These technological developments have resulted in a significant decrease in the emission factor. In the 20th century, this factor was only 1% and later, with the introduction of further modifications, it became a mere 0.25% (Hong et al., 1996a,b).

Since Roman times, the Cu/Al ratio has increased in ice samples, which indicates that considerable copper pollution occurred in the troposphere over the Arctic in this period. This copper might have originated during the high-temperature section of the processing as small-sized aerosol particles, and entered the atmosphere. These aerosols can easily reach the Arctic region, leaving the middle latitudes where they originated from (in the Roman times: mainly the Mediterranean basin, especially Spain; in the medieval ages: China).

Changes to the Cu/Al ratio in ice samples seem to correspond to the estimated changes of anthropogenic copper emission. Data derived from the ice cores in Greenland indicate low values until 2500 years ago, medium values from the Roman times until the Industrial Revolution, and suddenly increasing values near the recent period. Data from the Roman times show high variability. This can probably be traced back to the fact that in this period the production of copper occurred in short periods, with regard to how many copper coins were needed (Hong et al., 1996b).

According to the ice samples in Greenland, comparing production data with emission factors, atmospheric copper emission culminated twice before the Industrial Revolution. The first peak occurred in the golden age of the Roman Empire about 2000 years ago, with a mass of some 2300 MT/year, when the use of metal coins spread in the ancient Mediterranean. The second peak appeared in the golden age of the northern Sung dynasty (960–1279 AD) in China, at about 1080 AD, with a mass of some 2100 MT/year, when the Chinese economy was extensively developing and copper production increased. Since the smelting technology was primitive at that time, about 15% of the smelted copper got into the atmosphere. Though the total copper emissions of the Roman and Sung times was about a tenth of that in the 1990s, copper production did not reach even a hundredth of that in the recent period. Hemispheric copper pollution caused by copper emissions has a more than 2500-year-old history and copper emissions from the Roman and Sung times were as high as any time before 1750 (Hong et al., 1996a).

18.4 ENVIRONMENTAL AWARENESS IN ANCIENT ISRAEL

The environment is a natural issue of concern for Judaism. Much of the discussions center around the biblical commandment of "bal taschit," i.e., not to destroy, without purpose, any object from which someone might derive pleasure. Trees, fields, and rivers belonged to this circle. Jewish people knew that trees were very important and, for this reason, they prohibited the cutting down of trees around cities. Furthermore, trees were required to be watered and the environment to be cared for. Any form of luxury was prohibited, because luxury itself is a kind of waste. Beyond the prohibition of actual destruction, an entire series of laws deals with maintaining the general environmental quality of life. The Talmud requires:

1. that one must not open a shop in a courtyard if the noise pollution of customers will disturb his neighbors' sleep;
2. that one must put the pigeon cotes at least 50 cubit from the town walls, so that the poops would not damage the town's vegetable gardens;
3. that threshing floors must also be kept at this distance, to prevent the chaff from creating an air pollution problem for the city. So too carrion, graves, and tanneries also have this same distance requirement because of the odors they produce . . .

The fifth book of Moses is the basis of Jewish ecology. It specified, among other things, that soldiers were prohibited from relieving themselves on the field of the camp. They should leave the camp, then dig a hole, and when they are finished, they should bury it. It was prohibited to build a latrine near houses because they were malodorous. Since stench distributes in a different way in winter than in summer sewer channels were prohibited from opening during the summer. If anybody suffered from the sewage of another person, he/she could claim compensation. Sewage was not allowed to be released near kitchen gardens because it decreased the yields. Nature is natural and basic for Jewish people and they believe that one should live the same way even recently, as it was written in the Bible, Torah, and Talmud. The biblical cities in Israel were surrounded by a *"migrash"*—an area of 1000 cubit left for public enjoyment in which nothing may intrude. For this reason trees must be kept 25–50 cubit (depending on the species of tree and the amount of shade each species provides) from the city wall. Furthermore, according to the rabbis, the migrash may not be turned into a field, as it destroys the beauty of the city. Interestingly, a field cannot be made into a migrash, as it will diminish the crops. In the Temple service, olive wood and wood from grape vines are prohibited to use on the altar. One opinion holds that the reason is also of concern for the settlement and cultivation of the land of Israel. The second opinion is much more specific. These types of woods burn with a great deal of smoke, and air pollution is to be avoided. Jerusalem, as the holiest of cities, also had special environmental legislation designed to protect its unique environment for the enjoyment of its inhabitants and visitors. In that regard, all garbage was removed from the city, dunghills were prohibited from the city area, and no kilns were allowed to operate within its border. In this way, vermin and smoke were kept out of the city and the quality of life was improved (Mészáros, 2001). Tanning manufacturers should be settled at least 60 cubit from the city wall, because they were extremely smelly. They were prescribed to be built to the eastern side of the city, since in Israel, northerly and westerly winds were generally the most frequent, and this way, the stench would not return to the city. Mills were only allowed to be built at least 50 cubit from the city wall, because when they were operating much dust got into the air, which was harmful to humans when inhaled. Furthermore, wheat powder was described not only as unhealthy, but also as harmful for the fields. Therefore, mills should possibly be built far away from fields. On the other hand, one can read in the Talmud that smoke is not only bad and harmful, but destroys the Garden of Eden of God as well. Hence, the relationship between God and humans becomes worse and they draw away from each other. The Talmud also says that the soul of God lives in everything: in animals, plants, and stones, etc. Therefore, He must not be offended, because if God had wanted a smoky world he would have created it. In the law-book of Tosefta it was written that it was prohibited to wash in drinking water. Each well should be covered by a roof so that snakes, insects, and vicious souls could not attack the water in it. Sewage holes were not allowed to be dug near a neighbor's well. The Jewish lawbooks dealt with noise pollution, too. Millstones caused a lot of noise and vibration during work. For this reason, mills were not allowed to be established near the city. The operation of a school (it was considered to be a big one if it had at least 50 students) depended on the inhabitants of the neighboring houses. Namely, the noise of children could disturb the inhabitants (Bal Tashchit, 5th Book of Moses).

This raises the question of why rabbis dealt so much with the environment in the past and why this environmental -sensitivity was later pushed into the background. The answer might be that Jewish people didn't have farms for a long time and thus they didn't feel themselves close to the nature. Therefore, they didn't appreciate the value of the field as much as their ancestors did. But now that they can go back to their real homes, they perhaps will listen to the sounds of nature and their environment more carefully. According to the proverb, *"The clean Jewish people take better care of their environment, than the dirty Romans"* (Bal Tashchit, 5th Book of Moses).

Acknowledgments

The authors would like to express their gratitude to Claude F. Boutron (Laboratoire de Glaciologie et Géophysique de l'Environment du Centre National de la Recherche Scientifique, Unité de Formation et de Recherche de Mécanique Université Joseph Fourier, Grenoble, France) for the exceptionally comprehensive contribution, moreover thanks are given to Noa Feller, Ian Strachan, and Keith Boucher for their useful hints on the topic.

References

Archaometallurgy Group, Beijing University of Iron and Steel Technology, 1978. A Brief History of Metallurgy in China. Science Press, Beijing.

Bal Tashchit, 5th Book of Moses (Deuteronomy).

Boutron, C.F., 1995. Historical reconstruction of the Earth's past atmospheric environment from Greenland and Antarctic snow and ice cores. Environ. Rev. 3, 1−28.

Boutron, C.F., Patterson, C.C., 1986. Lead concentration changes in Antarctic ice during the Wisconsin/Holocene transition. Nature 323, 222−225.

Boutron, C.F., Görlach, U., Candelone, J.P., Bolshov, M.A., Delmas, R.J., 1991. Decrease in anthropogenic lead, cadmium and zinc in Greenland snows since the late 1960s. Nature 353, 153−156.

Boutron, C.F., Rudniev, S.N., Bolshov, M.A., Koloshnikov, V.G., Patterson, C.C., Barkov, N.I., 1993. Changes in cadmium concentrations in Antarctic ice and snow during the past 155,000 years. Earth Planet. Sci. Lett. 117, 431−444.

Boutron, C.F., Candelone, J.P., Hong, S., 1994. Past and recent changes in the large scale tropospheric cycles of Pb and other heavy metals as documented in Antarctic and Greenland snow and ice: a review. Geochim. Cosmochim. Acta 58, 3217–3225.

Brimblecombe, P., 1987a. The Big Smoke. A History of Air Pollution in London Since Medieval Times. Methuen, London, New York.

Brimblecombe, P., 1987b. The antiquity of smokeless zones. Atmos. Environ. 21 (11), 2485.

Brimblecombe, P., 1995. History of air pollution. In: Singh, H.B. (Ed.), Composition, Chemistry and Climate of the Atmosphere. Van Nostrand Reinhold, New York, NY, pp. 1–18.

Capasso, L., 2000. Indoor pollution and respiratory diseases in ancient Rome. Lancet 356, 1774.

Chun, Y., 2000. The yellow-sand phenomenon recorded in the Chosunwangjosilok. J. Meteorol. Soc. Korea 36, 285–292.

Colbeck, I., Nasir, Z.A., 2010. Indoor air pollution. In: Lazaridis, M., Colbeck, I. (Eds.), Human Exposure to Pollutants Via Dermal Absorption and Inhalation, 17. Springer, Dordrecht, pp. 41–72.

Costa, C.D.N., 1997. Dialogues and Letters: Seneca. Penguin Books, London.

Driver, G.R., Miles, J.C., 1952. The Babylonian Laws Legal Commentary. Clarendon Press, Oxford.

Eisinger, J., 1984. Lead in history and history in lead. Nature 307 (5951), 573.

Eisinger, J., 1991. Early consumer protection legislation—a 17th C law prohibiting lead adulteration of wines. Interdiscip. Sci. Rev. 16, 61–68.

Epstien, R., 1999. Pollution and the environment: some radically new ancient views. *Vajra Bodhi Sea: A Monthly Journal of Orthodox Buddhism.* 1, 30. pp. 36, 12; Pt. 2, 30 (2000), pp. 37–38; Pt. 3, 30 (2000), pp. 36–38, 46.

Epstein, R., 1992. Pollution and the environment. Vajra Bodhi Sea 30 (Pt. 1), 36. 12.

Fouache, E., 2013. Using the geo-aerchological approach to explain past urban hazards. In: Serre, D., Barroca, B., Laganier, R. (Eds.), Resilience and Urban Risk Management. Taylor & Francis Group, London, pp. 15–20.

Goldstein, E. (Ed.), 1988. Pollution. Social Issues Resources Series, Inc, Boca Raton, FL.

Hartwell, R., 1967. A cycle of economic change in Imperial China: coal and iron in northeast China, 750–1350. J. Econ. Soc. Hist. Orient 10, 102–159.

Healy, J.F., 1988. Mining and Metallurgy in the Greek and Roman World. Thames and Hudson, London.

Heidorn, K.C., 1978. A chronology of important events in the history of air pollution meteorology to 1970. Bull. Am. Met. Soc. 59, 1589–1597.

Hong, S., Candelon, J.P., Patterson, C.C., Boutron, C.F., 1994. Greenland ice evidence of hemispheric lead pollution two millennia ago by Greek and Roman civilizations. Science 265, 1841–1843.

Hong, S., Candelone, J.P., Soutif, M., Boutron, C.F., 1996a. A reconstruction of changes in copper production and copper emissions to the atmosphere during the past 7000 years. Sci. Total Environ. 188, 183–193.

Hong, S., Candelone, J.P., Patterson, C.C., Boutron, C.F., 1996b. History of ancient copper smelting pollution during Roman and medieval times recorded in Greenland ice. Science 272, 246–249.

Hughes, J.D., 1993. Pan's Travail. John Hopkins University Press, Baltimore, MD.

Karatzas, K., 2000. Preservation of environmental characteristics as witnessed in classic and modern literature: the case of Greece. Sci. Total Environ. 257, 213–218.

Makra, L., 2000. Wandering in China. Változó Világ 37. Press Publica Kiadó, Budapest (in Hungarian).

Mamane, Y., 1987. Air-pollution control in israel during the 1st and 2nd century. Atmos. Environ. 21 (8), 1861–1863.

Markham, A., 1994. A Brief History of Pollution. Earthscan, London.

McNeill, J.R., 2001. Something new under the sun. An Environmental History of the Twentieth-Century World. W.W. Norton & Company, New York, London.

Mészáros, E., 2001. A Short History of the Earth. Vince Publisher Ltd, Budapest (in Hungarian).

Mészáros, E., 2002. The mankind and the environment before the industrial revolution. História 5−6, 21−24 (in Hungarian).

Morin, B.J., 2009. Reflection, Refraction, and Rejection: Copper Smelting Heritage and the Execution of Environmental Policy. Ph.D. dissertation. Michigan Technological University.

Morin, B.J., 2013. The Legacy of American Copper Smelting: Industrial Heritage Versus Environmental Policy. University of Tennessee Press, Knoxville, TN.

Nriagu, J.O., 1983a. Lead and Lead Poisoning in Antiquity. Wiley, New York, NY.

Nriagu, J.O., 1983b. Occupational exposure to lead in ancient times. Sci. Total Environ. 31, 105−116.

Nriagu, J.O., 1990. Global metal pollution: poisoning the biosphere. Environment 32 (7), 8.

Pounds, N.J.G., 1990. An Historical Geography of Europe. Cambridge University Press, Cambridge, London.

Renberg, I., Persson, M.W., Emteryd, O., 1994. Pre-industrial atmospheric lead contamination detected in Swedish lake sediments. Nature 368, 323−326.

Rosman, K.J.R., Chisholm, W., Boutron, C.F., Candelone, J.P., Görlach, U., 1993. Isotopic evidence for the sources of lead in Greenland snows since the late 1960s. Nature 362, 333−335.

Satake, K., 2001. New eyes for looking back to the past and thinking of the future. Water Air Soil Pollut. 130 (1−4), 31−42.

Schäfer, E.H., 1962. The conservation of nature under the tang dynasty. J. Econ. Soc. Hist. Orient. 5, 299−300.

Scheidel, W., 2009. In search of Roman economic growth. J. Roman Archaeol. 22 (1), 46−70.

Strabo, 1988. The Geography of Strabo, vol. 5. Harvard University Press, 9780434992119p. 541.

Turco, R.P., 1997. Earth and Seige: From Air Pollution to Global Change. Oxford University Press, Oxford.

Tylecote, R.F., 1976. A Hystory of Metallurgy. Mid-County, London.

Waldron, H.A., 1973. Lead poisoning in the ancient world. Med. Hist. 17, 391−399.

Waldron, H.A., 1985. Med. Hist. 29 (1), 107−108. Available from: https://doi.org/10.1017/S0025727300043878.

Weeber, K.W., 1990. Smog über Attika: Umveltverhalten im Altertum. Artemis, Zürich.

Zeek, W.C., 1986. Technology and culture. In: Nriagu, J.O. (Ed.), Lead and Lead Poisoning in Antiquity, 1983. Wiley, New York, NY, pp. 129−130.

Poisoning in Ancient Rome: Images and Rules[*]

Evelyn Höbenreich[1] and Giunio Rizzelli[2]

[1]Institute of the Foundations of Law/Department of Roman Law,
University of Graz, Graz, Austria [2]Chair of Roman Law,
University of Foggia, Foggia, Italy

O U T L I N E

19.1 VENEFICIUM AND LEGAL TERMINOLOGY

Poisoning (*veneficium*) is a crime which is committed by the administration of *venena*: i.e., by substances or practices which may also belong to the sphere of magic and are apt to alter whatever they come into contact with (which includes both the organism of a person and his or her

[*]The ancient Greek and Latin sources have been translated into English by the authors of this article. For the translation of the text from Italian into English we are obliged to Sebastian Puchas and Marlene Peinhopf. We thank Aglaia McClintock for checking of the abstract.

mind). The activity of using *venena* is considered to be a *veneficium*, penalized, and, accordingly, punished, if—dependent on the period and the context—the effects caused by the *venena* are deemed harmful or censurable from an ethical point of view. Thus, a legislative enactment dating from the age of Sulla, the *lex Cornelia de sicariis et veneficis*, passed in 81 BCE, apart from containing provisions against persons going about armed with the intention of killing or thieving (*sicarii*), persecutes the *venefici*, i.e., the perpetrators of *veneficium*, an activity which consisted of the preparation or administration of *venena*. Its fifth chapter—as Cicero asserts (*Cluent.* 148)—orders the punishment of anybody who prepared, sold, bought, kept, or administered a noxious poison (*venenum malum*). It almost seems as if the legislator has enumerated the single criminal activities in the chronological order in which they succeed each other and which are, respectively, destined for the preparation, selling, acquisition, and hence possession, and, eventually, administration of the substance for the purpose of killing (Fig. 19.1).

The legal literature confirms the reference found in the speech of Cicero in defense of Cluentius. Indeed, references of a similar purport are to be found in the work of the jurist Marcianus at the beginning of the 3rd century CE, and in a later legal text, the so-called *Pauli sententiae* (PS) Marcian. 14 *inst.* D. 48.8.1.1: "Someone is also liable who confections ⟨and⟩ administers poison for the purpose of killing a man." Marcian. 14 *inst.* D. 48.8.3pr.-1: "Under chapter five of the same *lex Cornelia* on armed 'gangsters' or murderers and poisoners, someone is punished who makes, sells, or possesses a drug (*venenum*) for the purpose of homicide. § 1. The person who sells baneful medicines (*mala*

FIGURE 19.1   Lucius Cornelius Sulla Felix, approx. 54 BCE. *From https://www.coinarc-hives.com/e38c3b73fae7d2d88e347367085e500d/img/elsen/135/image00151.jpg (01.08.2018)*

medicamenta) to the public or possesses them for the purpose of homicide is liable to the penalty of the same law." And, eventually, PS 5.23.1: "The lex Cornelia imposes the penalty of deportation on a person who has kept, sold, prepared a poison (venenum) destined to procure the death of a man."

The assessment of the venenum as "baneful" (malum), which can be found in the Ciceronian text, concurs with the description found in the Marcianian text which asserts that the penalized behavior is calculated to cause the death of a (hu)man: although the formulations of Cicero and Marcianus do not coincide in every point, both indicate that a certain behavior comes within the scope of criminal law, if the preparation, possession, and administration of venenum is aimed at killing somebody. Furthermore, the jurist also attests to an extension of the law, because the punishment inflicted upon those who sell drugs to the public which have proved to be lethal or those who keep substances which can be used to cure, with the intention to kill, was the same as that affixed by the lex Cornelia with reference to the sicarii and venefici. Therefore, Marcianus does not use "malum," i.e. "baneful" with regard to venenum, because that adjective is reserved for "medicamenta" (mala medicamenta), in order to indicate that the substances or concoctions aimed at healing somebody can also have lethal effects: if someone intends to kill by their use, their possession is punished; if this intention does not subsist, but the person who has consumed them nevertheless perishes, the person who has sold them to the public is punished.

19.2 PERPETRATORS, TRIALS, STEREOTYPES

In the sources, the charge of poisoning is frequently brought against females. These cases may concern a woman, in general one of low social status, who professionally devotes herself to the preparation of venena, like Locusta, who provided the poison for Agrippina in order to kill Claudius, and for Nero to murder Britannicus: Tacitus (ann. 12.66 and 13.15) mentions her as notorious for her many crimes and as being held among the instruments of the realm. This charge, however, may also concern a person of higher status, even an empress, like Livia Drusilla, the wife of Augustus, who was suspected, rightly or wrongly, of being responsible for the deaths of Marcellus, Gaius, and Lucius, and even of the princeps i.e. the Roman emperor, himself, in order to assure the succession of her son, Tiberius Claudius Nero (cf. Tac. ann. 1.3, 5; Cass. Dio 53.33.4, 55.10a.8−10): in these passages there is no explicit reference to the act of poisoning, however, the instances of suspicion which are mentioned can only be connected with such deeds.

It has to be underlined that for a long time the notion of *veneficium* continued to be related to the sphere of magic and, consequently, the woman concocting or administering the poison was somehow seen as a sorceress. Besides, it is a commonplace in antiquity that all that is related to the female universe is generally entangled with sorcery, with the consequence that the way of seducing and "bewitching" a man could be conceived and described simply by recourse to magic arts. A case in point may be found in the passage in which Cassius Dio depicts Cleopatra while she is reducing Mark Antony to her slave by availing herself properly of those arts, in combination, however, with *eros* (Cass. Dio 49.33.4).

Long prior to the passing of the *lex Cornelia de sicariis et veneficis* a trial was documented which took place in front of the *comitia* (i.e., the Roman legislative assembly) in 331 BCE and ended with the conviction of 170 matrons. In the short account written by Valerius Maximus (2.5.3) at the beginning of the Principate, the case seems to be emblematic because it was destined to lead to a series of trials in connection with various kinds of acts considered to be cases of *veneficium*.

As Maximus explains, the victims of these crimes seem to have been the husbands of the female perpetrators. They were killed by stealth with poison, i.e., in a particularly insidious and detestable way, since they had no chance to defend themselves. Crimes of that kind were thought to be particular to feeble persons, incapable of confronting their opponent openly and directly. Indeed, what appears in this passage as a faint echo is commonly accepted in Greek and Latin literature, i.e., the opinion of woman's relatively weak physiology that plays a role in her presumed fragile psychology. As a short treatise on physiognomy attributed by the ancients to Aristotle (and which, at any rate, perfectly fits into the Peripatetic tradition) expounds, female animals, which are less strong and courageous than their male counterparts, are therefore all the more insidious (i.e., there exists a strong connection between weakness, cowardice, and a disposition to injustice: *Physiogn.* 809a, 26-810a, 8). On the other hand, in the above-mentioned treatise the connection between the occult character of the practice, and the malicious intention of those who perform it, appears to be a lesser misconduct with regard to that which the Latin author seems to suggest.

In the course of the centuries, however, this motive seems to have lost none of its strength, as indicated by the fact that at the end of the 19th century an Italian forensic doctor (Giuseppe Ziino) used it in commenting on a statistic which showed that there were more females than males among the accused, to explain that "the innate weakness of women, even more than their perfidiousness, induces them to use a weapon which kills insidiously and does not force the (female) murderer to fight openly with the victim." The topic is also much diffused among works by other famous authors, including Cesare Lombroso and Richard von Krafft-Ebing, where instead of scientific arguments we encounter simply commonly accepted assertions. Thus, in an influential physiological-criminological manual for physicians, jurists and sociologists, dating from 1902 and dedicated to the celebrated psychiatrist Emil Kräpelin by the author, Gustav Aschaffenburg, one reads: "All in all, the female crime is characterized rather by baseness, the male crime rather by brutality."

19.3 TRAINING FOR THE COURTS

These incidents do certainly not exhaust the cases of *veneficium* committed by women against their husbands. Here, however, another motive to which the sources refer fits in: that of the insatiable female erotic desire as a reason for committing *veneficium* in connection with adultery, which basically can also be described as a kind of *veneficium*: a *veneficium* which the adulteress commits against her own organism by introducing into it the blood of a male stranger, and which therefore has a corruptive influence. Thus, the corruption in this case is also an ethical one (shamelessness), which necessarily is connected with *veneficium* which, in turn, can be ascribed to shamelessness, according to a *circulus in probando* (i.e., circular reasoning) involving the body and mind. This can be further illustrated by an example of *ratiocinatio*, i.e., reasoning by positing a question, to be found in a compendium of rhetoric compiled in the first decades of the 1st century BCE, the *Rhetorica ad Herennium* (*Rhet. Her.* 4.23): "When our ancestors condemned a woman for one crime, they considered that by this single judgment she was convicted of many transgressions. How so? Judged unchaste, she was also deemed guilty of poisoning. Why? Because, having sold her body to the basest passion, she had to live in fear of many persons. Who are these? Her husband, her parents, and the others involved, as she sees, in the infamy of her dishonor. And what then? Those whom she fears so much she would inevitably wish to destroy. Why inevitably? Because no honorable motive can restrain a woman who is terrified by the enormity of her crime, emboldened by her lawlessness, and made heedless by the nature of her sex. Well now, what did they think of a woman found guilty of poisoning? That she was necessarily also unchaste. Why? Because no motive could more easily have led her to this crime than base love and unbridled lust. Furthermore, if a woman's soul had been corrupted, they did not consider her body chaste. Now then, did they observe this same principle with respect to men? Not at all. And why? Because men are driven to each separate crime by a different passion, whereas a woman is led into all crimes by one sole

FIGURE 19.2   Marcus Tullius Cicero, Musei Capitolini. https://de.wikipedia.org/wiki/
Marcus_Tullius_Cicero#/media/File:M._Tullius_Cicero_IMG_2414_B1.jpg (01.08.2018).
Cicero - *By Freud [GFDL (http://www.gnu.org/copyleft/fdl.html), from Wikimedia Commons.*

passion." That is what a young contemporary Roman preparing to
embark on a forensic career learned (Fig. 19.2).

The stereotype of the adulteress-*venefica*-sorceress also seems to
assume concrete characteristics in some women who were accused in
famous trials held at the beginning of the Principate. Thus, under the
reign of Tiberius, among the charges against Lepida, sentenced for
having simulated a birth with the intention of giving her husband a
false heir, are those of having committed adultery, *veneficium*, and of
having consulted Chaldean astrologers on the destiny of the imperial
family (Tac. *ann.* 3.22–3.23). However, Claudia Pulchra is persecuted
for having committed adultery and *veneficia* against the *princeps*, and for
having performed *devotiones* or magical arts (Tac. *ann.* 4.52.1 and 3).

Naturally, men may also commit the crime of poisoning, and the
ancient authors confirm that fact in abundance, but, as far as the cases
of *veneficium* are concerned, it seems nevertheless more probable that
they have been committed by a woman. This has been observed by
Quintilian in his work on the training of orators, where he deals with
probative arguments, and what factors should be used in their presentation, such as the sex of the perpetrator. Indeed, he asserts, it seems to be
easier, for example, to believe that an act of banditry (*latrocinium*) was
committed by a man and an act of *veneficium* by a woman (Quint. *inst.*
5.10.24). It is again Quintilian who returns to the argument of the
adulteress who kills by poison. When claiming the authority of wise

men and famous citizens as particularly efficient witnesses, he quotes a sentence by Cato (Quint. *inst.* 5.11.39): "If an adulteress is on her trial for poisoning, is she not already to be regarded as condemned by the judgment of Marcus Cato, who asserted that every adulteress was as good as a poisoner?" Indeed, the orators elaborate this motive by investigating its different outcomes, as shown by several Latin declamations (exercises which were considered indispensable for those who, between the end of the Republic and the first years of the Principate, decided to dedicate themselves to the art of forensic oratory) which have come down to us.

19.4 JURISTS AND THE INTERPRETATION OF LAWS

Even if not intended, poisoning is often caused by females. This fact is recorded by the jurisprudential literature which deals with situations arising in connection with the storage, handling and provision of *venena*. Marcianus mentions a *senatus consultum*, probably dating from the 1st century CE, concerning the case of a woman guilty of having administered a medication to another woman, which was intended to facilitate her conception (the jurist speaks of a drug *"ad conceptionem"*), but which, however, eventually caused her death. The woman is found guilty by the senate and, in consequence of the sentence which concludes the trial, condemned to relegation (antique punishment, a milder form of exile or death penalty).

The passage is the following (14 *inst.* D. 48.8.3.2): "But the addition of 'baneful drug' indicates that there exist certain drugs which are not baneful. The term (*venenum*) is therefore neutral, covering as much a drug prepared for the purpose of healing as one for the purpose of killing, as also that which is called an aphrodisiac. However, only that (kind of drug), which is possessed for the purpose of killing a man, is disapproved by the law. It is, however, ordered by *senatus consultum* that a woman who, not admittedly with malicious intent but giving a bad example has administered a fertility drug from which the recipient had died, shall be relegated."

The senate punishes the woman's deportment because it has caused the death of a person. This punishment, however, according to Marcianus—although it seems probable that the senate wanted to make a clear statement against practices deemed socially dangerous—is based upon the fact that the action is considered a "bad example," certainly not because it was aimed at killing her who consumed the medication: therefore it goes beyond the original scope of the Sullan law, which contained provisions against all who made, sold, bought, kept, or administered poison for the purpose of killing.

For a full understanding of the jurist's discourse it is, however, necessary to keep in mind what he states immediately prior to that, namely, that for the *lex Cornelia* to be applied, the concocted potion must qualify as *venenum malum*. In the case in question, indeed, the medication is not considered as *malum*, i.e., baneful and aimed at causing death. This clearly emerges from the fact that, as Marcianus argues, *"venenum"* is a neutral designation which may refer either to substances for the purpose of healing or to those for the purpose of killing. As the jurist points out, also the *amatorium*, an aphrodisiac draught, is considered to be a *venenum*: an explanation which seems to indicate that the use of such drugs was common at that time. This is, by the way, confirmed by another *senatus consultum* mentioned in the subsequent paragraph of the same fragment, which extends the *lex Cornelia* to the careless administration (which, therefore, probably, had caused some harm) of substances like the mandragora, and the salamander, which are also known for being aphrodisiacs, by dealers in cosmetics, spices, and unguents— the *pigmentarii*.

The *senatus consultum*, as can be read in Marcianus' fragment (14 *inst.* D. 48.8.3.3), supplements and generalizes the provision in § 2 by extending the penalty of the *lex Cornelia* to the careless (*temere*) administration of substances which in principle are apt to cure but which, due to their dangerousness, may also have baneful effects. The jurist's enumeration is not to be seen as exhaustive: "It is laid down by another *senatus consultum* that dealers in cosmetics, spices and unguents are liable to the penalty of this law if they recklessly hand over to anyone hemlock (*cicuta*), salamander (*salamandra*), monkshood (*aconitum*), pinegrubs (*pituocampis*), a beetle (*buprestis*), mandragora (*mandragora*), Spanish fly (*cantharis*) and whatever is prepared to cure a person." (Marcian. 14 *inst.* D. 48.8.3.3). As all of the mentioned substances can be used as medicines as well as poison (some also as aphrodisiac), and their handing over is punished only if done carelessly, it must be presumed that they have been given as *venena bona*, i.e., therapeutically. However, since the administration is not effected with the intent to kill, how can one verify whether it was performed in a careless way? The answer is: only if it has lethal consequences or causes a serious deterioration in someone's health.

Marcianus' observation that apart from baneful *venena*, there do exist other *venena* which are not baneful (which serves to explain that one may also be liable for the preparation or the administration of a drug without the incriminated action being premeditated) is confirmed by Roman legal literature, in an explanation by Gaius, at some prior time, in his commentary on the XII Tables (4 *ad l. XII tab.* D. 50.16.236pr.).

The *venenum*, explains the jurist, must be qualified as "benign/inoffensive" (*bonum*) or "baneful/harmful" (*malum*), because the term *venena* (drugs) embraces also *medicamenta* (medical substances, medicaments). Gaius continues that *"venenum"* is, in general, something which is capable of changing the nature of that which it comes into contact with. Similarly, he notes, also in the Greek language *"phármakon"* refers to both, substances which may cure, and those which may harm, i.e., to the medication as well as to the poison. In order to support his own argument, Gaius calls to witness a very respectable source, namely, a Homeric verse (*Od.* 4.230), taken from the description of the drugs Helen was taught to prepare by the Egyptian Polydamas, a woman, hailing from a country where a great variety of such substances were produced. The double function which the *phármaka* may exert indeed seems to be a motive which, in the period in which Gaius writes, appears to have been popular in Greek literature for centuries. An example in case is to be found precisely in connection with Helen, in the *Encomium*, written towards the end of the 5th century BCE by the sophist Gorgias of Leontini, in which the author, maybe with reference to the medical knowledge of the epoch, distinguishes among *phármaka* which cure a sickness and those which kill (§ 14) (Fig. 19.3).

The jurists' efforts to define the notion of *venenum*, which is of importance for the purpose of the *lex Cornelia de veneficis*, is justified by the circumstance that there exists an obvious connection between the nature of the administered (concocted, sold, kept) substance and the criminal intent of the person who intends to commit *veneficium*, in the sense that

FIGURE 19.3 "Witches", Mosaic from the Villa Cicerone in Pompeii, III-I BCE, Museo Archeologico Nazionale Napoli. https://commons.wikimedia.org/wiki/File: Pompeii_-_Villa_del_Cicerone_-_Mosaic_-_MAN.jpg (01.08.2018). Villa del Cicerone Mosaic - *By WolfgangRieger, via Wikimedia Commons.*

the intent of the perpetrator seems to be more obvious if that substance is a poison, i.e., a *venenum malum*. On the other hand, one and the same *venenum* is basically apt to be used for different purposes, be they legal (aimed at healing) or illegal (aimed at impairing the health of a person, potentially also causing her/his death), dependent on the administered quantity, the type of ingestion, etc. The mandragora, for example, can be used as a soporific, but, if consumed in an abusive way, may also have lethal consequences. This double function of the *venena* presents an impetus for the orators to introduce a series of complex facts which may provoke sophisticated arguments, illustrating how a substance meant to cure has lethal effects or a poison has healing effects, like the border case of the consumption of icy water which caused the death of the person to whom it was administered. Thus it seems obvious that on the one hand the criminal intent (i.e., the intention to kill) of the perpetrator must be proved in order to be able to declare her/him liable in terms of the *lex Cornelia*, taking into account a whole range of facts which do not limit themselves to the question of simply determining the possession of the *venenum*. However, on the other hand, the necessity emerges to limit the diffusion of life-threatening practices by persecuting the use of *venena* which, contrary to the intention of the person who administered the substance, has proved to (potentially) have lethal effects: a hypothesis which does not pertain to the original wording of the Cornelian law.

In any case, the legal provisions in relation to the careless administration of certain medications, as described in the fragment by Marcianus, have been subjected to several modifications over the course of time. At least that is the case with an *amatorium*, the administration of which, together with that of an abortifacient, is penalized (beginning at a certain point in history which can no longer be determined), if it does not cause the death of the person who consumed it, because it is considered a "bad example." That is confirmed in the PS (5.23.14 = Paul. 5 *sent.* D. 48.19.38.5): "Those who administer an abortifacient or aphrodisiac draught, even if they don't do it with bad intention, are still condemned, because the deed sets a bad example, if of lower rank to the mines, if of higher status to relegation to an island with the forfeiture of part of their property. But if for that reason a man or woman died, they are punished with an atrocious form of death penalty."

Thus, in this case—contrary to what occurred at the time of Marcianus—the mere administration of abortifacients and aphrodisiac draughts was penalized despite not resulting in death, because they were considered *mali exempli res*. The inflicted punishment varied, however, depending on whether the perpetrators were persons of lower rank (*humiliores*) or persons of higher position (*honestiores*)—categories which were very prominent in the PS. If in consequence of such an administration of poison somebody dies, the perpetrator is destined to be subjected to the *summum supplicium*, the death penalty. It is unknown whether this applies also to fertility drugs, *ad conceptionem*, which are mentioned by the Severan jurist. Maybe the precise reference to the abortifacient or to the *amatorium* can be considered as an indicator. This may be due to the fact that the Pseudo-Paulinian text was inspired by a normative provision concerning a particular case (but it is also possible that the abortifacient and the aphrodisiac draught were just mentioned as examples). At any rate, the case in point seems indicative of a legal policy intending to offer severe responses to a more efficient social control of activities—so it might be presumed—of those professionally preparing and dealing in drugs and unguents of this mentioned type.

However, we must not forget that as background to these events significant to the reconstruction of the legal practice in ancient Rome, ghosts have been conjured up by poets, which have given rise to fears and anxieties haunting the collective imagination for centuries. Consider, for example, the tragic figure of Phaedra, who—in the description of Propertius (2.1.51−2.1.54)—devotes herself in vain to the preparation of potions in order to make her stepson fall in love with her, or that of Medea, a passionate woman, but also a dangerous sorceress, furious and fatal in her striving for revenge.

Further Reading

Aschaffenburg, G., 1923. Das Verbrechen und seine Bekämpfung. Einleitung in die Kriminalpsychologie für Mediziner, Juristen und Soziologen; ein Beitrag zur Reform der Strafgesetzgebung, 3. Aufl. C. Winter, Heidelberg.

Botta, F., 2014. Osservazioni in tema di criteri di imputazione soggettiva dell'*homicidium* in diritto romano classico. Diritto@Storia 12, http://www.dirittoestoria.it/12/tradizione-romana/Botta-Imputazione-soggettiva-omicidio-diritto-romano.htm [03.07.2018].

Cavaggioni, F., 2004. *Mulier rea*. Dinamiche politico-sociali nei processi a donne nella Roma repubblicana. Istituto Veneto di Scienze, Lettere ed Arti, Venezia.

Cherubini, L., 2010. *Strix*. La strega nella cultura romana. UTET Libreria, Torino.

Cilliers, L., Retief, F.P., 2000. Poisons, poisoning and the drug trade in ancient Rome. Akroterion 45, 88−100. http://akroterion.journals.ac.za/pub/article/view/166 [03.07.2018].

Cilliers, L., Retief, F.P., 2015. Poisons, poisoning, and poisoners in Rome. Med. Antiqua http://www.ucl.ac.uk/ ~ ucgajpd/medicina%20antiqua/sa_poisons.html [03.07.2018].

Cloud, J.D., 1969. The primary purpose of the Lex Cornelia de sicariis et veneficis. Zeitschrift der Savigny-Stiftung für Rechtsgeschichte. Romanistische Abteilung (ZSS.RA) 86, 260−283.

Cloud, J.D., 2009. Leges de sicariis: the first chapter of Sulla's lex de sicariis. Zeitschrift der Savigny-Stiftung für Rechtsgeschichte. Romanistische Abteilung (ZSS.RA) 126, 114−155.

Crawford, M.H. (Ed.), 1996. Roman Statutes, vol. II. Institute of Classical Studies, School of Advanced Study, University of London, London.

Ferrary, J.-L., 1991. Lex Cornelia de sicariis et veneficis. Athenaeum 79, 417−434.

Höbenreich, E., 1988. Due senatoconsulti in tema di veneficio (Marcian. 14 *inst.* D. 48.8.3,2 e 3). Archivio Giuridico "Filippo Serafini" (AG) 208, 75−97.

Höbenreich, E., 1990. Überlegungen zur Verfolgung unbeabsichtigter Tötungen von Sulla bis Hadrian. Zeitschrift der Savigny-Stiftung für Rechtsgeschichte. Romanistische Abteilung (ZSS.RA) 107, 249–314.

Kaufman, D.B., 1932. Poisons and poisoning among the Romans. Classical Philol. 27, 161–164.

von Krafft-Ebing, R., 1906. Psychopathia Sexualis. With Special Reference to the Antipathic Sexual Instinct. A Medico-Forensic Study. Rebman, London–New York, NY.

Lombroso, C., Ferrero, W., 1895. The Female Offender. D. Appleton & Co., New York, NY.

Longo, G. (Ed.), 2008. [Quintiliano], La pozione dell'odio (Declamazioni maggiori, 14–15). Edizioni Università di Cassino, Cassino.

Nardi, E., 1971. Il procurato aborto nel mondo greco romano. Giuffrè, Milano.

Nörr, D., 1986. Causa Mortis. Auf den Spuren einer Redewendung. C.H. Beck, München.

Querzoli, S., 2011. About "raptus" and "veneficium" in Marcian's Institutions. Ostraka 20, 83–94.

Querzoli, S., 2012. Se i veleni non sempre uccidono: veneficia e lex Cornelia nelle Istituzioni di Marciano. Annaeus 5, 55–72 (= Ostraka 21, 165–176).

Redl, G., 2005. Die fahrlässige Tötung durch Verabreichung schädigender Substanzen im römischen Strafrecht der Prinzipatszeit. Revue Internationale des Droits de l'Antiquité (RIDA) 52, 309–324.

Rizzelli, G., 2013. In: Rodríguez López, R., Bravo Bosch, M.J. (Eds.), Note sul veneficium. Mulier. Algunas Historias e Instituciones de Derecho Romano. Editorial Dykinson, Madrid, pp. 1–20.

Wacke, A., 1979. Fahrlässige Vergehen im römischen Strafrecht. Revue Internationale des Droits de l'Antiquité (RIDA) 26, 505–566.

Ziino, G., 1890. Compendio di medicina legale e giurisprudenza medica secondo le leggi dello Stato ed i più recenti progressi della Scienza ad uso de' medici e giuristi, third ed. Società editrice libraria, Milano.

"Gleaming and Deadly White": Toxic Cosmetics in the Roman World[*]

Susan Stewart
Independent Librarian, Edinburgh, United Kingdom

The range of cosmetic products available to the ancient Romans was extensive and included foundations, face powders, antiwrinkle creams, hair dyes, eye make-up, rouge, breath fresheners, deodorants, and hair removers.[a] Cosmetics were used largely, though not exclusively, by women who, in applying these products, endeavored to attain the perfection embodied in the literary descriptions and visual representations of fictional women, imagined goddesses, and often idealized members of the social elite. In accordance with contemporary standards, women

[*]See Nicander *Alexipharmica* 2.74ff.

[a]There is no conclusive evidence that the Romans used lipstick.

strove to make their faces look pale with just a hint of pink, their eyes seem large, their hair an attractive color, their skin smooth, and their bodies free from unwanted hair. In reality, some of the make-up applied to achieve this desired "look" not only had the potential to enhance, alter, conceal, or remove but could also be poisonous. The general vocabulary relating to make-up reflects this ambiguity; *medicamentum*, meaning cosmetics in Latin, refers not only to cosmetics and to medicines but also to poisons and even enchantments. Similarly, the Greek word *pharmakon* pertains to medicines or substances either taken inwardly in the form of oils or draughts or topically in the form of ointments but can also refer to a poisonous drug, charm, or spell.[b]

The main toxic substances used in make-up, either as beauty products in their own right or as ingredients in beauty products, were the various minerals; in particular, lead, antimony, mercury, and arsenic. In this chapter, I examine the facts, and the fiction, that surround the use of these minerals concentrating on the period of the Roman Empire, when the expansion of Roman domination brought flourishing trade, even with countries beyond the borders of the Roman Empire, and encouraged familiarity not only with actual ingredients or products themselves but also with the ideas and practices of other cultures in respect of cosmetics.

A good part of our scientific or quasiscientific information relating to the Romans' use and understanding of these dangerous substances comes from the *Historia Naturalis*, the encyclopedic work of Pliny the Elder; note that, in turn, some of his knowledge was gleaned from the Materia Medica of Dioscurides. There are also a number of medical texts, written in Greek (as much medical learning was Greek in origin and many in the profession were Greek) but also in Latin, that make reference to make-up; for example, the works of Galen, Scribonius Largus, and Celsus. Other writers, from the Satirists (Juvenal, for instance) to the love poets (among them, Ovid and Propertius) and indeed the dramatists (Plautus and Terence) include reference to make-up as a matter of social comment and a vehicle for humor. That is to say, nonmedical writers often included references to cosmetics in their text because this topic was a useful rhetorical tool for the purposes of defining the feminine and at the same time defaming the character of men. Moreover, to all intents and purposes, the literary evidence we have for cosmetics was penned by men, which contributed to the gendered bias and denied women, who used cosmetics most, a voice. In tackling the subject of ancient cosmetics scholars must appreciate the importance of the literary rhetoric while, at same time, endeavoring to separate the fact

[b]Petr.126, Ov.*A.A.*3.205(cosmetic), Cic.*Pis* 6.13 (medicine) Suet. Calig. 50 (enchanted potion) Varr.*Ap* Non.345,23; Liv 8.18. (poison). For *pharmakon* see Plat. Phaed.

FIGURE 20.1   Toilette scene funerary relief from Neumagen, Germany, 3rd century AD; Rheinische Museum, Trier. *From Wikimedia commons.*

(where that is possible) from the fiction. On the plus side however, there is plenty of written material to work with indicating familiarity both on the part of the writer, and of his audience, with make-up.

Aside from the literary evidence, an abundance of visual images and archeological remains that pertain to the matter of cosmetics survive. Contemporary artwork, including paintings, mosaics, sculptures, and funerary reliefs also reflect familiarity with beauty products. Note, however, that there are plenty of rhetorical messages (relating to wealth, status, and gender, for example) here too. Women are depicted at their toilette with all the paraphernalia that making-up entailed; servants, mirrors, boxes, bottles, palettes, mixing spoons, and the rest (see Fig. 20.1). There is, however, virtually nothing among the surviving images that can be said categorically to show women actually wearing make-up. Furthermore, men do not appear anywhere in the visual record associated with cosmetics. Turning to the archeological evidence, the pots (*pyxides*) and bottles (*unguentaria*) that contained beauty products in use at this time survive in abundance. Where we are lucky enough to find some residue inside these vessels, this can now be analyzed using modern noninvasive techniques including synchrotron radiation and mass spectrometry.[c] Ancient cosmetics is an area of study that, until relatively recently, merited scant attention, being given little

[c]*See* Ribechini et al. (2011).

if any space in scholarly books where the topic might have been relevant and somewhat more attention in books for the general reader where inaccurate information was often reiterated by different authors, thereby allowing fiction to appear as fact. However, due in part at least to the influence of feminism and an increasing interest in gender studies some valuable work has been done on make-up in the last few decades: I refer to the work of Olson, Richlin, and Wyke.[d] Having made these points as regards research into ancient cosmetics in general, I move on to discuss toxic cosmetics specifically.

20.1 A FAIR COMPLEXION

Psmithium, *also that is* cerussa, *or lead acetate is produced at lead works . . .it is useful for giving women a fair complexion.* **Pliny the Elder NH.34.176**

White lead (*cerussa* in Latin, in Greek *pysmithion*) was manufactured, in the classical period, by steeping lead shavings in vinegar. Women applied the resultant white powder to make their faces appear pale, a matter not only of fashion, but of status. A pale complexion could suggest to the onlooker that the individual upon whom he or she had settled their gaze did not spend too much time outside; that is, she was not a working class woman but instead belonged to the upper echelons of Roman society. Equally, a pale complexion could imply that the woman sporting it was healthy, fertile, and therefore likely to make a good marriage partner; while exposure to lead has a cumulative effect corroding the surface of the skin as well as causing potential damage to the central nervous system and to the main internal organs, lead poisoning can also result in infertility; so much for the assumption that a pale complexion might indicate success in producing a healthy son and heir.

Although there were certainly safer foundations on the market including, for example, kaolin (*creta*) and "white earths" such as *chia terra*, white lead (*cerussa*) is most popularly referred to in the written texts. However, to what extent white lead was used as make-up compared with any other substance applied for the same purpose in the Roman world remains unknown. In the archeological record, finds of white lead are not uncommon. Lumps of white lead have been found in cosmetics boxes from excavated graves from the Hellenistic period for example (see Fig. 20.2).[e] However, samples found at Pompeii (and

[d]For a general introduction to the topic of Roman cosmetic see Stewart (2007). For work on the nature of cosmetics themselves form a historians point of view see Olson (2009) and Stewart (2012). For gender matters and the discourse of rhetoric and reality in respect of cosmetics see Wyke (1994) and Shear (1936).

[e]See Shear (1936).

FIGURE 20.2   Cosmetic container with tablets of white lead found in a grave in Athens dating from 5th century BC. *From Wikimedia commons.*

therefore dating from the 1st century AD), as well as samples from other sites in the west from the Roman period, show a prevalence of more natural chalk-based alternatives.[f] Although these data are fascinating, there is little we can draw from them in terms of the scale of use at any particular time. Nor indeed can we establish any preference for applying a particular product. In short, we cannot assume, on the basis of the analysis of some samples whose survival is, after all, pure chance, that by the time of the Roman Empire white lead was being rejected in favor of other substances.

What we do know is that those in the medical profession, including Celsus and Galen, along with Pliny the Elder, and others with some medical knowledge, were aware of the dangers of lead. While Celsus expounds on the healing properties of this substance, as a treatment for wounds, headaches, and joint pain, for example, he also includes remedies to counteract its poison too.[g] Pliny also acknowledges that while *cerussa* "is useful for giving women a fair complexion... it is a deadly

[f]Welcomme et al. (2006).

[g]2.42.166 (wounds); 1.272, 1.458 (headache and joint pain) and requiring an antidote 2.122.

poison" (Plin. *HN* 34.176). Vitruvius, writing about the use of lead in a wider context as a building material, also notes that "cerussite in particular is said to be injurious to the human system" (Vit. *De Arch.* 8.6.10−11).

Although awareness of these hazards is clearly expressed in some of the written texts, we cannot be sure how far the dangers of lead were understood by the general public. Poets and playwrights, Martial and Plautus, for example, concentrate on putting across the rhetorical message that wearing make-up reflects the immorality of the wearer and their lack of status, rather than referring to the dangerous nature of particular cosmetics. In love poetry, drama, and satirical verse, it is often old women who are noted as wearing white lead to conceal their wrinkles, the telltale signs of aging; part of the rhetoric portraying women as devious characters while also stressing the unattractiveness of old age. It is interesting to speculate that, in the case of some women at least, the appearance of age might have been as much the result of the toxic effects of wearing lead make-up as it was a true reflection of advancing years, though this is of course impossible to verify.

Early in the 20th century scholars argued that lead poisoning was a significant factor in the collapse of the Roman Empire. This theory was based on the belief that exposure to lead resulted in infertility and a falling birth-rate among the Roman aristocracy. The argument in favor of this theory reemerged in 1983 expounded by geochemist Jerome Nriagu.[h] It is certainly true that the people of the Roman world, in particular those living in the cities across the Empire, would have been exposed to lead in many areas of their lives. Aside from applying lead as a cosmetic, their water supply flowed through lead pipes, they cooked in lead vessels, applied plasters containing lead for medical purposes, and even ingested lead in their wine using it as a preservative. However, with the exception of one possible rather earlier account of lead poisoning noted by the Greek poet Nicander, writing in the 2nd century BC, together with evidence of relatively high lead levels found in exhumed bones from the Roman period, we have no definite record of lead poisoning until the 7th century AD.[i] The theory that lead poisoning contributed significantly to the collapse of the Empire is largely discredited by scholars today and the scale of the health impact from the use of lead argued for in earlier research is believed to have been overestimated. Indeed, when Nriagu raised these theories in the 1980s, they were quickly refuted by Scarborough, an eminent classicist and pharmacologist.[j] In dismissing this argument here, I point out that exposure to

[h]Nriagu (1983).

[i]Nicander *Alexipharmica* 1.600. Paulus of Aegina 3.64.

[j]Scarborough (1984).

lead through cosmetics was but a small part of the ancients' contact with this potentially dangerous substance.

20.2 ROUGE

She's blushing!—Yes, modesty suits a pale skin but it is better put on. The real thing can be a nuisance *Ov.* **Am.3.7.7**

Not only white but red pigments, such as red lead or lead tetroxide (*minium*), were used as make-up. This was applied as rouge. Cinnabar (*cinnibaris*), or red mercuric sulfide, another brilliant red, was also used to heighten the color of the cheeks and create a complexion that could be compared to the subtle colors of nature. The poet Ovid describes the ideal female complexion as follows; "In her face the lily and the rose are glowing still—snow white pale red." (Ov. *Am* 3.35−6).

Both red lead and cinnabar were known poisons at this time. Indeed inhaling the dust or powder from these beauty products may have been a potential health hazard not only for the women wearing these cosmetics but also for the maidservants tasked with applying them. Cheaper and safer alternatives did exist; for example, the dregs of red wine (*faex*), the red dye extracted from the roots of alkanet (*anchusa*) a type of borage, and the juice of the mulberry (*morum*). According to Ovid, one woman even rubbed her cheeks with poppies steeped in cold water.[k]

What then was the appeal of known dangerous substances, such as red lead and cinnabar? As with white lead, we do not know how far the general populace were aware of the dangers of these substances; some may have used them unaware of the health hazards. However, I offer an alternative explanation. Red lead was imported (largely from Spain) and cinnabar brought to the cities of the Mediterranean from Spain and India. Red pigments such as these, sometimes coming from beyond the boundaries of the known world in antiquity, were commonly believed to be the congealed blood of dragons. The far-flung origins of these rouges and the stories that built up around them gave these products considerable exotic appeal. Their exclusivity, even scarcity when compared with the rather more readily available supply of left-over wine or mulberries, bestowed on the owner and user both glamor and status. The pursuit of this illusion of grandeur, if you like, imparted by the enthusiastic use of such mysterious products as red lead and cinnabar might have encouraged the use of these substances, overriding any consideration of the dangers.

[k]Ov. *De Med. Fac.* 100.

20.3 EYE MAKE-UP

In the same mines as silver there is found what is properly to be described as a stone, made of white and shiny but not transparent froth; several names are used for it, stimi, stibi, alabastrum *and sometimes* larbasis **Plin. HN. 33.101.**

Powdered antimony sulfide, another toxic substance, toxic not only when ingested but also dangerous when absorbed through the skin, had been a popular black eye make-up used for brows and lashes and to define eyes since Ancient Egyptian times; the Egyptians called this *mesdemet*. Both Dioscorides and Pliny the Elder describe antimony (*stimmi* or *stibium*) as an eye cosmetic. According to the former, *stimmi* was "a good paste of stibnite is a cosmetic... enlarger of the eyes" (Dios.1. 555). Pliny the Elder noted that: "Antimony has astringent and cooling properties but is chiefly used for the eyes ...in beauty washes for women's eyebrows it has the property of magnifying the eyes." (Plin. *NH* 33.102). An eye make-up known as *fuligo* consisting, in safe form, of either soot or lampblack, could also be manufactured, in a potentially harmful variety made from powdered antimony. *Galena*, another eye make-up also used from ancient Egyptian times was poisonous even to the touch. Made from malachite, a lead ore, mixed with silver, this was also applied in powdered form. Soot (*favilla*) no doubt a cheaper and much less prestigious option in terms of eye make-up, was easily obtainable from the spitting oil lamps common to the city brothels and, by implication, and on account of its ready availability, probably used by lower-class women. Eye make-up, whether powdered antimony or soot, or galena, was applied much like modern-day eyeliner or mascara and, maybe for ease of application could be mixed with water or perhaps scented oil such as oil of roses. In the archeological record, eye make-up containers often consist of two tubes joined together. One would have held the powdered antimony or lampblack and the other the liquid, whether water or oil, for mixing. The addition of scented oil might have added to the attractiveness of eye make-up (in terms of smell), though of course if it were based on antimony, it was no less toxic.[1]

The frequency with which archeologists come across oculists' stamps attests to the proliferation of eye diseases at this time. Eye infections were caused by a number of factors including heat and dust as well as a lack of understanding as to how bacteria spreads. Some of treatments prescribed for eye complaints contained some of the same toxic ingredients or were the same toxic products that were applied as make-up.

---

[1]Juvenal gives us an acutely observed description of a transvestite applying soot with a damp pin. Juv.2.93.

A lead sulfide mixture known as *collyrium* was recognized as both an eye salve with healing properties as well as being used as a general term for eye liner. Recent research has established that in fact eye make-up containing lead, mercury, or lead compounds was not all bad news.[m] Indeed these products may have had an antibacterial effect and the ancients seem to have understood these benefits.

20.4 HAIR REMOVERS

How nearly was I recommending to you that there should be no shocking goat in the armpits and that your legs should not be rough with harsh hair

Ov *Ars. Am.* 3.194

At this time, underarm hair was taboo for both sexes, being associated with bad odor and poor hygiene. Pumice stone was used for removing unwanted hair. Tweezers too served this purpose and are common site-finds for this period. As these tools are even excavated in numbers at army camps where the community would have been predominantly male, we can safely conclude, especially given the accepted code regarding body hair at this time, that men did indeed use these toilette instruments to remove hair. According to Seneca, hair plucking was a service available to men when they visited the public baths.[n] Ideally, the entire surface of women's bodies should be hairless. The use of depilatory creams and pastes was associated predominantly with the removal of hair from the female body. However, Pliny the Elder does claim to be "ashamed to confess that the chief value now set on resin is for a depilatory for men" (Plin. HN 24. 124).[o] While using a pumice stone or pulling your hair out with a set of tweezers was essentially safe and effective even if it might irritate the surface of the skin, depilatory pastes included some toxic concoctions, in particular arsenic (*arsenicum*). There was also the option of orpiment (*auripigmentum*), yellow arsenic, the stuff, when it came to murder by poisoning, of the modern detective novel. Furthermore, one might apply *psilothrum*, a toxic mixture of arsenic and quicklime, green in color.[p] Antimony was also an ingredient in hair removers. Some, like orpiment, were believed

[m]Cf. Murube (2013). See also Tapsoba et al. (2010).

[n]Sen. Ep.56.2.

[o]Note also his remark that "Depilatories I myself indeed regard them as a woman's cosmetic, but now today men also use them" (Plin. *HN* 26.164).

[p]Mart. 6.93.

to have a cleansing effect.[q] Pliny confuses us when it comes to how some of these substances might have worked, stating that "before using any depilatory the hairs must first be pulled out." (Plin.32.137). It is possible that these concoctions were intended to stop the hair growing in again rather than simply to remove the surface hair there already. Certainly, most of these products were not only toxic but caustic; the astringent qualities of such products would have stripped off skin as well as hair, if left on for any length of time.

To conclude, some ancient cosmetics were indeed potentially dangerous, especially if used regularly over a prolonged length of time. In practical terms, without any record of extensive poisoning as a result of using these products we cannot draw any real conclusions as to the how these risks materialized. We do know that the ancients continued to use such products despite at least some knowledge of the dangers. Perhaps we can liken this usage to our own predisposition to smoke despite being aware of the damage this can cause to our health. Ideologically, the exclusive status bestowed on the individual female through the use and ownership of specific beauty products, may have been encouragement enough to take the risk. Although the flip side of this argument might be that, rhetorically speaking, these substances could be seen to poison woman as a gender, corrupting her physical morality in their application and mirroring her innermost flaws. However, as those who condemn cosmetics outright (satirists, some poets, and playwrights) do not dwell on the poisonous nature of these substances this argument is not really expressed. Certainly, in order to gain any sense of reality vis-a-vis the use of harmful beauty products by the Romans we need to treat the evidence very carefully, teasing any suggestion of real life from the rhetorical message to which this subject matter is so firmly attached. However, it is the rhetoric that in a sense is a clue to the reality surrounding the use of toxic make-up at this time. That is to say, the scholar must understand that the rhetoric surrounding cosmetics is also among the reasons for their use.

References

Murube, J., 2013. Ocular cosmetics in ancient times. Ocul. Surf. 11 (1), 2—7.

Nriagu, J., 1983. Lead and lead poisoning in antiquity. N. Engl. J. Med. 308, 660—663.

Olson, K., 2009. Cosmetics in Roman antiquity: substance, remedy, poison. Classical World 102 (3), 291—310.

Ribechini, E., Modugno, F., Perez-Arantegui, J., Columbini, M.P., 2011. Discovering the composition of ancient cosmetics and remedies: analytical techniques and materials. Anal. Bioanal. Chem. 401 (6), 1727—1738.

[q]Celsus 5.5.

Scarborough, J., 1984. The myth of lead poisoning among the Romans: an essay review. J. Hist. Med. 39, 469–475.

Shear, T.L., 1936. Psmythion. In: Shear, T.L. (Ed.), Classical Studies Presented to Edward Capps. Princeton University Press, Princeton, NJ, pp. 314–317.

Stewart, S., 2007. Cosmetics and Perfumes in the Roman World. Tempus, Stroud.

Stewart, S., 2012. Cosmetics and perfumes in the Roman world: a glossary. In: Harlow, M. (Ed.), Dress and Identity. British Archaeological Reports, Oxford.

Tapsoba, I., Arbault, S., Walter, P., Amatore, C., 2010. Finding out Egyptian gods' secret using analytical chemistry biomedical properties of Egyptian black makeup revealed by Amperometry at single cells'. J. Anal. Chem. 82 (2), 457–460.

Welcomme, E., Walter, P., Van Elslande, E., Tsoucaris, G., 2006. Investigation of white pigments used as make-up during the Greco-Roman period. Appl. Phys. A 83 (4), 551–556.

Wyke, M., 1994. Woman in a mirror: the rhetoric of adornment in the Roman world. In: Archer, L., Fichler, S., Wyke, M. (Eds.), Women in Ancient Societies: An Illusion of the Night. Macmillan, London/New York, NY, pp. 134–151.

Further Reading

Richlin, A., 1995. Making up a woman: the face of Roman gender. In: Eilberg Schwartz, H., Doniger, W. (Eds.), Off with Her Head: The Denial of Women's Identity in Myth Religion and Culture. University of California Press, California/Berkeley.

21

Cherchez la Femme: Three Infamous Poisoners of Ancient Rome

Louise Cilliers

Honorary Research Associate, Department of Greek, Latin and Classical Studies, University of the Free State, Bloemfontein, South Africa

OUTLINE

Toxicology in Antiquity
DOI: https://doi.org/10.1016/B978-0-12-815339-0.00021-4

The reign of the Julio-Claudian emperors in the 1st century CE was characterized by political instability, tension, and intrigues in Rome. The upper class had yet to become accustomed to the transition from Republic to Empire, and to the absolute power of an emperor. Cases of murder and poisoning were frequent. Poisons played a prominent role in the imperial court; to accept an invitation for dinner (especially by the emperor) was dangerous, about as good as signing one's own death warrant. Murders were, however, committed among all classes of society to procure an inheritance, to eliminate a husband or stepson, or to get rid of one's enemies (Kaufman, 1932: 158). The satirist Juvenal, describing the moral decay of the elite, claimed that poisoning for personal benefit became a status symbol (*Sat.* 1 73–76).

21.1 STEREOTYPES

During this era three women gained notoriety for their expertise in poisoning: Locusta, Martina, and Canidia. It seems an anomaly that the perpetrators of such an insidious type of crime were often women—supposedly the weaker sex—who killed for hire (Juvenal, *Sat.* 6.659–661). However, the perception that *veneficium* (poisoning) was particular to weak people who cannot defend themselves (such as women)[1] was commonplace in Greek and Latin literature: "it is rooted in the cultural representation of the female physiology and psychology" (Höbenreich and Rizzelli, 2015: 44). This is already the view proposed in a short treatise attributed to Aristotle (*Physiognomy* 809a. 26-810a8). Poisoning is also related to women in the sense that the poisoner, who had the knowledge to concoct the poison, was regarded as a kind of sorceress; this correlates with the general perception of the female universe as a magical sphere.

21.2 SOURCES

Our knowledge of poisons available during Roman times is derived from the writings of various authors living at the time. Theophrastus, associate of and successor to Aristotle as head of the Lyceum (4th century CE), was the first to identify plants with medicinal (and poisonous) properties (Smith, 1952: 154). In the 1st century CE Dioscorides wrote his famous *De materia medica*, which superseded all existing literature in classifying remedies and drugs from the animal, vegetable, and mineral kingdoms. This work, which dealt with close to 1,000 drugs, became the standard text for centuries to come. Information on poisons can also be gleaned from the

---

[1]Cf. Rudyard Kipling's poem "The female of the species is more deadly than the male."

writings of Scribonius Largus (1–50 CE), Pliny the Elder (23–79 CE), and the poet Nicander of Colophon (2nd century BCE).

Incidents of poisoning reached a peak during the reign of the Julio-Claudian emperors, in particular that of Claudius (reigned 41–54) and Nero (reigned 54–69). For this period we are fortunate to have detailed descriptions by the historians Tacitus, Suetonius, and Dio Cassius of several deaths by poisoning. In some cases we also have an indication of the poison used and its effect.

21.3 INCIDENTS OF POISONING IN WHICH THE THREE INFAMOUS WOMEN WERE INVOLVED: LOCUSTA

Locusta was the most infamous of the women under discussion. Apparently of Gallic origin, she was convicted of many crimes during Claudius' reign, but not immediately executed. She prepared the poison which killed Claudius as well as Britannicus, and we can also assume that the murder of Silanus, the popular governor of Asia, was due to Locusta's poisons with Agrippina acting as the prime instigator (Tac. *Ann.* xiii.1; Dio Cassius lxi. 6.4; RE 26:769). Ironically, she also supplied a poison which Nero wanted to use to commit suicide after learning that the Senate was calling for his death. However, it was either removed from the casket by his household slaves, or he did not have the courage to take the poison and instead instructed one of his slaves to run him through with a sword (Suet. *Nero* 47). Tacitus reports, regarding Locusta, that "she was skilled in such matters...had lately been condemned for poisoning and had long been retained as one of the tools of despotism" (*Ann.* xii. 66). She continued to ply her trade throughout Nero's rule (Suet. *Nero* cc. 33 and 47) and was condemned to death by Nero's successor, the Emperor Galba (Dio Cassius 64.3.4).

21.3.1 The Murder of Claudius

It is believed that when Agrippina, the second wife of Claudius, wanted to dispatch her husband to enable her son, Nero, to become emperor, she approached Locusta to prepare a poison for him (the Jewish historian, Josephus, however, only mentions a rumor of poisoning in *Jewish War* 20.8). Tacitus states that Agrippina decided that Locusta should prepare "some rare compound which might derange his mind (i.e., to make him unaware that he was being poisoned) and to delay death" (*Ann.* xii.66). Claudius was, at that stage, already 65 years old, sickly, and a heavy drinker. There is agreement among the sources

FIGURE 21.1   A bust of the emperor Claudius.

that the emperor was to attend a priestly banquet on the October 13, 1954, and, after that, a dinner in the palace, at which the eunuch, Halotus was the taster or *praegustator* whose task it was to taste all food offered to the emperor.[2] Mushrooms, Claudius' favorite dish, were on the menu that evening (Fig. 21.1).

Hereafter the historians' versions differ. According to Dio Cassius (lxi.34) Claudius was carried out of the banqueting hall after the dinner, apparently drunk (which seems to have happened often), and then poisoned by Agrippina in the imperial quarters in the palace. He died in the course of the night.

---

[2]Tasters [slaves or freedmen] had become so common in this time that they even formed a *collegium* guaranteeing benefits for widows after the taster's death.

Tacitus, however, relates (*Ann.* xii. 66–7) that the poison was infused into or poured over the mushrooms (hence Nero's sick joke later that mushrooms are the "food of the gods"). This would of course have involved Halotus. The effect of the poison was not immediately perceptible, due to the emperor's intoxicated condition. His bowels were relieved, which seems to have saved him. Agrippina was dismayed and called on the emperor's trusted physician, C. Stertinius Xenophon, who must have been drawn into the conspiracy only at this late stage. Under the pretense of helping the emperor to vomit, he introduced into his throat a feather smeared with a rapid poison, after which Claudius died. Aconite, which causes immediate death, could have been used, but there are scholars who believe that Claudius was killed by a poison inherent in the mushroom itself, the *Amanita phalloides* (Grimm-Samuel in Horstmanshoff, 1999: 40).

Suetonius (*Claud.* c. 44) mentions various versions of the incident, among them that it was the emperor's taster, Halotus, who, at the banquet with the priests, allowed the poisoned mushrooms to be served to the emperor. Another version is that Agrippina herself, at a family dinner, mixed the poisons into the mushrooms and served them to Claudius. Reports as to what followed also differ: some say that he "became speechless as soon as he swallowed the poison, and after suffering excruciating pain all night, he died just before dawn." Others say that "he fell into a stupor, then vomited up the whole contents of his overloaded stomach and was given a second dose, perhaps in a gruel...or from a syringe..." to evacuate his stomach.

21.3.2 The Poisoning of Britannicus

Nero immediately succeeded Claudius as emperor although it was really his half-brother, Britannicus, Claudius' own son, who was the legitimate heir to the throne and thus a threat to Nero's position. Suetonius and Dio Cassius describe the poisoning of Brittanicus, a crime among several more in which Locusta played a role.

Dio Cassius (lxi.7.4) and Tacitus (*Ann.* xiii.17) both recount that Nero, who was a (bad) artist, was jealous of Britannicus' beautiful voice and greater popularity, and that he had on several occasions tried to seduce Britannicus to humiliate him, but that it was Agrippina who forcefully reminded Nero that Britannicus was the rightful heir and a threat, and urged Nero to take some action against him. Tacitus further relates (*Ann.* xiii. 15) that Nero, "having no charge against his brother and not daring openly to order his murder," directed that poison be prepared through the agency of the tribune Julius Pollio, "who had in his custody a woman under sentence for poisoning, Locusta by name, with a vast reputation for crime." The poison was administered to Britannicus by his tutors, but it was either too weak, or he eliminated it through his bowels. Locusta was flogged by Nero himself and charged "that she had prepared a medicine instead of a poison" (Suet. *Nero* c.33). She thus prepared, before his eyes, "a rapid poison of previously tested ingredients." Suetonius tells us that Nero first tried it on a goat, but as it lingered for five hours, he had her strengthen the mixture and then gave it to a pig, which instantly fell dead (*Nero* c.33). This, then, was the dosage administered to Britannicus at the second attempt. While dining with Nero and other aristocratic friends, Britannicus' food was tested by the *praegustator* as usual, and thereafter a hot drink made of wine and water (already tasted), was served to him, but since it was too hot, the young prince refused it; cold water in which the poison had been mixed was then poured in. Tacitus states that "this so penetrated his entire frame that he lost alike voice and breath" (*Ann.* xiii.16). Nero, remaining unperturbed among the ensuing commotion, casually remarked that this was a common experience—Britannicus often had such seizures since he suffered from epilepsy since his infancy—thus after a brief pause the company resumed their mirth. Agrippina, however, who was clearly unaware of the intrigue, was terrified, and fearing for her own life, realized that "here was a precedent for matricide" (*Ann.* xiii.16). Britannicus' lifeless body was then quickly prepared for the funeral, and buried in the Campus Martius "amid storms so violent that in popular belief they portended the wrath of heaven" (*Ann.* xiii. 17). A story is told by Dio Cassius (lxvi.7.4) that Nero had the body covered with plaster of Paris to hide the livid effect of poisoning, but that the rain washed it off.

Suetonius (*Nero* c.34) concludes the story of this crime by stating that Nero rewarded Locusta for "her eminent services with a pardon and large estates in the country, and actually sent her pupils",[3] presumably to learn her trade. She was eventually put to death by the emperor Galba (Dio Cassius 64.3.4).

21.4 MARTINA

The infamous Syrian poisoner, Martina, was involved in the poisoning (if poisoning it was) of Germanicus, the son of the emperor Tiberius' brother, Drusus. In this case we do not have as many details about the poisoning itself—apart from the fact that it happened in Syria

[3]Cf. Juvenal *Sat.* 1.69–72: "Do you see that distinguished lady? She has the perfect dose/ for her husband—old wine with a dash of parching toad's blood./Locusta's a child to her: she trains her untutored neighbors/to bury their blackened husbands. Ignore the gossip"

where Germanicus was posted, and there were no direct witnesses. Our main source, Tacitus, "who admired Germanicus and disliked Tiberius" (Scullard, 1976: 280), obviously used this incident to highlight Tiberius' jealousy of his popular nephew, Germanicus[4] and to highlight the increasing unpopularity of Tiberius.

Germanicus was recalled from Germany after three campaigns there, and sent to the Eastern provinces where there was unrest. At the same time, Tiberius sent a new governor to Syria, Gnaeus Calpurnius Piso (who was on good terms with the emperor), inter alia to keep an eye on the young prince—Tiberius did not want ambition for military glory to lure Germanicus into a war with Parthia. While in Syria, Germanicus visited Egypt, probably only to view the antiquities, but thereby transgressing the ruling that no senator was allowed to visit Egypt without the emperor's permission. On his return to Syria, Germanicus found that Piso had canceled most of his arrangements about ruling the province efficiently; this led to savage insults from both sides; Germanicus sent Piso a formal letter renouncing their friendship, and ordered him to leave the province. In this period Germanicus became ill, and died (19 CE) soon after Piso had left the province, all the while believing that he had been poisoned by Piso (Tac. *Ann.* ii.69). Germanicus' widow, another Agrippina, sailed to Rome with her husband's ashes; Piso, who lingered near the coast to stay informed about Germanicus' health, unwisely reentered Syria after he heard about his death, ostensibly to take over the command of what he thought was a vacant province; however, the newly appointed acting legate forced him to leave. Tacitus relates (*Ann.* ii.74) that the poisoner, Martina, who was believed to be implicated in Germanicus' death, was sent to Rome to stand trial there, but that she had suddenly died at Brundisium; "that poison was concealed in a knot of her hair, and that no symptoms of suicide were discovered on her person" (*Ann.* iii.7). It was thus believed that she had been murdered to prevent her from giving any information on who had presumably hired her to poison Germanicus (Fig. 21.2).

In addition to Germanicus and Piso being at odds as described above, Agrippina and Plancina, Piso's wife (and an intimate friend of Livia, the mother of Tiberius), were not on good terms either. Tacitus tells us that Martina, "a woman infamous for poisonings in the province," was "a special favorite of Plancina." How and where the poisoning took place we are not told, but it is clear that if it was poisoning, it caused a lingering illness in which ill intent can only with great difficulty be proved. We are further told that Germanicus' body, "before it was burned, lay

[4]Cf. Suetonius, *Tiberius* c. 52: "As to Germanicus, he (sc. Tiberius) was so far from appreciating him, that he made light of his illustrious deeds as unimportant, and railed at his brilliant victories as ruinous to his country"

FIGURE 21.2   Busts of Agrippina and Germanicus.

bare in the forum at Antioch," and that "it is doubtful whether it exhibited the marks of poisoning" (Tac. *Ann.* ii.73), for men gave conflicting opinions, depending upon whether they pitied Germanicus or were biased toward Piso.

In Rome, Tiberius, due to the difficulty of the inquiry and all the rumors, referred the case to the senate. From the accusation of poisoning Piso was acquitted, since the senate stated that they could not be convinced that there had been treachery about the death of Germanicus (*Ann.* iii.14). However, he was found guilty of unlawfully entering a province. He committed suicide before being sentenced, only entreating Tiberius that his son be saved. Plancina, equally detested, was indeed saved, thanks to her friendship with Livia (*Ann.* iii.15).

Suetonius has little to say about the death of Germanicus, but adds (*Tib.* cc.52–53) that when Germanicus' wife, Agrippina, who was as popular with the people as her husband, complained about her husband's death, she was exiled to the island Pandateria, where she continued her complaints, and was then on the order of Tiberius "beaten until one of her eyes was destroyed." She subsequently perished of self-inflicted starvation (*Tib.* c.53). Her sons, Nero and Drusus, who were equally popular with the people, were also, after many false allegations, banished, and starved to death (*Tib.* c.54).

21.5 CANIDIA

Canidia is not connected with any specific poisoning and is mentioned only in a few of Horace's Epodes and Satires, where she is
depicted as an evil woman who was involved in magic and poisoning.
The scholiasts say that Canidia was really called Gratidia, and that the
name Canidia was derived from her gray hair (Latin *a canitie*), and that
she was an old flame of Horace's with whom he had quarreled, but
Page (1973:473) disagrees and states that such guesses are futile.

In Epode 5 Canidia ("a mannish lustful hag...with her locks and
disheveled head entwined with tiny adders") is preparing a charm
("gnawing her uncut nails with her bile-black teeth") with which to
secure the affections of the aged Varus, and for this charm a young boy
is to be killed ("his marrow and liver, excised and dried, will serve for
an aphrodisiac potion"). Three other witches help her to prepare the
other ingredients ("a nocturnal screech-owl's feathers and eggs smeared
with the blood of a nasty toad") and to dig the pit in which the boy is
to be buried. When Varus, the adulterous old man, angry because the
original charm did not work, does not come to her door, Canidia starts
to prepare a more potent philtre to fire his passion. Then the boy, panic-
stricken because of the threat to kill him, and seeing that prayers are
idle, curses the witches and threatens them with solemn execrations
and the vengeance of his ghost: that the witches are to be flung out
beyond the city's boundaries and left unburied for carrion-birds to feed
on. Page (1973: 473), who clearly does not have much appreciation for
this Epode, states that it "is hardly more than an immature attempt to
depict one of those scenes which were popular with ancient readers."

In Epode 17 Horace, by focusing on Canidia, presents us with some
of the atrocities committed by those practicing magic. The epode is in
the form of a dialog in which Horace admits his defeat in his constant
battle against Canidia, and sarcastically entreats Canidia's pity; he even
recants his denial of the power of witchcraft. She replies that his entreat-
ies are in vain, since he publicly disclosed her mysteries and made her
the talk of the town. She states that her workshop is aglow with poisons
and magic spells and that she is burning certain objects so that, as they
are consumed, Horace may also be consumed (burning an image of the
victim or something that belongs to him, is typical in witchcraft). She
boasts that she can prepare "very swift poisons," but that Horace
must not imagine that she will use them on him, for whom there awaits
"a more lingering doom."

We meet Canidia again in Satire 1.8. The old burial ground below
the Esquiline Hill has been converted into lovely gardens by
Maecenas, but the witches, "those hags who are forever plagueing the

souls of men with their spells and potions," still visit it to gather bones and deadly plants. Horace states that he sees Canidia, "walking barefoot, her black robe tucked up and her hair streaming free, shrieking with the elder witch Sagana, their faces both made hidious by a deathly pallor." The nearby statue of Priapus describes how he/it witnessed the nocturnal witchwork of the two hags: they "scrape away the earth with their nails, then, taking a black lamb, they set about tearing it to pieces with their teeth, letting the blood trickle into the trench, from where they mean to summon the spirits of the dead to answer their questions." The statue of Priapus suddenly lets out a fart, which sends the two witches scattering, leaving behind their false teeth, hair, and charms.

In Satire II.1 56 there is a reference to hemlock, which could be the poison preferred by Canidia, mentioned a few lines earlier: a scoundrel will not get rid of his elderly mother by cutting her throat or strangling her, but "a cocktail of honey and hemlock will finish the old girl off" (line 56).

Satire II.8 contains a short reference to Canidia in a description of a disastrous dinner party when the guests dash off without tasting a thing, "as if the banquet had been blighted by Canidia, whose breath is more deadly than that of an African snake"—the breath of certain serpents was supposed to be deadly (Columella 8.5.18).

In the prescientific world of Horace these hair-raising descriptions of magic would have found appreciative readers—whether the poet himself believed them or not.

21.6 IDENTIFICATION OF POISONS

The question arises as to what kind of poisons Locusta used. Three kinds of poisons were distinguished in antiquity in terms of their effect:

1. A slow poison, which is applied over time and affects the brain;
2. A slow poison which affects the body; and
3. A rapid poison which is intended to bring about the death of the victim as quickly as possible (Horstmanshoff, 1999: 39).

Our knowledge of poisonous substances known to the Romans is derived from the writings of Dioscorides, who differentiates between drugs from the animal, vegetable, and mineral kingdoms. Vegetable poisons were best known and most frequently used: those most mentioned in ancient sources are poison hemlock (*Conium maculatum*, the poison which killed Socrates), monk's hood (*Aconitum*), deadly nightshade (*Atropa belladonna*), thorn apple (*Datura stramonium*), autumn crocus (*Colchicum autumnale*), black henbane (*Hyoscyamus niger*), mandrake (*Atropa mandragora*), and hellebore (*Helleborus niger*). Strychnine was unknown to the Romans, and there is no evidence that cyanide was extracted from the kernels of fruits. The ancients could not really distinguish between edible and poisonous mushrooms. The commonest poisonous mushrooms were the *Amanita muscaria*, *A. pantherina*, and *A. phalloides*.[5]

However, in a culture of high superstition when scientific ignorance about its effects made the detection of poisoning virtually impossible, it was very difficult to prove incidents of poisoning. It is likely that many poisoners went unpunished, while innocent citizens were wrongly condemned. Furthermore, many of the "poisons" were also used as medications, differing only in the dose given.

21.7 A FORENSIC INVESTIGATION

In Agrippina's attempt to end Claudius' life, she ordered Locusta to prepare a slow poison which would "derange his mind and delay death" (see above). We do not hear about this slow poison again, which was supposed to take gradual effect, but which, administered in the first attempt, poured over a *delectabilis boletus* (a delicious mushroom), was ineffective either because of Claudius' intoxicated condition, or because he voided his bowels. Thereafter, his physician smeared a feather in his throat, with the excuse that it would make him vomit any poison, but which, since it was dipped in a rapid poison, killed him instantly. It could have been aconite, a cardiotoxin which causes paralysis and kills rapidly without discomfort or pain. This was probably also the poison Cleopatra used, hidden in a ring or hairpin, to commit suicide (Retief and Cilliers, 1999: 102). Theophrastus considered aconite (also known as wolfsbane) the deadliest of all the poisons; Ovid referred to it as "the stepmother's poison."

This was also the pattern of the poisoning of Britannicus—diarrhea thwarted the first attempt. When Nero raged at Locusta for preparing "a medicine instead of a poison," she said in excuse "that she had given a smaller dose to shield him of the odium of the crime. He replied: 'It is likely that I am afraid of the Julian law'" (Suet. *Nero* c.33).[6] It is interesting to note that Locusta here reveals some knowledge of the

---

[5]Dioscorides thought that an unhealthy environment could make harmless mushrooms toxic (iv.82), whilst Pliny (NH 22.92–99) said that they became poisonous once a snake had breathed on them.

[6]This would have referred to the Lex Cornelia de sicariis et veneficis of 82 BC, enacted by the dictator L. Cornelius Sulla to protect the citizens against assaults by terrorists (*sicarii*), poisoning and black magic.

dose–response effect—the dividing line between poisoning and healing was a thin one.[7] Recall Nero and Locusta's experiments, mentioned above, in which they gradually increased the dose of a poison until they reached an amount which killed a pig immediately.

Can the information that the ancient sources provide about Britannicus' symptoms help us to decide what poison was used? Tacitus states that "this so penetrated his (sc. Britannicus') entire frame that he lost alike voice and breath" (*Ann.* xiii.16). He must also have had seizures, since Nero stated Britannicus suffered from epilepsy since his infancy. Which of the poisons previously mentioned produce such symptoms? The symptoms of henbane are "rapid onset of dry mouth, tachycardia and a progression of neurological symptoms varying from sedation to delirium, hallucination, mania, paralysis, coma and death" (Cilliers and Retief, 2001: 91–93). Thorn apple and deadly nightshade present basically the same symptoms, but the symptoms of aconite seem to come closest to those of Britannicus: "extremely poisonous, even in small doses. Rapid onset of numbness and paresthesia of the mouth and throat which spreads over the rest of the body; pain and twitching of the muscles, progressing to general weakness, cold and clammy extremities, arrhythmia and hypotension, respiratory paralysis, drowsiness, occasionally convulsions, stupor and death" (Cilliers and Retief, 2001: 92).

Tacitus mentions that the cremation of Britannicus took place in great haste, on the same night. The excuse given by Nero, that mourning should be kept to a minimum for those who died prematurely, seems plausible enough at first sight, but Britannicus was already 14 and had assumed the *toga virilis* (i.e., a white toga presented to boys to indicate their transition to manhood). Dio Cassius records that the *communis opinio* (i.e., the commonly held opinion) about the hasty funeral was fear of the discovery of poisoning[8]—it was commonly believed that poisoning would cause the skin of the victim to develop dark patches and turn a hue the color of lead.[9] To obscure this livid coloration suggestive of poisoning, Nero ordered the body covered with plaster of Paris. There was a heavy rain, though, when Britannicus' corpse was carried to the pyre and it was washed off (Dio Cassius 61.7.4). However, modern science

---

[7]Arsenic, for instance, was a byproduct of lead, silver, and goldmines; it was found that, if heated, it became a beautiful red pigment which was used by artisans. But arsenic was also used extensively as medicine for various ailments (Cilliers and Retief, 2001: 95).

[8]Cf. also that Germanicus' body was exposed in the forum in Antioch before cremation. "Whether it bore marks of poisoning was disputable: for the indications were variously read, as pity and preconceived suspicion swayed the spectator to the side of Germanicus, or his predilections to that of Piso" (*Ann.* ii.73.5). "Indications of black magic in Germanicus' house (*Ann.* ii.69) aroused suspicion."

[9]Juv. *Sat.* i. 73–76; Dio Cassius lxi.7.4; Galen, *De locis affectis* vi.5, viii.422–3K.

does not support the ancients' view that poisoning would cause the skin of a corpse to turn livid (Lewin in Horstmanshoff, 1999: 41 n. 9).

21.8 CONCLUSION

We have already observed that the sociopolitical circumstances in the 1st century CE with the transition of Rome from Republic to Empire, when ambition and political intrigues were the order of the day and emperors who abused their absolute power came to the throne, could have played a significant role in the increase in cases of poisoning and suicides. This period is very well documented, and Tacitus' and Suetonius' dramatic descriptions of the poisoning of members of the ruling class like the emperor Claudius, Germanicus, and Britannicus, made a lasting impression. In the 2nd century CE during the reign of the so-called Five Good Emperors, when the Empire reached its peak, very few cases of poisoning are reported—perhaps because this period was not so well documented, or because the stress and fear characteristic of the previous century have made place for peaceful circumstances under able emperors. However, in the 6th century, during the reign of Justinian (518–602), further steps had to be taken to suppress assassination and poisoning, and the aid of a physician was required in the investigation of all cases.

References

Cilliers, L., Retief, F.P., 2001. Poisons, poisoning and the drug trade in ancient Rome. Akroterion 45, 88–100.

Höbenreich, E., Rizzelli, G., 2015. Poisoning in ancient Rome: the legal framework, the nature of poisons, and gender stereotypes. In: Wexler, P. (Ed.), History of Toxicology and Environmental Health. Toxicology in Antiquity, Volume II. Elsevier, Amsterdam.

Horstmanshoff, M., 1999. Ancient medicine between hope and fear: medicament, magic and poison in the Roman Empire. Eur. Rev. 7 (1), 37–51.

Kaufman, D.B., 1932. Poisons and poisoning among the Romans. Class. Philol. 27, 156–167.

Page, T.E. (Ed.), 1973. Q. Horatii Flacci. Carminum Libri IV. Epodon Liber. Macmillan Education Ltd, Basingstoke & London.

Retief, F.P., Cilliers, L., 1999. Die dood van Kleopatra. Suid-Afrikaanse Tydskrif vir Natuurwetenskap en Tegnologie 18 (3), 100–103.

Smith, S., 1952. Poisons and poisoners through the ages. Medico-Legal J. 20 (4), 153–167.

Further Reading

Beck, L.Y., Trans., 2005. Pedanius Dioscorides of Anazarbus. De Materia Medica. Olms-Weidman, Hildesheim.

Cary, E., 1925. Dio Cassius. Roman History, Vol. VIII. Loeb Classical Library, Harvard University Press, Cambridge, MA (Books 61–70).

Church-Brodribb, 1966. (Trans.). Tacitus. The Annals and the Histories. The New English Library, Macmillan & Co., New York.

Furneaux, H., 1897. The Annals of Tacitus, Vol. II. Clarendon Press, Oxford.

Gensel, Canidia, 1899. In: Band VI., J.B. (Ed.), RE: Paulys Realencyclopädie der classischen Altertumswissenschaft. Metzlersche Verlagsbuchhandlung, Stuttgart.

Green, P., 1967/1998. Juvenal. The Sixteen Satires. Penguin Classics.

Hughes-Hallett, L., 1999. Cleopatra. Histories, Dreams and Distortion. Pimlico, London.

Jones, W.H.S., 1951. Pliny. Natural History, Loeb Classical Library, Vol. VI. Harvard University Press, Cambridge, MA (Books 20–23).

Kipling, R., 1919. Rudyard Kipling's Verse. Inclusive Edition, 1885–1918. Hcton, London.

Lucusta, K.Z., 1927. In: Band XXVI., J.B. (Ed.), RE: Paulys Realencyclopädie der classischen Altertumswissenschaft. Neue Bearbeitung. Metzlersche Verlagsbuchhandlung, Stuttgart.

Palmer, A. (Ed.), 1971. The Satires of Horace. Macmillan, London.

Retief, F.P., Cilliers, L., 2000. Poisons, poisoning and the drug trade in ancient Rome. Akroterion 45, 88–100.

Rolfe, J.C., Trans., 1965/1979. Suetonius, Volume I and Vol. II. Loeb Classical Library, Harvard University Press, Cambridge, MA.

Rudd, N., 1979. Horace. Penguin Classics, Satires and Epistles.

Rutten, A.M.G., 1997. Ondergang in Bedwelming. Drugs en Giften in Het West-Romeinse Rijk. Erasmus Publishing, Rotterdam.

Scullard, H.H., 1959/1976. From the Gracchi to Nero. A History of Rome from 133 BC to AD 68. Methuen & Co., London.

Shepherd, W.G., 1983. Horace. The complete Odes and Epodes. Penguin Books.

Stein, Martina, 1930. In: Band XXVIII, J.B. (Ed.), RE: Paulys Realencyclopädie der Classischen Altertumswissenschaft. Neue Bearbeitung. Metzlersche Verlagsbuchhandlung, Stuttgart.

Whiston, W., 1987. The Works of Josephus. Complete and Unabridged. Hendrickson Publishers, Peabody, MA.

22

Did Hannibal Really Poison Himself?

Francesco M. Galassi

Archaeology, College of Humanities, Arts and Social Sciences,
Flinders University, Adelaide, SA, Australia

Hannibal of Carthage (247−183/181 BC) (Fig. 22.1) is still remembered as one of the world's greatest military leaders, his brilliant tactics and maneuvers having been studied for centuries in military academies. Raised by his father Hamilcar Barca (c. 275−228 BC) in the deepest contempt and hatred for the Roman Republic victorious against the Carthaginians in the First Punic War (264−241 BC). After conquering a vast territory in the Iberian Peninsula in 219 BC, by capturing Saguntum, an ally of Rome, he set off on a military adventure which would make him Rome's worst nightmare. *"Hannibal ad portas!"*, "Hannibal at the gates!" would become a popular saying which speaks for the extreme terror generated in the collective psyche of the Roman population during the so-called Hannibalic War. Having crossed the Alps with an army including elephants, he utterly crushed a series of mighty armies sent against him by the Roman Senate, mainly at the battles of the Trebbia River (218 BC), Lake Trasimene (217 BC) and Cannae (216 BC), a carnage in which the Roman army led by Consuls Gaius Terentius Varro and Lucius Aemilius Paulus was slaughtered. Having won most of the southern Italian peninsula to his cause and seeking an alliance with the kingdom of Macedonia and the Seleucid Empire, Hannibal was only one step from absolute victory.

FIGURE 22.1 Hannibal's presumed bust. *From Mommsen's Römische Geschichte, gekürzte Ausgabe (1932) by Phaidon Verlag (Wien-Leipzig). Image in the public domain. Credits: https://it. wikipedia.org/wiki/Annibale#/media/File:Mommsen_p265.jpg.*

Missing, however, the opportunity to take Rome when she was on her knees, and failing to receive adequate support both from his Hellenistic allies and his own motherland, he eventually allowed Rome to recover and reconquer the lost territories. A new figure had emerged in the meantime from the fires of war, Publius Cornelius Scipio (later named "Africanus," 236–183 BC), a young and brilliant military leader for the Roman Republic who rose to prominence by defeating the Carthaginians in Spain and then by moving the theatre of war to Africa, finally liquidating Hannibal's army at the Battle of Zama (202 BC). Although the war had been lost, Hannibal, sent into exile by his own countrymen, did not concede defeat and pursued a kind of personal struggle against the Roman enemy by offering his own expertise and services to Rome's next enemies, the Near Eastern kingdoms. Nonetheless, Rome's push towards the east seemed unstoppable and her advance meant fewer shelters for Hannibal. Aged, yet still vigorous—he only lamented the loss of his sight at one eye during the Italian Campaign—he found in King Prusias of Bithynia (c. 250–c. 182 BC) a last friend and protector. The eastern king, however, could not resist diplomatic pressure from Rome to hand in Hannibal and ultimately agreed to break the sacred laws of hospitality (Culican and Hunt, 2018).

Historical research agrees on the fact that the old Punic general, realizing that his time had come and seeing the Bithynian troops closing in on his fortified castle near the town of Lybissa (modern-day Gebze), took his own life through a poison in order to avoid being taken prisoner by his mortal enemy and dragged as a captive into the streets of Rome. These conclusions are drawn from later accounts of Hannibal's demise and burial, since no coeval primary sources—either Roman or by one of the Carthaginian's associates—have been preserved for our contemporary scrutiny. In addition, these sources appear to focus mainly on the transient nature of glory and destiny's last irony in that Hannibal's death place is named Lybissa, reminiscent of Lybia (or Africa according to ancient conventions), his homeland.

At first glance, it can be noticed that several versions of Hannibal's death exist. From a paleopathological and historico-medical perspective it is thus important to examine them in detail in trying to reconstruct the most likely final scenario and cause of death for the great Carthaginian commander. To achieve this goal, as with other prominent historical characters such as Julius Caesar (Galassi and Ashrafian, 2016), Alaric I (Galassi et al., 2016), and the Roman Emperor Hadrian (Petrakis, 1980), the principles of retrospective diagnostics are best implemented through a philologico-clinical approach to the existing sources arranged in chronological order of composition (Mitchell, 2017; Galassi and Gelsi, 2015).

Cornelius Nepos (c. 110–c. 25 BC) in his *De Viris Illustribus* [XII] gives a rather vivid description of Hannibal's last moments of life:

> "Hannibal indeed confined himself to one place, living in a fortress which had been given him by the king; and this he had so constructed that it had outlets on every side of the building, always fearing lest that should happen which eventually came to pass. When the Roman ambassadors had gone thither, and had surrounded his house with a number of men, a slave, looking out at a gate, told Hannibal that several armed men were to be seen, contrary to what was usual. Hannibal desired him to go round to all the gates of the castle, and bring him word immediately whether it was beset in the same way on all sides. The slave having soon reported how it was, and informed him, that all the passages were secured [*omnisque exitus occupatos*], he felt certain that it was no accidental occurrence, but that his person was menaced, and that his life was no longer to be preserved [*neque sibi diutius vitam esse retinendam*]. That he might not part with it, however, at the pleasure of another, and dwelling on the remembrance of his past honours, **he took poison, which he had been accustomed always to carry with him** [*quod semper secum habere consuerat, sumpsit*]." *Nepos (1886)*

Livy (59/64 BC–17 AD) in his monumental *History of Rome* [Liv. 39.51] gives a more detailed and romanticized account of the Carthaginian's demise:

> "Having regard to the dangers which were all around him, in order that he might always have some way of escape in readiness, he had made seven exits from his house, and some of these were secret, lest he might be hemmed in by guards. But the dread power of kings leaves nothing unexplored when they want it traced down. They surrounded the whole area about the house with guards, so that no one could escape from it. When the word was brought to him that the king's troops were in the vestibule, Hannibal attempted to escape by a side door which was out of the way and especially adapted to a stealthy departure, and when he found that this too was blocked by guards stationed around it [*sensit et omnia circa clausa custodiis dispositis esse*], **he called for the poison which he had long kept ready for such emergencies** [*venenum, quod multo ante praeparatum ad tales habebat casus, poposcit*]. "Let us," he said, "relieve the Roman people of their long anxiety, since they find it tedious to wait for the death of an old man. [...]" Then, cursing the person and the kingdom of Prusias, **he drained the cup** [*poculum exhausit*]." *Livy (1936)*

Plutarch (c. 46–120 AD) in his *Life of Flamininus* [20.5], after a premise on the impossibility of an escape: "[...] he set out to make his escape by way of the underground passages, but encountered guards of the king, and therefore determined to take his own life."
gives three alternative versions of Hannibal's death:

1. "Some say that he **wound his cloak about his neck** [ἱμάτιον τῷ τραχήλῳ περιβαλὼν] and then **ordered a servant to plant his knee in the small of his back, pull the rope towards him with all his might until it was twisted tight** [κελεύσας οἰκέτην ὄπισθεν ἐρείσαντα κατὰ τοῦ ἰσχίου τὸ γόνυ καὶ σφοδρῶς ἀνακλάσαντα συντεῖναι καὶ περιστρέψαι], and so to **choke** [μέχρι ἂν ἐκθλίψαι τὸ πνεῦμα] **and kill him** [διαφθείρειεν αὐτόν]";
2. "some, too, say that he **drank bull's blood** [αἷμα ταύρειον πιεῖν] in imitation of Themistocles and Midas";
3. "but Livy says that **he had poison** [φάρμακον ἔχοντα] which he ordered to be mixed, and **took the cup** with these words: 'Let us now at last put an end to the great anxiety of the Romans, who have thought it too long and hard a task to wait for the death of a hated old man'." (Plutarch, 1921).

Appian (c. 95–c. 165 AD), in his work *The Syrian Wars* [2.11], cursorily touches upon the Carthaginian's demise, suggesting that poison was the *instrumentum mortis*:

"he caused Prusias to put him to death **by poison** [ἔκτεινε διὰ τοῦ Προυσίου **φαρμάκῳ**]" (Appian, 1899).

Pausanias (c. 110–c. 180 AD), in his *Description of Greece* [8.11], in sharp contrast with the sources seen above, gives a radically different interpretation of the way in which Hannibal lost his life:

"Hannibal received an oracle from Ammon that when he died he would be buried in Libyan earth. So he hoped to destroy the Roman empire, to return to his home in Lybia, and there to die of old age. But when Flamininus the Roman was anxious to take him alive, Hannibal came to Prusias as a suppliant. Repulsed by Prusias he jumped upon his horse, **but was wounded in the finger by his drawn sword. When he had proceeded only a few stades his wound caused a fever, and he died on the third day** [ἀπωσθεὶς ὑπ' αὐτοῦ ἀνεπήδα τε ἐπὶ τὸν ἵππον καὶ γυμνωθέντος τοῦ ξίφους τιτρώσκεται τὸν δάκτυλον. Προελθόντι δέ οἱ στάδια οὐ πολλὰ πυρετός τε ἀπὸ τοῦ τραύματος καὶ ἡ τελευτὴ τριταίῳ συνέβη]. The place where he died is called Libyssa by the Nicomedians" (Pausanias, 1918).

A much later Latin-language author, Eutropius (latter half of the 4th century AD), in his *Breviarium ab urbe condita* [4.5], favors the previously examined version of the poisoning story:

"And, since he was to be handed in to the Romans, **he drank a poison** [*venenum bibit*] [. . .]." *Eutropius (1853)*

Pseudo-Aurelius Victor (4th century AD), in *De viris illustribus urbis Romae* [42.6], adds an interesting particular on the location of the poison:

"[. . .] **he took the poison which he kept beneath the gem of his ring** [*quod sub gemma anuli habebat, veneno absumptus est*] [. . .]" *Victor (1911)*

Six of seven ancient sources (Nepos, Livy, Plutarch, Appian, Eutropius, and Pseudo-Aurelius Victor) accept that Hannibal died because of poison and they all agree that his taking it was a voluntary act that occurred as a consequence of Prusias' betrayal. Of these six sources, three (Nepos, Livy, and Pseudo-Aurelius Victor) mention the fact that Hannibal used to carry some poison on his person (Pseudo-Aurelius Victor even adds that he kept it concealed in his ring), as if he was expecting that he might find himself in the necessity to use it to avoid being captured by his mortal enemies; three sources (Livy, Plutarch, and Eutropius) additionally specify that he *drank* the poison. None of the authors says what kind of poison this was. Two of seven sources, instead, either give alternative interpretations for Hannibal's death (Plutarch) or even propose a totally different explanation, as Pausanias does. His version of the great general death should, however, be looked at with a certain degree of scepticism. While it is unquestionable that even a trivial wound in the preantibiotic era might have resulted in a life-threatening infection, Hannibal's story is close, among others, to that of Epaminondas in the context of a discussion on oracles and how people can often misunderstand their meaning. Since the oracle had predicted that Hannibal would die on Lybian soil, he thought that he would really perish in Lybia (i.e., Africa), while his Lybia would instead be Lybissa, in Anatolia. Furthermore, a story like that—featuring a once-upon-a-time mighty conqueror who dies because of a self-inflicted wound—might be read as a way of minimizing the aura of mythology surrounding Hannibal's name, although the former interpretation still appears more likely, given the oracular context.

Plutarch's first option for Hannibal's death (i.e., a slave choking him) offers some very interesting details, particularly considering the limited time Hannibal had to commit suicide, since the house had already been surrounded by Prusias' soldiers. In order to avoid being captured still alive and conscious, whatever means would be implemented to take his own life, should have been both effective and, above all, fast-acting.

Several poisons were known in classical antiquity and their noxious effects were the subject of studies and medical observations. Among others, vegetal poisons such as *Atropa belladonna*, *Aconitum napellus*, *Conium maculatum* (hemlock), or *Ricinus communis* are good candidates for taking Hannibal's last breath, since they are somewhat fast in their action, starting immediately after digestion (about 30 minutes) (Cilliers and Retief, 2000).

In order to act more swiftly and avoid being taken alive, his servant's intervention might have been fundamental. By pulling a rope around Hannibal's neck, the slave would have interrupted the blood supply to his master's brain in the carotid arteries, since only 20 seconds are necessary to make somebody faint, while longer times (2–4 minutes) are enough to cause irreversible brain tissue damage and potentially even death itself. If the act of pulling, as Plutarch underlines, was indeed extremely forceful, it is not unthinkable that Hannibal's cervical vertebrae got fractured. If his mortal remains were still available for a forensic examination—which is no longer the case—such a more extreme eventuality could be proved by the identification of a fracture of the hyoid bone, which is pathognomonic for deaths by hanging, a situation not too dissimilar from the presented one in which an extreme force is exerted upon the neck by a pulled rope (Lyle, 2008).

Quite interestingly, such a particular mechanism of death (a combination of poison and physical killing) has an interesting parallel in the demise of one of history's most cruel characters, the Nazi dictator Adolf Hitler (1889–1945), who first ingested a cyanide capsule and ensured his death by shooting himself in the head in his bunker in Berlin when surrounded by the Red Army, on April 30, 1945.

In conclusion, while Hannibal can be reasonably considered to have committed suicide by taking a poison, one must also consider the limited time he had to succumb. Thus, the likelihood of a combined method of death, in which a mechanical injury of some kind ensured the hero's demise, should be seriously taken into account.

References

Appian, 1899. The Foreign Wars. Horace White. The Macmillan Company, New York, NY.

Cilliers, L., Retief, F.P., 2000. Poisons, poisoning and the trade in ancient Rome. Akroterion 45, 88—100.

Culican, W., Hunt, P., 2018. Hannibal. Encyclopaedia Britannica Website: https://www.britannica.com/biography/Hannibal-Carthaginian-general-247-183-BC (accessed 24th May 2018).

Eutropius, 1853. Abridgement of Roman History. Translated, with Notes, by the Rev. John Selby Watson. Henry G. Bohn, York Street, Covent Garden, London.

Galassi, F.M., Ashrafian, H., 2016. Julius Caesar's Disease: a New Diagnosis. Pen and Sword, Barnsley, South Yorkshire, UK.

Galassi, F.M., Gelsi, R., 2015. Methodological limitations of an etiological framing of Ariarathes goitre: response to Tekiner et al. J. Endocrinol. Invest. 38 (5), 569.

Galassi, F.M., Bianucci, R., Gorini, G., Paganotti, G.M., Habicht, M.E., Rühli, F.J., 2016. The sudden death of Alaric I (c. 370-410AD), the vanquisher of Rome: a tale of malaria and lacking immunity. Eur. J. Intern. Med. 31, 84—87.

Livy, Ltd, 1936. Books XXXVIII—XXXIX with an English Translation. Cambridge, Mass., Harvard University Press, Cambridge; William Heinemann, London. Published without copyright notice.

Lyle, D.P., 2008. Howdunit Forensics: A Guide for Writers. Writer's Digest Books, Cincinnati, Ohio, p. 161.

Mitchell, P.D., 2017. Improving the use of historical written sources in paleopathology. Int. J. Paleopathol 19, 88—95.

Nepos, C., 1886. Lives of Eminent Commanders (J. S. Watson, Trans.). Henry G. Bohn, London.

Pausanias, Ltd, 1918. Pausanias Description of Greece with an English Translation by W. H.S. Jones, Litt.D., and H.A. Ormerod, M.A., in 4 Volumes. Harvard University Press/ William Heinemann, Cambridge, MA/London.

Petrakis, N.L., 1980. Diagonal earlobe creases, type A behavior and the death of Emperor Hadrian. West J. Med. 132 (1), 87—91.

Plutarch, Ltd, 1921. Plutarch's Lives. With an English Translation by Bernadotte Perrin. Harvard University Press/William Heinemann, Cambridge, MA/London.

Victor, Pseudo-Aurelius, 1911. In: Pichlmayr, Franz (Ed.), De viris illustribus urbis Romae. Teubner, Leipzig [English translation of quotation by the author of this chapter].

23

Drugs, Suppositories, and Cult Worship in Antiquity

David Hillman

Independent Scholar, Specializes in Ancient Pharmacy and Medicine, Madison, WI, United States

23.1 INTRODUCTION

The oldest Greco-Roman medicines were plant-derived chemicals and animal toxins used predominantly in gynecology and obstetrics. [Our earliest Greek medical sources include treatises on the proper regulation of menstruation, conception, and delivery. The Hippocratic *Nature of Women* (Περί(Γυναικείησ Φύσιος)) contains lengthy instructions on the preparation and application of pessaries, douches, and other gynecological drugs. The names of gynecological drugs are among the most prominent within the corpus of Greco-Roman pharmacy. Some botanical components of these widely used drugs and drug concoctions derived their common names from the gynecological activity they promoted: e.g., the plant known as "aristolochia" means "best childbirth." The nexus of drugs, women, gynecology, and religion dates back to the foundation of Greco-Roman culture (see Riddle, 2010). Greek and Roman myth and cult rites actively promote the use of gynecological drugs in association with the worship of the earliest Mediterranean deities, including mother goddesses, witches, and oracles.] Female physicians, priestesses, and midwives belonged to the oldest pre-Hippocratic traditions of medicine and pharmacy in the Mediterranean and were largely responsible for the establishment of western medicine and pharmacy.

Scholars have closely examined drugs used as potables, edibles, and inhalations, but little is known of the vaginal and anal suppositories used in antiquity, their unique methods of administration, or their use in cultic sexual practices. Despite the under-researched nature of these drugs—due in large part to their direct association with practices and bodily functions considered taboo by modern cultures—a knowledge of their use helps to paint a more complete picture of the historical nexus of drugs and ancient cults as well as the gynecological sophistication of classical civilization.

23.2 DRUGS AND CULTS

Virgil's *Aeneid* preserves a mythic episode that aptly illustrates the predominant role of exotic botanicals and venoms in antiquity as well as their connection with traditional cults. Ancient mystery rites referenced in numerous Greek and Roman poets illustrate the use of potent drugs that affect human sexuality; many of these drugs were applied to the anus and vagina by means of medicated dildos and as pessaries.

Virgil's treatment of the goddess Allecto illustrates the use of drugs within the context of ancient cult ritual. Allecto, a Fury, was very much the quintessential sibyl. [The figure of Allecto as the virgo or κόρη is foreshadowed in the Etruscan maiden figure of Vanth, who is a combination of Hecate and the Furies. See *The Religion of the Etruscans*, ed. by Grummond and Simon (De Grummond and Simon, 2006). There is also an element of the Lasas in Allecto, who are known to carry the alabastron (ἀλάβαστρον), an overtly sexual device used to apply μύρον as seen in Aristophanes' *Lysistrata*. The alabastron was mentioned in the gospel of Mark as a perfume container possessed by Mary Magdalene, an alleged prostitute.] Like the goddess Justice, she was an eternally youthful dancer, possessed of the bloom of life, whose mind was immovably set on the "purgation" of the unjust; as an avenging Fury, she sang of the beauty of pain—a psychic pain she imposed as punishment upon the impious.

Virgil described Allecto as one of the daughters of Black Night; she was a Fury who exercised a power to overthrow the sanity of mortals irredeemably possessed by greed, ill-will, and hubris. When heaven needed brother to fight brother, it summoned Allecto; in the seventh book of the *Aeneid*, Juno called upon the teenage goddess to drive a queen crazy in order to start an uncivil war between the inhabitants of Italy and the Trojans. (Virgil, *Aeneid*, 7.341−7: "Then Allecto, saturated with the poisons of the Gorgons ... takes up a position at the most private portal of queen Amata ... into whom the goddess forcefully inserts a single measure of snake venom taken from her black hair; she applies the suppository in her body's natural pocket, at the place of her most intimate being." Author's translation.)

Allecto spread her maddening poison by applying a snake-derived drug to a mortal woman's biological "pocket," or in Virgil's words, the location of her most intimate sensations. Like other Bacchants, the goddess removed the venom from her own hair, which was traditionally depicted in both art and literature as being full of vipers. (As seen in Aeschylus' *Eumenides* and Euripides' *Bacchae*. Interestingly, most of the deities associated with ritual Bacchic worship were young girls who had not yet had children, yet each was able—like the Nymphs—to provide breast milk to their charges. These nurses are directly associated with the viper of Dionysus, and it is known that viper venom produces a prolactin-like response in mammary tissue.) After this venom was absorbed vaginally, Virgil says it coursed through the queen's body, intoxicated her, and sent her into an ecstatic frenzy.

23.3 BACCHANTS AND VIPER VENOM

Before ancient seers summoned spirits like Allecto from the underworld—an arcane practice known as necromancy—they invoked Bacchus, the god of ecstatic dance. The worship of this mystery cult divinity, known variously as Dionysus, Bromius, and Zagreus, was connected in classical literature and art with the handling of the European horned viper (*Vipera ammodytes*). This snake accompanied Maenads and Bacchants in their religious processions and was frequently associated with young, virgin girls of myth who nursed gods and heroes alike.

In antiquity, snakes like the horned viper were closely bound up with the practice of medicine and the exercises of cults. The Greeks, Romans, and Etruscans did not openly distinguish between the practice of medicine and religion, and even the goddess of health herself—

Hygeia—was sometimes portrayed with a snake positioned over an offering dish in a posture that appears to imply the act of milking the animal's venom.

The use of viper venom in ancient medicine and cults may explain how young girls associated with the cult of Bacchus, who were not previously pregnant, were able to function as wet nurses. Crotoxin is a phospholipase A2 neurotoxin produced by a South American pit viper (*Crotalus durissus terrificus*). This crotalid toxin has similar biological activities to the components of horned viper venom and acts on mammary epithelial cells to stimulate the secretion of casein using the same biochemical pathways as prolactin (see Ollivier-Bousquet et al., 1991). Stimulating the production of breast milk by the application of viper venom may have been responsible for the ability of these nonparturient girls to lactate.

Divinities like the Furies—one of whom was Allecto—and their sisters, the Death-Spirits (Keres), were represented as being infused with viper venom; the ancient world believed these female entities were particularly hearty and possessed significant stamina, longevity, and an apparent immunity to weakness or illness. Some of them are even called "dragonesses," and are involved with the "burning off" of human mortality. It appears that the priestesses of Hecate, Priapus, and Demeter/ Persephone were involved in the consumption of viper venom. [*Aeschylus was nearly executed for revealing the secrets of the Eleusinian mysteries in his plays on Orestes. One of these plays (The Cup-bearers) contains a dream of a "dragoness" in which her breast milk is injected with the venom of a snake.*] These priestesses were considered to be physically superior to ordinary mortals, and like the Furies were of a superior physical constitution—and somewhat immune to disease. Interestingly, crotoxin has shown itself to be a potential anticancer drug (see Cura et al., 2002).

23.4 ANCIENT VAGINAL SUPPOSITORIES

Allecto's application of a snake-venom-based drug to a woman's vagina is in direct step with ancient gynecological practices. According to the author of the Hippocratic treatise titled *The Nature of Women*, numerous compound drug mixtures in antiquity were used to produce abortion, control uterine bleeding, and treat disease. These drugs were applied directly to the vagina by means of cylindrical pessaries. [Hippocrates, *Nature of Women*, 71: "Irritating suppositories that draw blood: mix frankincense and myrrh with blister beetle, forming them as a big as an oak gall; make into an elongated shape, attach it around a feather with a flock of wool, tie it with a thread of fine linen, soak in white Egyptian unguent or rose unguent, and apply." Translated by Paul Potter.]

Frankincense, myrrh, and blister beetle were all common ingredients in gynecological drug mixtures. Aromatic botanicals were exceptionally common as basic components of Greco-Roman suppositories, unguents, and ointments.

Blister beetle was a very popular aphrodisiac in antiquity just as it has been in the modern world. Cantharidin, a terpenoid secreted by blister beetles, is an irritant or blistering agent that induces priapism. Cantharidin can cause gastrointestinal and renal dysfunction, but it has been used since classical antiquity as an effective aphrodisiac. [Cantharidin has also been used in China. For its history and toxicology see Moed et al. (2001).]

Rose oil, like other fragrant volatile oils, was most often used as a base for gynecological applications. The combination of a drug like cantharidin with aromatic oils may have been responsible for the abortifacient and/ or contraceptive qualities of these aphrodisiacs. The caustic properties of cantharidin would have irritated the lining of the uterus to such a degree as to induce sloughing and therefore the prevention of implantation. Rose oil concoctions were typically associated with the priestly followers of Aphrodite, who were typically associated with temple prostitution.

According to the aforementioned passage from the Hippocratic *The Nature of Women*, vaginal suppositories were compounded and then worked into a penis-like shape for the sake of application. These drugs were used to affect reproductive physiology, sexual activity, and disease. Greek and Latin vocabulary concerned with suppository usage was highly specific. For example, both cultures employed specific verbs denoting the act of applying of suppositories to the vagina and anus. ["Προστίθημι" in Greek and "subdo" in Latin denote the application of a pessary or a suppository. The Greek gynecological texts found in the Hippocratic corpus—which could not have been written by Hippocratic physicians due to their liberal use of abortifacients—are full of the nominal forms of this verb and associated nouns that indicate or denote a "pessary" or "suppository."]

23.5 DRUGS AND SEXUALITY

Dioscorides, a first century CE physician and compiler of botanical and animal-derived drugs, wrote at length about the use of oils, ointments, and unguents in sexual contexts. These drugs were typically divided into compounds that "harden" and "soften" the external genitalia—also known as "heating" and "cooling" drugs. The Greco-Roman world used these substances to prepare the genitalia for copulation, to prevent disease and to regulate menstruation; these drugs, despite their numerous varied uses, were known collectively as "aphrodisiacs."

Opobalsamum, known to biblical scholars as the "balm of Gilead," was one such valuable aphrodisiac. According to Dioscorides, opobalsamum was the hard-to-collect exudate of a Middle Eastern tree—a substance produced in response to structural injury. [Dioscorides, *De Materia Medica*, 1.19 (Dioscorides, 1958).] Like the "tears" of myrrh, frankincense, and even the opium poppy, this drug was a potent secondary metabolite that was harvested, packaged, and sold in markets throughout the Mediterranean. (Many of the exudates harvested from injury are referred to as "tears" due to their appearance and the fact that they drip slowly from incisions made in the botanicals whence they are derived.) Dioscorides reveals that expensive drugs like opobalsamum were so valuable that they were frequently adulterated by drug sellers. [According to Dioscorides, opobalsamum was so valuable that it was worth twice its weight in silver. 1.19. Because of this, he speaks extensively about the best means of testing the drug in order to determine its purity (Dioscorides, 1958).]

Opobalsamum—among other things—was a diuretic, an abortifacient, and used to treat abrasions of the vulva [Dioscorides, *De Materia Medica*, 1.19 (Dioscorides, 1958)]. Like other secondary metabolites, opobalsamum possessed distinct antifungal and bacteriostatic properties that may have been responsible for its use as a means of decreasing the prevalence of certain sexually transmitted diseases. It is also significant that an ointment compounded of opobalsamum was also believed to counteract snake venom—something to which ancient priestesses, like those who worshiped Bacchus, Hecate, and Allecto, would have been exposed.

23.6 APHRODISIAC SUPPOSITORIES AND MAGIC

Priestesses who served divinities like the infernal Furies, or their leader, the witch goddess Hecate, were known in antiquity to exercise powers over demons; such manipulation of the natural world was considered to be the practice of magic. Just as ancient magic was not distinguished from the practice of religion, the use of drugs and the act of sex were not considered to be strictly secular or nonreligious operations; sex, drugs, and religion were a single entity in the ancient world.

According to Petronius, the first century CE author of the *Satyricon*, Roman priestesses were known for their drug-induced sexual practices. In one particular episode of his novel, Petronius allows one of his characters to illustrate the power of the priestess-sorceress to control and manipulate human sexuality by means of drugs:

"Oenoethea, drawing out a leathern prick, dipped it in a medley of oil, small pepper, and the bruised seed of nettles, and proceeded by degrees to direct its passage through my hinder parts.... with this mixture the old woman laboriously sprinkled my thighs; ...and with the juice of cresses and southern-wood washing my loins, she took a bunch of green nettles and began to strike gently all the vale below my navel." (Petronius, Satyricon, 138. Translated by W.H.D. Rouse; Warmington, 1987)

In actuality, the sorceress Oenoethea was acting as a priestess-physician in this passage; she was attempting to cure Petronius' character of his impotency. What is most curious about the passage is that Petronius does not seem to question the reality that men were readily penetrated anally in medico-religious acts in antiquity. In other words, it was acceptable for men to have drugs applied anally by means of medicated dildos.

The use of nettles both externally and internally as a promoter of penile erection may be the result of the presence of 3,4-divanillyltetrahydrofuran, a lignan that appears to liberate available testosterone and thus may influence sexual drive. [For a closer look at lignans, see Touré and Xueming (2010).]

Medico-religious ceremonies also employed the use of drugs with the intent of producing magical "fluids." Bodily secretions like semen, vaginal ejaculate, saliva, and even breast milk were all important ingredients of magical ceremonies—including healing acts, exorcisms, and "purgations." [The roots of these practices are found in Egypt. See *The Mechanics of Ancient Egyptian Magical Practice* by Ritner (1993).] In another episode found in the *Satyricon*, Petronius causes his characters to employ magic in order to induce erection and ejaculation:

Then the old woman took a twist of threads of different colors out of her dress, and tied it around my neck. Then she mixed some dust with spittle, and took it on her middle finger, and made a mark on my forehead despite my protest.... after this chant she ordered me to spit three times and throw stones into my bosom three times, after she had said a spell over them and wrapped them in purple and laid her hands on me and proceeded to try the force of their charm on the powers of my groin. Before you could say a word, my sinews obeyed her comment and filled the old woman's hands with a huge upstir (Petronius, Satyricon, 131. Translated by W.H.D. Rouse; Warmington, 1987).

Like the Fury Allecto, the sorceress in this passage employs magical elements including drugs, incantations, the manipulation of the sexual organs, and the use of body fluids to affect a change in her subject. And once again, the manipulation of sexuality by means of drugs and ritual magic is forced. In other words, Virgil's Allecto inspires madness by means of violently forceful sexual drug-magic, just as Petronius' sorceresses aggressively apply medicines in order to manipulate sexual experience in a religious setting.

23.7 CONCLUSION

The ultimate purpose of using drugs and magical ritual to affect human sexual experience was the veneration of deities. Greek and Roman sorceresses worked with one goal in mind; they desired to celebrate the gods by using magic to create oracular vision. Vaginal suppositories that maddened women, anally inserted medicated dildos that mesmerized men, and drugs used to force the expulsion of ejaculate were all elements of the creation of religious song.

Religious ecstasy was enhanced by drugs and the rituals of young priestesses, but the climactic moment of ancient medico-religious experience was always accompanied by a divine utterance or shout. This shout became the holy voice of the deity overseeing the performance of such mysteries; the god was directly served by means of this ecstatic vocalization. In other words, when Allecto forcefully drugged her victims by penetrating their most private body parts, and thus infused them with fury, their pained screams of lunatic agony served Justice.

References

Cura, J.E., Blanzaco, D.P., Brisson, C., 2002. Phase I and pharmacokinetics study of crotoxin (cytotoxic PLA2, NSC-624244) in patients with advanced cancer. Clin. Cancer Res. 8, 1033.

De Grummond, N., Simon, E., 2006. The religion of the Etruscans. In: De Grummond, N., Simon, E. (Eds.), The Religion of the Etruscans. University of Texas Press, Austin, TX.

Dioscorides, P., 1958. De Materia Medica. August Raabe, Berlin.

Moed, L., Shwayder, T.A., Chang, M.W., 2001. Cantharidin revisited: a blistering defense of an ancient medicine. Arch. Dermatol. 137, 1357–1360.

Ollivier-Bousquet, M., Radvanyi, F., Bon, C., 1991. Crotoxin, a phospholipase A2 neurotoxin from snake venom, interacts with epithelial mammary cells, is internalized and induces secretion. Mol. Cell Endocrinol. 82 (1), 41–50.

Riddle, J., 2010. Goddesses, Elixirs, and Witches: Plants and Sexuality throughout Human History. Palgrave Macmillan, New York, NY.

Ritner, R.K., 1993. The Mechanics of Ancient Egyptian Magical Practice. University of Chicago, Chicago, IL.

Touré, A., Xueming, X., 2010. Flaxseed lignans: source, biosynthesis, metabolism, antioxidant activity, bio-active components, and health benefits. Compr. Rev. Food Sci. Food Saf. 9 (3), 261–269.

Warmington, E.H. (Ed.), 1987. Petronius: Satyricon; Seneca: Apocolocyntosis [Rouse W.H. D., Trans.]. Harvard University Press, Cambridge.

Entheogens in Ancient Times: Wine and the Rituals of Dionysus

Carl A.P. Ruck

Department of Classical Studies, Boston University, Boston, MA,
United States

O U T L I N E

The abuse of psychoactive substances triggered by R. Gordon Wasson's revelations in his *Life* magazine article of the May 13, 1957 about drug-induced shamanic rites among the indigenous peoples of the New World and the subsequent popularization of Lysergic acid diethylamide (LSD), which Albert Hofmann had discovered in 1943, made it necessary to create the neologism "entheogen" to discuss the religious role of such substances, divorced from the drug culture of the so-called psychedelic revolution, with its excessive examples of self-indulgence and addiction. An "entheogen" is a mind-altering substance that, as its Greek roots indicate, induces the experience of being *entheos*, of communion with deity, of sharing an identity with the deity, of having the god dwell within (Ruck et al., 1979). Entheogens are central to the historical record of humankind's spiritual quest for the meaning of existence, documented in rock paintings as early as the Paleolithic and continuing through all periods in rituals of secret societies and among the ecclesiastical elite of most religions until the present.

Ironically, neither mushrooms nor LSD are, in fact, addictive, nor are many psychoactive agents listed in the Controlled Substances Act of

1970 and subsequent amendments, classified with a high potential for abuse. The threat that they pose is as a stimulus for political and theological innovation since there is no established context or etiquette for their use, whereas when incorporated into traditional ritual enactments, they can serve the opposite purpose of confirming societal norms or group identities through communal experience of shared metaphysical expectations. The latter was the situation in antiquity (Fig. 24.1).

Alcohol as a substance was unknown in antiquity, discovered only in the mid-14th century CE, when the Spanish Franciscan monk Johannes de Rupescissa succeeded in isolating ethanol from distilled wine, employing Arabic procedures for processing various metallic substances, such as antimony sulfide, called *koh'l* (kohl) in Arabic and used as an eye cosmetic since antiquity. In the 8th and 9th centuries, Muslim experimenters had used distillation for extracting volatile vapors which they could not capture or contain but which burned upon release as *aqua ardens*, "burning water." Rupescissa named his discovery aqua vitae, the "water of life." With the Arabic article *al* it became "alcohol." Rupescissa considered it the *quinta essentia* or "fifth essence," which Aristotle had theorized existed as the elusive substance that permeated and bound together the four elements of fire, air, water, and earth, that the 5th-century Empedocles of Acragas had posited as the fundamental materials from which the universe was constructed. Ethanol was immediately greeted as the "spirit molecule" and it quickly spread through philosophical and theological networks as a major advance in chemistry, which later disowned the mystical implications of its origin by separating itself as a science from alchemy (Fig. 24.2).

Early humankind employed a wide variety of toxic substances from botanical and animal sources to alter consciousness in both religious and recreational contexts. Among these were the ferments of sugary liquids, in particular honey, whose inebriating drink was called mead, providing the basic term for intoxicant in Greek and yielding the word "mad" in English. With the discovery of wild grapes with higher sugar contents, fermentation was applied to the crushed juices from the fruit of the vine to produce wine. All ferments, mead, beer, and wine, however, are limited in their content of alcohol by the fact that the natural yeasts that engender fermentation die when the concentration of ethanol makes the liquid too inhospitable for further growth. This occurs at the highest limit of around fourteen percent, which would be termed seven-proof by the modern nomenclature, a drink of moderate intoxicating potential (Fig. 24.3).

Nevertheless, it was customary in antiquity to drink wine diluted with three or four parts water, yielding a drink of around only two proof, or of quite low alcoholic content. A legendary Thracian wine required twenty parts dilution, and in actuality, a wine that went by that name in the Roman period required eight parts of water to be drunk safely. Despite the low alcoholic content, wine had a mind-altering potency that could result in complete mental derangement or narcosis after only four cups drunk over the period of several hours. The conclusion of a drinking party often continued out into the streets with public rowdiness and not infrequently violent and aggressive behavior (Rinella, 2010). Without dilution, wine might result in perma-nent brain damage or, as reported on one occasion, even death after only ten cups for a group of young athletes in their physical prime. A Greek drinking cup held about 100 mL. Ten cups would be a liter.

FIGURE 24.1   Mixing the potion: a woman displays a mushroom to a priest with staff (right); a young male with staff instructs a woman, who is adding a vine plant to the wine mixture in the *pithos* urn, partially buried in the ground: Greek 5th-century red-figure *hydria*, from a cemetery of ancient Ainos (modern Enez, Turkey), Museum of Edirne (ancient Adrianopolis, Thrace).

The virulence of the drink far exceeded its alcoholic content, but was derived from the various fortifying toxins added to it. These included venoms extracted from poisonous animals and insects and from toxic plants, including lethal poisons at dangerously threshold dosages (Hillman, 2008). Some drugs, like henbane, were considered age-appropriate and unseemly in an adult, and some were reserved for religious contexts, and profane recreational use was sacrilegious and punishable by death. Among the toxins added to the wine, identi-fiable agents include serpent and salamander venoms, hemlock, jimsonweed, aconite, cannabis, wormwood, ergot, and probably *N,N*-Dimethyltryptamine (DMT) from acacia and similar plants, as well as psychoactive resins and incenses. Other toxins are masked under names that probably refer to special compounded mixtures. In 2013, a storeroom of still extant wine dating from the mid-second millennium BCE unearthed in Canaan confirms the presence of psychoactive additives, which previously was documented only in literary texts (Associated Press Article, 2013).

Since opium was among the additives, addiction was inevitable, and persons were satirized for drunkenness and overindulgence. The low alcoholic content of diluted wine probably precluded the possibility of actual alcoholism in most cases, and the perceived dependency was due to the toxic additives. This is probably true even of those individuals who were said to have indulged in drinking their wine unmixed, in the Thracian fashion, to the point of permanent derangement or madness. Insanity is the outcome of only the most severe alcoholic addiction. Several prominent figures, especially philosophers, chose to end their lives by drinking. A cup of purely alcoholic wine is hardly appropriate as a means for intentional suicide. Wine was also the customary way for administering medicinal preparations.

The vessels for the drink involved in certain religious initiations each had a characteristic design, differing from the wine cups of the symposium, and were appropriately larger to hold the complete dosage required for the experience and the potion was probably not drunk over an extended period of conviviality.

Wine was associated with a particular deity and his analogs in other cultures. By the Classical period, which was the 5th century BCE in Athens, this was the god Dionysus, also known as Bacchus, and there was a complex etiquette or social norm surrounding his worship and the consumption of his drink. The civilized product resulting from the controlled recognizably fungal growth of the fermenting yeasts was contrasted with the wild naturally occurring toxins, among which mushrooms, containing psychoactive psilocybin and muscimol, and ergot of grain containing Lysergic acid amide, played a fundamental role as similarly fungal. The mycelium of the ergot fungus fruits with recognizable mushrooms, visible to the unaided eye. The wine drink with its toxic additives was seen as mediating between the dichotomy of natural or wild toxins and their evolution or taming through the art of viticulture. A similar civilizing fungal growth was recognized in the leavening yeasts of bread, linking Dionysus and his analogs with the goddesses of the grain, Demeter and her daughter Persephone, and their analogs, and with the netherworld realm of Hades and the afterlife, either chthonic in the blessed fields of Elysium or celestial in the empyrean beyond the solar disc (Ruck, 2006). The entheogen derived from ergot figured in the visionary potion of the Eleusinian Mystery, which was a religious initiation that began in the mid second millennium and lasted until the Christians persecuted and destroyed the sanctuary in the 4th century. It was an experience that afforded meaning to human existence in the context of spiritual dimensions and something that most of the great thinkers, writers, artists, and politicians of the Greco-Roman world had undergone.

The etiquette of drinking involved different scenarios for men and women and different locales, either within the city or without. For the men, the ritual was the symposium, a "communal drinking." The best-documented exemplar, apart from the numerous depictions on the mixing vessels and cups for the event, is Plato's *Symposium* dialog. The men gathered at the house of the host and drank a succession of toasts, while engaged in discussion of a proposed topic. The host determined the admixtures, the rate of dilution, and the frequency and number of rounds. Throughout the increasing inebriation, the guests were challenged to maintain sobriety, although even in Plato's symposium, where the men have decided to drink abstemiously since they were still drunk from the previous day, all the guests except Socrates, the comedian Aristophanes, and the tragedian Agathon end up unconscious until dawn.

The characteristic cup for the symposium was the *kylix*, and its design as a broad saucer supported by a narrow stem challenged the drinkers to maintain decorous sobriety since it would be unstable in drunken hands. The god himself never drank from such a vessel, but from a double-handled mug called a *kantharos*, holding about a half liter. In addition to its depiction on vase paintings, several fine exemplars survive, probably intended as dedicatory offerings to the deity.

Customarily, a hired female, who sang, danced erotically, and played music, provided entertainment, which often included engaging in sexual activity with the guests. The men, moreover, were usually seated with their male lovers and engaged with each other, as well as with the female professional companion.

The symposium celebrated Dionysus in his avatar as the wine. The women's celebration honored the manifestations of the deity that predated viticulture. For them, he was Bacchus and they became bacchants. Although customarily secluded in the women's quarters of their homes and subjected to the dominance of their husbands, they left their houses and assembled outside the city in a locale symbolic of the wilderness, typically a mountainside, not a place for viticulture, especially in the winter season of the ritual, but where wild herbs abound, particularly mushrooms. They were also called madwomen or maenads, for the extreme state of ecstasy they experienced. This involved delusional fantasies of sexual abandon to the amorous onslaught of hybrid creatures imagined as ithyphallic goat-men or satyrs, who represented the spirits of the wilderness and its herbal toxins. Often the god himself was thought to materialize among them as their divine lover, cross-dressing as a female. This hallucinatory reality is well documented in both literary sources and in numerous vase paintings. The women also beat tambours, played instruments, and danced, creating a terrifying din of movement and sound. On one occasion documented in a historical record, they descended from the mountain in such a state of derangement that they fell unconscious in the center of a village and had to be protected through the night from sexual molestation by a garrison of soldiers.

FIGURE 24.2   Return of Persephone: Persephone, carrying a crossbar torch, stands in a chariot drawn by four white horses, with her mother Demeter; behind the chariot is Artemis/Bendis with Thracian hunting boots and hound; in front is Hermes with spotted fawn above him; below are three water nymphs (Danaids, Oceanids?); above is the divine bride (deceased) with Eros and Aphrodite; inner surround of florets (mushroom caps?); outer surround of tendril vine blossoming in two female heads wearing the Phrygian cap: handled broad dish with four mushroom knobs, red-figure broad funerary *patera* from Apulia, c. 330–320 BCE, attributed to the Baltimore Painter, Art Institute of Chicago.

The bacchants are never depicted drinking wine, since that was not the source of their intoxication. The emblem of their mountain celebration indicates the nature of their ritual activity. This is the thyrsus, the long hollow reed of the giant fennel, into which leaves of ivy have been stuffed, seen protruding from its top. It was customary for herb gatherers to use such an implement as the container for the wild plants that they gathered. The maenads were enacting a mimesis of herb gathering, and the object of their quest was symbolized as the wild ivy. It was a plant seen as the primordial vine from which the grapevine was hybridized. Its leaves and berries induced mental derangement in their natural state, without the civilizing intervention of viticulture to tend and prune the vine as necessary to induce it to fruit with the grapes from which the new intoxicant could be manufactured through the process of fermentation. The thyrsus was also called a narthex, which is the name of the giant fennel reed, but it also has a transparently obvious etymology as the "container" (*thex*) for the "narcotic" or drug/entheogen.

The prototypic primordial plant symbolized by the leaves of ivy and similar wild vines like bryony (cucumber) and smilax (bindweed) was the uncultivable psychoactive mushroom. Thus, in ordinary culinary nomenclature, the stipe or stem of a mushroom was called its thyrsus. This is a terminology that appears to have been in use for at least a millennium. The fungal cap spreading above the thyrsus represents the toxic herbs gathered into its container. In the case of the Amanita mushrooms that yield the visionary muscimol, the psychoactive toxin is primarily confined to the rind of its cap. Both bryony and smilax contain tropane alkaloids that yield LSA (Lysergic acid amide, a precursor to LSD).

It is well documented as a folkloric motif that magical plants require special procedures to address or appease their indwelling spirits as rituals for their gathering. The scenario for the bacchant revel indicates the delusional fantasies enacted by the women. They marshaled themselves into generic groupings according to age, as pubescent, maternal, or postmenopausal, and sexually seduced the plant, which in the case of the mushroom was a common metaphor for the penis. The dominant theme is the hunt for the animate manifestation of the plant, typically an animal like the bunny with prolific sexual connotations or the bull with its aggressive masculinity or the wolf for the shamanic motif of the werewolf transmogrification. They are never depicted, however, with any implements like a net or a trap for the hunt, and they had never received instruction for the activity, which was not an ordinary female pursuit. The spirit of the plants materialized and sexually engaged with them. As the primitive antecedent to the grapevine, its proper role was to yield to its civilized successor. This was enacted as a sacrificial victim, probably in earlier times an actual human offering, whose raw flesh they tasted. The manner of the slaughter was termed a rendering apart or *sparagmos* with their bare hands, hardly likely to be anything more than a metaphor in the case of so powerful and dangerous a beast as a bull. The bull must be interpreted, like all the other supposed victims, as a fantasized zoomorphic persona of the gathered plant. Thus, the plant was also mothered as their baby, which they tore to pieces, although there is no evidence that they ever took their children to the mountain revel. They did, however, implausibly nurse wild animals as their babies. Obviously, the women could not have hunted and caught these animals except as zoomorphic animating personae of the magical plants.

FIGURE 24.3   Maenad: Attic white-ground kylix depicting a maenad holding a thyrsus and a leopard, 490–480 BCE, Staatliche Antikensammlungen, Munich.

The thyrsus in the lore of the mythological tradition was also the container in which the titan Prometheus concealed the fire that he stole from the celestial gods and entrusted to mankind as the drug that set humans upon their path to consciousness, making them aware of mystical dimensions of existence beyond reality. This was the primordial entheogen. Not surprisingly, the bacchants were also said to handle fire in their hands, and were not burnt, since the fire is also a metaphor.

The actual source of the toxins that induced their extreme delusional state and ecstasy may have been the serpents that they were frequently depicted handling. One historical account describes stroking the cheeks of the serpents (Ritter, 2013), which would indicate the method for milking the venom, which would then probably be applied as a dermal agent in sexual mimesis since ingestion in most cases would destroy the toxins.

By the Roman period, the bacchant revel had broadened to include men. A notorious event in the year 186 BCE is recorded both in the historical record and in the surviving inscription that preserves the decree of the senate known as the *senatus consultum de Bacchanalibus*, which attempts to limit the rites which had spread from the Greek colonies in the south of the Italian peninsula and had reached as far as the city of Rome. The allegations indicate that women and their knowledge of drugs were originally the dominant source, but that the rites, in addition to actual sexual profligacy, involved the sodomizing of young men and even their sacrificial murder.

In Greece of the Classical age, the symposium of the men was countered by the bacchanalia of the girls and women. The god of wine was the ultimate mediator between realms, bringing the wild into balance with civilization. Thus a further ritual of etiquette for the drinking of his intoxicant was celebrated as the February festival of the Anthesteria. This enacted the opening of the fermented vats of new wine. Here the drinking was from individual pitchers containing about 700 mL and apparently intended as the probable daylong dosage. As the god returned from his apparent sojourn in the subterranean world, he brought with him all the ghosts of the departed to join in a communal feasting with their still-living relatives. This was a reunion that brought the whole family together, not only the living and the dead, but also husbands with wives, together with their children, the nubile daughters, exposed to public view, and even the babies. The latter were given their first taste of wine at the tender age of three or four, and they were indoctrinated into the metaphysical symbolism of the drink. The children are depicted on the characteristic toy-sized drinking pitchers for the festival playing in the vicinity of the gravestones or impersonating their elders in performing some of the most sacred rituals in honor of the god (Wasson et al., 1978).

The festival lasted three days and the drinking induced a state of mind that was parodied on the comic stage as delusional. Despite the drunkenness, the experience was accorded a positive potential. It included humorous attempts to demonstrate sobriety by standing upon a greased wineskin, and in the mythological lore, it coincided with the cure of the Orestes from the pursuit of the demons who maddened him, pursuing him for the murder of his mother. It set the precedent for the superior claim of the father over the mother as the parent of the child.

The supreme gift of the deity of intoxication was his patronage of the theater. In the 6th century under the tyrant Pisistratus, what had begun as a mushroom cult celebrated in the rural mountain villages was imported into the city of Athens. It eventually developed into the theatrical festivals of comedy and tragedy, which made the city into the cultural icon that has assured its role as the fountainhead of the Classical tradition. The hallucinatory realities of the stage enactments were a prime instrument in indoctrinating the populace into the lore of their mythological heritage. Throughout the several days of day-long performances, the audience drank a specially doctored wine, facilitating the pretense of thespian impersonation and blurring the boundaries of imagined and real.

Of the two types of performances, tragedies enacted the motif of the necessary demise of the primitive as fundament for the civilized, essentially the theme of the bacchant revel. The comedies, on the other hand, took a different view. They held a finger up to the world and imagined a paradise where baser instincts had their way. Reality could be molded with the fickleness of the phallus and the inexhaustible metaphors it traditionally inspires. Mediating between these two extremes was still another genre of the theater called the satyr play after the costuming of its dancers. Here the theme was the stories of tragedy, but treated as parody with comic intent.

Within the context of the entire cycle of Dionysian events, the intoxication served to reinforce societal norms and cultural identity, but the aggressive rowdiness of drunken symposiasts also roused suspicion of seditious intent aimed at the established political authority.

References

Associated Press Article, Brandeis University, November 22, 2013. Archaeologists discover largest, oldest wine cellar in Near East. Science Daily.

Hillman, D., 2008. The Chemical Muse: Drug Use and the Roots of Western Civilization. St. Martin's Press, New York, NY.

Rinella, Michael, 2010. Pharmakon: Plato, Drug Culture, and Identity in Ancient Athens. Lexington Books, Lantham, MD.

Ritter, M. 2013. Ancient Wine Cellar Unearthed in Israel Shows Canaanites Enjoyed a Sophisticated Drink. AP, New York. < http://www.huffingtonpost.com/2013/11/22/ancient-wine-cellar-israel_n_4323848.html > (accessed 03.12.15).

Ruck, C., 2006. Sacred Mushrooms of the Goddess: Secrets of Eleusis. Ronin Publishing, Berkeley, CA.

Ruck, C., Bigwood, J., Staples, D., Ott, J., Wasson, R., 1979. Entheogens. J. Psychedelic Drugs 11 (1—2), 4—5.

Wasson, R., Hofmann, A., Ruck, C., 1978. The Road to Eleusis: Unveiling the Secret of the Mysteries. Harcourt Brace Jovanovich, Inc, New York, NY.

Entheogens (Psychedelic Drugs) and the Ancient Mystery Religions

Mark A. Hoffman
Independent Researcher and Author, entheomedia.org

25.1 PHARMACOLOGICAL ROOTS OF RELIGION

The Mystery Religions of the ancient world frequently, if not always, employed the use of psychoactive drugs or "entheogens" to induce

Toxicology in Antiquity
DOI: https://doi.org/10.1016/B978-0-12-815339-0.00025-1

profound altered states of consciousness. Such experiences were indispensible to the initiation of members, their rituals relating to spiritual development, and ultimately the attainment of the "peak" experience(s) that represented the apotheosis and fulfillment of their theological and spiritual aspirations.

Thus, more than any other principle, the entheogenic—i.e., pharmacological—induction of nonordinary states defines the theology and practice of the ancient Mystery Religions.

A significant body of scholarly data has been applied to attempts to identify specific entheogens used throughout the ages in the Classical Mystery Religions as well as among the early developmental stages of the historical religions (i.e., those thought to be based in some part, at least, upon historical events). Thus, the mythology, art history, and canonical, apocryphal, and historical literary traditions of religions including Judaism, Christianity, Islam, Buddhism, and Hinduism have been shown to contain evidence for entheogenic practices.

Drawing upon ancient sources, religious-comparative evidence, and modern scientific data, it has been possible to identify, with varying degrees of certainty, a range of substances that are likely to have been ritually used to induce altered states of consciousness. The following is a necessarily cursory survey of some of the best documented and historically significant of these religions and their probable use of specific entheogens in a toxicological context, extrapolating likely scenarios relating to the symptomatology of practical entheogenic initiation and how such effects were interpreted by the initiates and those facilitating the experience.

The original dominance of ancient entheogenic sacramentalism has survived as deeply entrenched and culturally celebrated mythopoetic remnants regarding magical foods and beverages. Christianity has the Eucharist, referencing the blood and body of Christ; Judeo-Christianity, the Forbidden Fruit of the Tree of Knowledge, and manna of heaven; while Buddhism and Sikhism preserve a tradition of imbibing *amrita* (from Sanskrit "immortality") and Taoism has maintained an alchemical tradition regarding an elixir (or mushroom, or fruit) of immortality. The Vedic roots of Hinduism (and Buddhism) and the Iranian/Persian religions, including Islam and Sufism, emerged largely from a shared Indo-European sacramental culture steeped in the sacraments, *soma* and *haoma* (from Proto-Indo-European "pressed" or "pounded") respectively, which contributed to the survival of entheogenic themes in their practices and mythologies.

Prior to the historical religions, and within the cultural and mythopoetic contexts from which they derived, the mythologies of the mystery traditions abound with evidence of magical herbs capable of inducing "immortality" and frequently sought by gods and heroes.

25.2 HERMENEUTICS AND A DEFINITION OF TERMS

"The Mystery Religions" is a somewhat inadequate general phrase that usually denotes the dominant yet diverse religious practices of the ancient Greek and Roman periods. Similar confusion results from the use of the phrase "Classical civilizations," which may either refer specifically to the ancient Greco-Roman world, or to any of the highly developed cultures of the ancient world. For the purposes of this chapter, we extend the meaning to include any of the popular ancient cults that were characterized by ritual secrecy and initiation.

As a side note, it is important to clarify that although this chapter may be a bit beyond the scope of mainstream Western toxicology, it is still quite relevant. Psychoactive substances capable of inducing rarified states of consciousness, which some today might refer to as "intoxication," have a role to play in our historical understanding of toxic agents.

We use the word *entheogen*, first coined in 1976 (Ruck et al., 1976), to describe substances that would be inadequately termed "psychoactive" or "hallucinogenic." While "psychoactive" encompasses all substances and means of "affecting the mind," entheogens are specifically substances that have been used to induce experiences that are considered "spiritual" by those who imbibe them. They previously were called "hallucinogens" in the scientific literature, but many specialists have chosen to abandon this woefully inadequate term in favor of "psychedelic" and/or "entheogenic" in order to avoid the explicit connotation of "illusory" or "false or deluded perceptions" carried by "hallucinogen."

"Entheogen" overcame the inability of other misnomers and neologisms to properly convey "transcendent and beatific states of communion with deity," and more accurately describe these states within the cultural contexts in which they were and are used. An entheogen has also been defined as "any substance that, when ingested, catalyzes or generates an altered state of consciousness deemed to have spiritual significance." The entheogenic epiphany often involves the experience of the dissolution of distinctions or boundaries between the individual and the mystical or supernatural dimensions of the universe, and a sense of direct communion with pure and primal consciousness or divinity (Hoffman, 2004). Thus, in addition to its theological implications, "entheogen" also carries a distinctly gnostic or deist connotation that implies a direct, unmediated experience of deity, and shares many attributes in common with shamanic practices and belief systems. A general effect of entheogens is high-voltage, slow-wave, synchronous brain activity that is said to increase the connectivity between the emotional and behavioral brain centers, and between lower-brain areas and the frontal cortex, which occurs primarily via serotonin, dopamine, and mesolimbic disinhibition (Winkelman and Hoffman, forthcoming).

These effects on the brain provide an observable basis for the therapeutic use of entheogens and the universality of ecstatic shamanic phenomenology. Therapeutic effects include the activation of emotional processes of the limbic system and paleomammalian brain relating to attachment, emotive responses, and social cohesion (Winkelman, 2001).

Another important aspect of entheogen-inspired experience, and one especially relevant to ancient Mystery Religions, is that of infusing the initiates' mythological framework with a tangibly new, more immediate, and more profound meaning. One author describes this process as the way in which the "supernatural becomes natural" (Winkelman and Baker, 2008), and Sigmund Freud described the psychiatric use of psychedelics as a way to "make conscious the unconscious" (Hanson et al., 2009, p. 314). Perhaps the eminent chemist Alexander Shulgin puts it most simply when he wrote that these (psychedelic) compounds, "have given me ... a personal understanding of just who I am and why I am" (Shulgin and Perry, 2002).

Those who ingest entheogenic substances often describe the effects in terms of experiencing as if through a "spectator ego," "where he or she experiences a bond with nature and society, a sense of overwhelming revelation and truth, and a vivid awareness of his or her surroundings" (http://informahealthcare.com/doi/abs/10.3109/9781420092264.016).

As scholars of mysticism and comparative religious studies have noted for some decades, the descriptions of mystical union and spiritual ecstasy are often indistinguishable from those induced by means of entheogens. While this fact alone is highly suggestive of a cross-cultural entheogenic component, in many (or even most) cases, an entheogen is directly or indirectly implicated within the history of a spiritual tradition itself. Though the identity of specific entheogens utilized in a given rite may be formulaically veiled, the prominence of place reserved for such sacraments within a given religious tradition follows from the fact that entheogenic experiences often have a profound and life-changing impact upon the initiate, affecting their beliefs and behaviors, and are frequently considered one of the peak experiences of initiates' lives (Fadiman, 2001).

In the case of the Eleusinian Mysteries of ancient Greece, a sacred potion was given to the initiates, and among the Vedic Indo-European peoples, the *soma* and *haoma* sacraments were consumed in order to gain spiritual insight, inspiration, and poetic prowess. In these cases, the identity of the specific sacrament was kept in highest secrecy, much like the alchemical Philosopher's Stone, as a cult secret to be closely guarded in order to avoid the profanation of their most sacred principles, and to not offend the divine spirit residing within the sacrament.

While a scientific, nonspeculative, toxicological assessment of the mystery sacraments is impossible given the dearth of knowledge regarding the specific compounds and combinations used in a given mystery rite, the use of certain sacramental plants and fungi is well documented in antiquity, and reasonable hypotheses can be extrapolated by means of multidisciplinary inquiries that employ data from fields as wide ranging as botany, chemistry, archaeology, ethnology, comparative religious studies, mythology, and art history.

In addition to perplexing pharmacological questions, we are confronted with problems related to the academic study of hermetic subjects, in this case Mystery Religions that employed "holy" or "sacred" sacraments. Not only are the specific and unique rites veiled in mystery due to a characteristic and pervasive secrecy, but the mystical states of consciousness with which we are concerned can be highly variable and subjective, especially once removed from their "set and setting" within the cultic context.

25.3 TOXICOLOGY

In the context of the ancient Mystery Religions, scholarly consensus holds that the "abuse" and mundane "recreational" use of entheogens in antiquity was, with few exceptions, impracticable, a fact largely due to the cultural context in which they were set, and the often-extreme secrecy regarding their plant sources and the specifics having to do with their preparation. Certainly, this perspective is valid in the sense that the cultic "set and setting," which included ritual preparations and proscriptions, was antithetical to the development of a profane "drug culture," but this is not to say the use of entheogens was limited to the cultic setting. In fact, the evidence for the widespread use of all manner of intoxicants in the ancient world (Hillman, 2008) contributes to a better understanding of the likely roles they played in the Mysteries. Another important consideration is that, generally speaking, the primary entheogenic compounds simply aren't addictive, and therefore they are not substances of "abuse."

While some of the entheogens discussed herein are known to cause undesirable physiological side effects, the psychospiritual and therapeutic benefits attributed to these experiences far outweigh the relatively benign and fleeting "toxicity" that may, at times, accompany them. This was as true among ancient initiates as it is today among contemporary practitioners. Thus, for instance, transient nausea, cramps, vomiting, or diarrhea is considered a relatively minor inconvenience when compared to the profundity of experiencing what was considered a direct contact with one's deity or deities in an ecstatic communion—an experience that is thought by many to sacralize, heal, inform, and reintegrate the sacred into the profane, and the individual with group psychospiritual dynamics.

Difficult psychological experiences, usually also transitory but sometimes becoming a major or even dominant side effect of a given 'bad trip," can occur. These are relatively rare and are much more likely to take place in individuals with a history of mental instability, or by individuals with highly ordered, inflexible, or rigid personalities (Hanson et al., 2009, p. 315), an exaggerated or narcissistic ego, and moderate to severe psychological disorders.

Problems also occur more often when entheogens are taken outside of a ritual context, and without the supervision and support of an experienced facilitator. In such cases, self-preparation of the drug can result in an inferior preparation, and self-administration can result in overdose and severe psychological and physiological symptoms.

One side effect of entheogenic "toxicity" commonly mentioned is "panic attacks," for which benzodiazepines or other sedative drugs are often indicated in current medical protocol; it is often also suggested that the medical professional provide a "supportive environment." Such symptoms, usually transitory, can also include fear, depression/despair, anxiety, and tension. Along with the physiological discomforts which might occur, these comprise most of what is known as the "ordeal" aspect of initiations, which in the context of the whole experience would have been considered a valuable and therapeutic component of initiation that introduces humility and additional meaning, perspective, and emotional impact to the experience. Thus, mild to moderate symptomatic difficulties would not necessarily have been considered antithetical, nor would they likely have been treated. Special medical attention would have been relatively uncommon or even rare, and been reserved for the treatment of acute symptomatology, such as extreme pain, vomiting, diarrhea/dehydration, anaphylaxis, or other serious allergic reaction, serotonin syndrome, or the onset of a psychotic adverse reaction (otherwise known as a "freakout") where the initiate might hurt themselves or others and disrupt the ritual.

In the ancient context, as today, the facilitators of entheogenic initiations would have been intimately aware of the wide range of potential effects of these drugs and would have developed both psychological and pharmacological treatments to address especially problematic symptomatology. While the psychological "treatment" would have been consistent with the theological and ritual expectations of the ongoing initiation, physical treatment might have included proactive herbal preparations, massage, movement, emetics, counseling, supervision, and restraint and other methods.

The complexity and vagaries regarding the physiological and neurophysiological actions of specific and combined phytochemicals generally indicates an advanced practical knowledge of pharmacology, dosage, and toxicology. Though much can be learned about the theological characteristics of a religion's entheogenic-sacramental complex, as much can be gleaned by studying the surviving, and surprisingly widespread, shamanic uses of these and similar substances.

25.4 SOURCES, CHEMISTRY, AND EFFECTS

Although entheogenic substances are found in a wide variety of flora, fauna, and fungi, their effects derive from similar chemical compounds known as *indoleamines* that variously affect the brain's neurotransmitter systems. The major classes of indoleamines, tryptamines (e.g., DMT, psilocin, and psilocybin), and phenylethylamines (e.g., mescaline, MDA) exert similar influences on serotonergic neurons.

Thus entheogens can be classified and grouped by chemical structure, as well as the compounds from which they are derived. Chemically related substances tend to exhibit similar effects. Other entheogenic compounds are sometimes termed "pseudohallucinogens" in the medical literature because they are said to produce psychotic and delirious effects without the classic visual disturbances of true hallucinogens. This classification is of limited use for our purposes, however, as various entheogens were/are combined, and the entheogenic nature of an experience is determined within a subjective cultural context. The experience also differs qualitatively even within the same culture, by means of multiple initiatory rituals associated with different classes of entheogens. Thus, for instance, the Huichol Indians of Mexico employ various entheogens at various times for various purposes (Hoffman, 2002).

Below is a very cursory survey of some specific compounds and their natural sources that were thought to have been used as primary ingredients or admixtures in various mystery sacraments. Other important psychoactive and potentiating plants such as opium and cannabis are also strongly implicated, but will be omitted from the discussion for the sake of brevity.

25.4.1 *Amanita muscaria*: "Poison" Apple of the Inner Eye

The psychoactive Amanita mushrooms, specifically *Amanita muscaria* and *Amanita pantherina*, have a well-attested entheogenic use among Siberian, European, and Pan-American shamanic peoples and are specifically implicated in the Mysteries of ancient Greece (especially the Mysteries of Dionysus), Rome (Mithraic Mysteries) (Ruck et al., 2011), and as the original Vedic plant-god *soma* (Wasson, 1968), and Avestan *haoma* among the gnostic Manicheans and early and mystically inclined Christians of later periods (Hoffman et al., 2000).

The unique and striking morphology of these "fairytale" mushrooms can often be an important clue to deciphering veiled, yet sometimes obvious, literary or art-historical references to the use of these important entheogens within certain mystery traditions. The decarboxylation of the often "toxic" muscarine levels of these amanitas, which is largely achieved by drying the fruiting bodies, results in the production of muscimole, the desired nontoxic entheogenic compound. The chemical properties of these mushrooms thus serve as a metaphor for the Mysteries more generally, i.e., when the secret of their preparation is known and applied, the "mystery" contained within them is revealed.

25.4.2 Ergot Alkaloids: A Grail Quest

A preparation of ergot alkaloids is strongly implicated as the key ingredient in the *kykeon* potion of the Greater Eleusinian Mystery, but the safe and practical process for its preparation in ancient times is yet to be rediscovered (Wasson et al., 1978). Due to the dangerous toxicological effects of some ergot alkaloids and the uncertainties associated with applying chemical techniques available in antiquity, very few human bioassays of possible ancient "recipes" have occurred, and the sound theoretical research remains speculative.

25.4.3 Psilocybin and Psilocin (Mushrooms) and DMT, 5Meo DMT: Spirit Molecules

These closely related entheogenic compounds should be mentioned in a discussion of the Mystery Religions. While evidence for their use in this context has not been fully explored, it is extremely unlikely that the chemical properties of psychoactive mushrooms and the natural sources of DMT would have been overlooked by ancient herbalists and alchemists.

25.4.4 Tropane Alkaloids

While it is highly debatable whether, and to what extent, the tropane-containing plants were used as primary ingredients in mystery sacraments *per se*, they deserve a special mention in terms of toxicity as pharmacologically hazardous compounds. Used as a supplement by contemporary preparers of ayahuasca to their DMT/beta-carboline brew, the dosage tropane additives must be carefully controlled in order to avoid acute and long-term deleterious effects. Atropine, scopolamine, and hydrocine, and perhaps other tropane alkaloids, are said to enhance and intensify the effects of compounds such as opium, cannabis, and other entheogenic compounds. Tropane alkaloids were and are used as primary ingredients for shamanic initiations among a number of the shamanic peoples of the Americas and were key ingredients in the medieval witches' "flying ointments."

References

Fadiman, J., 2001. Psychedelic Explorer's Guide: Safe, Therapeutic and Sacred Journeys. Park Street Press, Rochester, VT.

Hanson, G.R., Venturelli, P.J., Fleckenstein, A.E., 2009. Drugs and Society, 10th ed. Jones and Bartlett, Sudbury, MA [cites Snyder 1974. P. 44], Jones and Bartlett, Sudbury, MA 2009.

Hillman, D.C.A., 2008. The Chemical Muse: Drug Use and the Roots of Western Civilization. Thomas Dunne Books, St. Martin's Press, New York, NY.

Hoffman, M., 2002. Huichol wolf shamanism and A. muscaria. In: Hoffman, M. (Ed.), Entheos: J. Psyched. Spirit. 1 (2), 43–48.

Hoffman, M., Ruck, C.A.P., 2004. Entheogens (Psychedelic Drugs) and Shamanism. ABC-CLIO, Santa Barbara, CA, pp. 111–117.

Hoffman, M., Ruck, C.A.P., Staples, D.B., 2000. Conjuring Eden: art and the entheogenic vision of paradise. In: Hoffman, M. (Ed.), Entheos: J. Psyched. Spirit. 1 (1), 13–50. ⟨http://informahealthcare.com/doi/abs/10.3109/9781420092264.016⟩.

Ruck, C.A.P., Bigwood, J., Staples, D., Ott, J., Wasson, G., 1976. Entheogens. J. Psychedelic Drugs 11 (1–2), 145–146.

Ruck, C.A.P., Hoffman, M., Celdran, J.G., 2011. Mushrooms, Myth and Mithras: The Drug Cult that Civilized Europe. City Lights, San Francisco, CA.

Shulgin, A., Perry, W., 2002. Simple Plant Isoquinolines. Transform Press, Berkeley, CA.

Wasson, G., 1968. Soma: Divine Mushroom of Immortality. Harcourt, Brace & Jovanovich, New York, NY.

Wasson, G., Ruck, C.A.P., Hofmann, A., 1978. The Road to Eleusis: Unveiling the Secret of the Mysteries. Harcourt, Brace & Jovanovich, New York, NY.

Winkelman, M., 2001. Psychointegrators: multidisciplinary perspectives on the therapeutic effects of hallucinogens. Complement. Health Pract. Rev 6 (3), 219–237.

Winkelman, M., Baker, J., 2008. Supernatural as Natural: A Biocultural Approach to Religion. Pearson North America, Upper Saddle River, NJ.

Winkelman, M., Hoffman, M., forthcoming. Hallucinogens and entheogens. In: Segal, R., von Stuckrad, K. (Eds.), Vocabulary of Religion. Brill, Amsterdam.

Further Reading

Barile, F., 2010. Clinical Toxicology, Principles and Mechanisms, 2nd ed. CRC Press, Boca Raton, FL.

Dobkin, M.D., 1984. Hallucinogens: Cross-Cultural Perspectives. University of New Mexico Press, Albuquerque, NM.

Fantegrossi, W., Mernane, K., Reissig, C., 2008. The behavioral pharmacology of hallucinogens. Behav. Pharmacol. 75, 17–33.

Griffiths, R.R., Richards, W.A., McCann, U., Jesse, R., 2006. Psilocybin can occasion mystical-type experiences having substantial, sustained personal meaning and spiritual significance. Psychopharmacology 187 (3), 268–283.

Harner, M.J., 1973. Hallucinogens and shamanism. In: Harner, M.J. (Ed.), Hallucinogens and Shamanism. Oxford University Press, New York, NY.

McKenna, T., 1992. Food of the Gods: The Search for the Original Tree of Knowledge. Bantam Books, New York, NY.

Nichols, D., 2004. Hallucinogens. Pharmacol. Ther. 101, 131–181.

Nichols, D., Chemel, B., McNamara, P., 2006. The neuropharmacology of religious experience. In: McNamara, P. (Ed.), Where God and Science Meet: How Brain and Evolutionary Studies Alter our Understanding of Religion. Praeger, Westport, CT.

Ott J. Pharmacotheon: entheogenic drugs, their plant sources and history. 1993.

Ruck, C.A.P., Staples, B.D., Heinrich, C., 2001. The Apples of Apollo: Pagan and Christian Mysteries of the Eucharist. Carolina Academic Press, Durham, NC.

Ruck, C., Staples, B., Celdran, J., Hoffman, M., 2007. The Hidden World: Survival of Pagan Shamanic Themes in European Fairytales. Carolina Academic Press, Durham, NC.

Schultes, R., Hofmann, A., 1979. Plants of the Gods. McGraw-Hill, New York, NY.

Wasson, G., 1980. The Wondrous Mushroom: Mycolatry in Mesoamerica. McGraw-Hill, New York, NY.

Wasson, R.G., Kramrisch, S., Ott, J., Ruck, C.A.P., 1986. Persephone's Quest: Entheogens and the Origins of Religion. Yale University Press, New Haven, CT.

Ancient Mystery Initiation: Toxic Priestesses and Vaginal Communion

David Hillman

Indepedent Scholar, Specializes in Ancient Pharmacy and Medicine, Madison, WI, United States

OUTLINE

Toxicology in Antiquity
DOI: https://doi.org/10.1016/B978-0-12-815339-0.00026-3

26.1 INTRODUCTION

Ancient Mediterranean religious rites preserve evidence for the practice of a unique polyvalent pharmacy in which the human body was used to process, store, and administer ritual communal substances along with their so-called antidotes. Colleges of priestesses—environments in which women were recognized as physicians, oracles, and experts in the art of drug preparation—promoted the development of compound mixtures of plant- and animal-derived toxins that were used in rites meant to induce observable psychosis in voluntary human subjects. Cult members participating in these mystery operations consumed the blood, breast milk, and vaginal excreta of priestesses who were themselves chronically exposed to viperid venoms by means of vaginal suppositories and self-induced dermal lacerations. The consumption of a priestess' medicated body fluids within the context of private and semipublic initiation rituals was considered the culmination of an initiate's personal religious experience, a form of divine marriage.

Communal substances, administered to initiates during ritual sex acts, formed an important and complex component of ancient mystery religion performances. Drug concoctions, compounded by specialist priestesses, contained numerous plant- and animal-derived toxins, creating a complex pharmacological process whereby communal drug mixtures induced a much-desired ecstatic psychosis in participants while concomitantly protecting these same subjects from the adverse side effects of venom and toxin exposure.

Using the technical language developed by Late Bronze Age (c. 1200–1000 BC) mystery religions preserved within classical Greek and Latin religious and medical texts, it is possible to trace these sophisticated pharmaceutical practices back to a single priesthood and its founding priestess. Medea (Μήδεια, Gk.), an historically prominent monarch of the eastern Black Sea kingdom of Colchis (modern Republic of Georgia) was considered the most drug-savvy priestess of her generation and the founder of what has come to be known by historians as the Orphic Mysteries.

Medea was credited with the development of unique snake-derived concoctions as instruments of religious practice and general health. Her research can only be described as a period-specific, uniquely pioneering combination of gynecology, pharmaceutical experimentation, and religious ritual. Medea was so popular with her contemporaries that the Medes claimed to have changed their cultural identity from Aryan to Medic in order to recognize her as their ultimate patroness. In addition to establishing numerous temples in the Hellenic world, Medea founded the famous eastern religious fraternity of drug-using philosopher-physicians known as the Magi.

Classical Greek and Roman literary sources considered the toxicological experimentation of ancient mystery priesthoods with potentially lethal venoms to be the foundational spark of all Western medicine. In this way, the Greco-Roman world believed pharmacy as a science, was developed directly from the invention of communal toxicological practices designed to balance the desired effects of venoms with unwanted side effects. Western pharmacy, as we know and practice it, developed from experimentation with drugs employed in sexually explicit mystery religion rituals.

26.2 TOXIC PRIESTESSES AND THEIR MYSTERY RITES

The Drakaina—or Dragon Priestess—was a prominent presiding authority in the performance of ancient mystery initiations.[1] Considered to be masters-of-poisons, dragon priestesses performed necromantic initiation rituals meant to induce a resurrection-state in novitiates, a process called "new-birth." This metaphysical transformation was viewed as the soul's entrance into the *aion* (αἰών), a transdimensional space that simultaneously bounded and encompassed—in terms of ancient physics—the time stream of the material cosmos.[2]

After ancient mystery initiates received "aionic life" (a term translated into English as "eternal life"), they took up, as a brotherhood, the practice of a rigorous religious discipline known simply as "the Way."[3] This new-birth of mystery initiates was accomplished by the

---

[1]The Drakaina, or δράκαινα, was said to bring about her powers by mixing/preparing drugs in potable or consumable form. There are numerous named Drkakainai, among the most famous being Clytemnestra. Aristophanes was accused of profaning (revealing) the Mysteries in his trilogy on Orestes. In his *Libation Bearers*, Aeschylus records that the Drakaina produced and administered a mixture of blood and milk after being bitten in the breast by a snake/dragon (514 ff.) Orestes even describes his own "snakeification" or transformation into the Dragon. (Line 549: ἐκδρακοντωθεὶς δ᾽ ἐγώ, "I myself too have brought out the dragon.")

[2]On αἰών: The aion was not only the "space" surrounding the time stream, but also a divinity directly linked with the performance of mysteries. Aion was the Orphic Protogonid divinity known as Phanes and pictured as emerging from an egg wrapped in snakes. He was considered the guardian of the Oracle of Night, herself an Echidnaic deity.

[3]The earliest Orphic use of the term "the way" can be found in the *Hymns*, where the concept is linked to the functioning of Hecate and the Bacchic "third time" of the Aionic cycle. The first hymn is addressed to *Hecate* (#1) and opens with multiple epithets for goddess of "the way" (e.g., εἰνόδιος or "goddess of the way").

consumption of a compound drug mixture followed by the supplemental ingestion of blood, milk, and ejaculate derived from an "only-begotten savior" belonging to the "kingdom of God."[4]

The dragon-priestess, herself an immaculate virgin and "only-begotten" child of God, was responsible within her cult for "burning off the mortality" of newborns.[5] This process involved a complex magico-medical procedure and employed numerous botanically and zoologically derived pharmaceuticals.[6] Priestesses chronically exposed newborns to the fumes of compound incense mixtures while applying dermal salves. This process of subjecting a child "to the fires" became codified in numerous myths related to popular mystic figures.[7]

The specific rubrics of ancient mystery initiation ceremonies were intentionally preserved in the writings of Greek and Etruscan priests and priestesses in the form of mythic narratives characterized by the use of a highly specific ritual vocabulary (John Lydus, *De Ostentis. 2.6. B*). This cult *vox*, a religious "voice," was characteristic of the language of the mysteries.[8]

These practices involved the use of venoms of a wide variety of snake species. Milked venoms were added to compound healing ointments and applied to dermal lacerations made on human subjects during ritual cutting or whipping. Sometimes these subjects were

---

[4]A single technical vocabulary in Greek was used in the performance of pagan and Christian mysteries. Otto Kern in his edition of the *Orphic Fragments* recognized this as the *vox Orphica*. The Greek words for "savior" (σωτήρ) and "only-begotten" (μονογενής) are Orphic technical terms employed in pre-Classical religion.

[5]Apollonius, *Argonautica.* 4.869–72.

[6]Apollonius *Arg.* 4:1135 ff. refers to the Mystis (μύστις), the mystery priestess who nursed Dionysus; the infant god, like Achilles, was "brought through the fire." Achilles was reportedly the only one of his siblings to survive the "fire." Mystis (whose name means "initiator"), Medea's Cytaean maids, various nymphs, and other pharmacomagical practitioners are credited with the production and administration of a serous vaginal exudate that was mixed with sacramental compound drug concoctions. See Apollonius on λευκόν: ἄνθεα δέ σφιν νύμφαι ἀμεργόμεναι λευκοῖς ἐνὶ ποικίλα κόλποις. LSJ: λευκόν: 5. [select] τὰ λ. the menstrua alba of young girls.

[7]The most famous example being Demophoon's submission to the rite by the goddess Demeter as she searches for her abducted daughter Kore as seen in the Homeric *Hymn to Demeter*, 231ff. The episode of passing a child "through the fires" is prohibited in the book of Leviticus (Ch 18–20), where it is compared to an act of sexual adultery in meeting with the mystery standard of the "holy union" between the initiate and the god.

[8]The *vox Orphica* is a term used by Otto Kern, editor of the *Orphic Fragments*. It denotes a technical language employed by mystery religion participants.

infants who had been "devoted" to a temple or had been given up for adoption to oracular sisterhoods.[9]

The ancient prophetic voice was considered to be accessible only to those influenced by the "divine sprit" while following "the Way." Under the mental sway of a *Drakaina's* sexual "anointing," the initiate was said to become a guardian of an oracular voice; the bearer of the ambrosial word (λόγος, Gk.) of God.

26.3 COLLEGES OF ECHIDNAE

Dragon-priestesses from the Late Mycenaean Bronze Age (c. 1200–1000) as presented in Archaic, Classical, and Hellenistic literature (c. 900–100 BC), present all the characteristics of professional associations or colleges concerned with the perpetuation of the mysteries and their associated drug rites. Evidence of priestly colleges is reflected in both the mythic background of widespread religious rites as well as the specific drugs and drug-associated rites used by particular cults and the representation of these drugs in ancient pharmaceutical literature. For example, mystery-rite priestesses known as *Echidnae* (the Greek word for viper) ritually employed viperid venom along with numerous standardized "antidotes" needed to ameliorate the negative side effects these venoms produced. *Echidnae* priestesses were almost exclusively depicted in art and literature in association with serpents, pharmaceuticals, and ritual magic.

The titles used by *Echidnae* priestesses derived from their own peculiar duties within the mystery rite itself; descriptive titles were given to women and girls involved with varying aspects of the rite. For example, the *Medusae* (Medusae, L.) were female temple guardians infamously equipped with arrows made lethal when drawn through the guardian's poisoned hair.[10] The serpentine locks of the *Drakaina* and her assistants were pretreated with concoctions containing numerous poisons.[11] These *Medusae* were said to be able to paralyze, or "turn to stone," anyone

[9]Adoption or "devotion" to a temple was a widespread practice in the ancient Mediterranean. Numerous literary figures are either adopted by priesthoods or spend time as children under the tutelage of religious organizations. For example, Ion (pagan) and St. Mary (Christian) were both devoted to temple service, while Achilles, Jason, and Hercules were all directly connected with pharmaceutically savvy mystic tutors.

[10]Echidnae priestesses are classically portrayed as having oiled hair, full of vipers. For example, see Lucan, *Civil War*. 6.490. Bacchus and his followers were also characterized by the wearing of vipers in the hair: Euripides, *Bacchae*. 99–104.

[11]The "arrow toxins" in ancient Greek literature represent a group of botanical and animal-derived toxins/venoms. They are among the sacramental "drinks" for which Scribonius Largus prescribes antidotes. *Compositiones* #188 ff.

who dared to approach a dragon priestess or oracle without her explicit consent.[12]

Medea, the founder of the poison-wielding Medo-Persian Magi, bore the same priestly linguistic root (*med-*) in her name, and was said to be both the first priestess of the Dragon-priesthood as well as "the most accomplished in drug use." The same root (*med-*) is found in Latin words connected to the practice of the "*medicus*" (physician).[13]

26.4 COMMUNAL IOS-RITE

Dragon priestesses entered an ecstatic or Bacchic state of mania, during which they claimed to experience and influence transdimensional forces or "daimones" (demons) by manipulating "waves" of time–space distortions produced by "darkened stars."[14] The Greeks used the word ἔνθεος to describe this strange star-induced ecstasy, and mystics claimed it was part of a process of manifesting the "god within the breast" of the initiate.[15] In antiquity, entheogenic drugs, administered in concert with sexual acts, were considered an avenue for facilitating a sacred mystery union of an initiate and a divinity; this union was considered a "divine marriage."[16]

Dragon priestesses used their own bodies to mix, store, and administer ritual drugs. The vagina (κόλπος) and breasts of priestesses became

[12]The Echidnaic Medusae were featured as arrow-wielding guardians of Diana's priestess in Euripides' *Iphigenia in Tauris*. Orestes, even unbound, is unable to approach the priestess because of their presence.

[13]On the significance and derivation of the Med-root, see: Thelma Karen (1951).

[14]κῦμα (G.) "waves" or botanical "sprout." The metaphysical mechanism of harnessing a "dark star" for interdimensional travel between the Orphic terrestrial realm and the realm of the underworld is featured in Claudian's *Rape of Proserpine*. Claudian mentions numerous mystery cult-related mythological themes, as well as a description of the use of arrow poisons (toxins) in the midst of a description of transdimensional travel (179–203). According to John Lydus, the Etruscans taught that myth was a means of conveying complex interdimensional physics to limited chronic beings. That is, myth was allegory invented by Etruscan (and early Roman) priests in order to encode their rites. See Lydus De Ostentis 2.6.B.

[15]ἔνθεος, to indicate an intoxicated state of "the god in the breast" is used as early as the Orphic hymns (pre-Classical) and as late as the epic poetry of Nonnos (late Roman Empire). It is an epithet of Bacchic divinities involved with ecstatic states, and was a technical term of the *vox Orphica*.

[16]A good example of the ios-rite "gamos" or "sacred union" can be found in Euripides' account of the union of Apollo and Creusa. *Ion* 887ff. In this passage, Creusa inserts meadow saffron into her vagina.

biological synthesizers of sacred communal substances.[17] Priestesses performed sexually explicit dances accompanied by ecstatic songs and incantations that culminated in the forced ejaculation of drugged, physically restrained, partially paralyzed initiates.[18] During the performance of these mysteries, initiates consumed the vaginal and mammary excretions of the *Echidnae* priestesses. Deprived of the bodily fluids of these priestesses, the rite was reportedly fatal.[19]

Ancient mystery initiation was an elegant feat of biochemical engineering staged as a profoundly influential religious performance; ancient communion was quite simply the ritual exchange of medicated bodily fluids between priestesses and novitiates.[20] The ecstasy experienced during these rites was notorious. The breast milk, ejaculate, and blood of participants contained seemingly innumerable physiologically active compounds directly credited with producing the profound physical and mental symptoms endured by religious initiates.[21] This interplay between drugs and the human body was the basis of the Western concept of communion.

26.5 THE SCIENCE BEHIND THE RITUAL

The central scientific premise of ancient ritual initiations was a recognition of the innate tendency of drugs to interact with each other and

---

[17]The Latin term for the body as a vessel used to prepare drugs was "officina" and was used by Horace to describe the power of the witch Canidia in her use of Colchian drugs (*Epodes* 17.35).

[18]The rite was satirized by Petronius: Satyricon, 131.

[19]Seneca (*Hercules on Oeta*) chooses the theme of death brought about by the use of the *ios*-rite drug bereft of its corresponding "antidote." Seneca elaborates on the specific symptoms of *ios*-poisoning found both in Galen and Nicander. In addition, Seneca relates Orphic mystery initiation to the action of the play: 299–306.

[20]Mystery initiation drugs were known abortifacients. Jerome laments the use of these abortifacients by "lost Christian virgins" who perform ritual acts involving intoxicants: Jerome, *Letters*. 22.

[21]See: Scholia on Theocritus' iunx: Scholiast directly asserts that the sexual union (i.e., Zeus and Maiden or Zeus and the Hera) is the female *pharmaceutizing* the Zeus. The Greek verb φαρμακεύω, "to apply the pharmakon" (φάρμακον) connotes the application of drugs in the process of sexual engagement. The Victorian minds behind the authoritative LSJ lexicon considered this combination of drugs and sexuality to be an act of the female "charming" the male. The pharmakon, or drug, then becomes the sexual excreta of the female partner; sometimes referred to as "ambrosia" or "honey." The process of applying/administering these drugs/sexual rites was called φαρμακεία (pharmakeia), or in less-than-modern parlance, "sorcery."

the human body in ways that potentiate or augment their observable physiological effects. This basic but sophisticated pharmaceutical observation formed much of the intellectual foundation of classical alchemy, magic, and women's medicine. Ancient priestesses saw the rational and explicable interaction of drugs with the human body as the material basis for genuine religious experience; ancient magic was considered a mastery of elemental transformation.

The mystery rites of the *Echidnae* priestesses centered on the use of "*ios*," a family of animal-derived poisons used widely in antiquity as lethal toxins applied to projectiles. Wounds inflicted by ancient "arrow poisons"—as they were known—were considered potentially lethal and oftentimes completely untreatable.[22] *Ios* (ἰός, Gk.), translated as "poison" or "venom," was associated with bodily fluids produced by "mania-inducing" creatures, including, among others, the venoms of numerous snake species.[23]

Ios venom was said to produce a maddening mental state the Greeks termed oistros (οἶστρος). Eye-witness accounts of physicians and priests depict people under the influence of these arrow poisons as suffering from an acute and potentially dangerous psychosis;[24] consumers of arrow poisons reportedly suffered horrifying visual hallucinations and severe delirium that sometimes led to acts of violence and even murder.[25] Those with access to *ios*-related drugs in extra-religious settings—like Roman taverns—could be seen roaming the streets of large cities, often crawling on-all-fours while vocalizing like herd animals; some religious zealots under the influence of arrow poisons even made a practice of occupying public squares and busy thoroughfares where they railed about invisible, greed-punishing demons and the coming apocalyptic doom of the greed-stricken "generation of iron."[26]

---

[22]Oenonoe, first wife of Paris, herself skilled in drugs (φαρμακουργός), recognized the lethality of the arrow poison that struck her unfaithful husband. Lycophron, *Alexandra*. 61–64.

[23]Arrows dipped in ios were considered "burning poisons." Heracles' was famous for his arrows covered in the burning ios. See Valerius Flaccus, *Argonautica*. 1.107 ff. Ios was also associated with "burning" sexual passion. Jerome addressed the impropriety of using intoxicants in ritual to enflame passion: Jerome, *Letters*. 52. Heracles death was attributed to the same poison: Diodorus Siculus, *Bibliotheca Historica* 4. 38. 1.

[24]Scribonius, *Compositiones* #194. Scribonius, a Roman military doctor (I CE), includes "Toxicum" a literal "arrow poison" used as a drink in ritual initiation in his discussion of the *mala medicamenta:* drugs developed from venoms that could be lethal or curative.

[25]Nicander, *Alexipharmaca*. 207 ff.

[26]See Nicander, *Alexipharmaca*. 1–259, where the priest/physician discusses the oracular uses of *ios*-related drugs. See also Scribonius, *Compositiones* #139.

26.6 A COMBINATION OF POISONS AND ANTIDOTES

The maddening effects of the arrow poisons were considered to be a form of Bacchic ecstasy, a type of religious possession; an inspiration or indwelling of the "divine spirit." The goal of this Bacchic initiation was the union of the initiate with the voice of the divinity possessing the priestess. The temporary, and oftentimes disorienting mania, was the desired outcome of these drug-induced mysteries.[27] Adverse effects of the *ios*-rite, like excessive sweating, painful burning sensations, spasms, and uncontrollable bleeding were all considered treatable or amenable with the application-specific compound "antidotes."[28]

The *Echidnae* used the colorless, serous breast exudate of prepubertal female attendants—under the influence of chronic viper venom administration—as one means of ameliorating the side effects of the *ios*-rite.[29] This so-called "virgin milk" acted as the rite's antidote to the administered arrow poison.

Priests and physicians viewed breast milk as blood transformed by mammary tissue into a fine, nourishing substance with the same physical properties as semen.[30] "Virgin-milk" was not considered to be simply the product of the female, developing breast, but a fortified drug capable of conferring upon its recipient the peculiar, healthy, inspired nature of its biologic source.[31]

---

[27]Olympias, the mother of Alexander the Great, infamously partook in these Bacchic, Echidnaic rites: "...all the women of these parts were addicted to the Orphic rites and the orgies of Dionysus from very ancient times (being called Klodones and Mimallones) and imitated in many ways the practices of the Edonian women and the Thracian women about Mount Haemus, from whom, as it would seem, the word "threskeuein" came to be applied to the celebration of extravagant and superstitious ceremonies. Now Olympias, who affected these divine possessions more zealously than other women, and carried out these divine inspirations in wilder fashion, used to provide the reveling companies with great tame serpents, which would often lift their heads from out the ivy and the mystic winnowing-baskets, or coil themselves about the wands and garlands of the women, thus terrifying the men." Plutarch, *Life of Alexander* 2ff. Translated by Bernadotte Perrin.

[28]Ancient medical literature on antidotes is detailed: see Galen, *Antodotes* K 14, Scribonius *Compositiones* #165–199. Antidotes for both animal- and plant-derived poisons/toxins were commonly grouped together in medical sources.

[29]The young age of participants in sexually explicit rites/ceremonies would not have been a foreign concept to the ancient Greek or Roman. See Ion. fr. eleg. 26., where the milking of children is a theme used in the context of the Bacchic chorus (revelry).

[30]Macrobius, *Saturnalia*. 5.11.15-17. The Greco-Roman world considered both masculine and feminine ejaculate as a single substance, like breast milk, capable of imprinting upon an infant the traits of the person who produced it. The words for blood and semen can be used interchangeably.

[31]For oracular, virgin milk, note Apollonius, *Argonautica* 4.1731 ff.

Suckling from a intoxicating nurse was a common theme in ancient literature.[32] Zeus was suckled by an ambrosia-delivering she-goat.[33] Athena herself was depicted as a paradigmatic example of these nursing, adolescent priestesses, whose milk was used as an antidote in venom-related rituals.[34] Apollo was associated with packs of wolves known to guard and breast feed infants.[35]

Because the *ios*-rite induced a potentially dangerous psychosis, officiating priestesses physically restrained initiates during the ritual administration of arrow poisons. Priestesses bound the feet and hands of novices and placed them in supine positions favorable to the rectal administration of drugs by means of penis-shaped applicators made of leather, glass, or polished stone.[36] Initiates were given an oral sedative and physically restrained before the anal application of *ios*-rite arrow poisons.[37] Once unwanted symptoms of the arrow poisons set in, priestesses breast-fed initiates, an act that reportedly furnished them with an ameliorating "antidote" capable of limiting any predictable, negative side effects.[38] Ancient authorities recognized the valued contribution of

---

[32]Scribonius Largus, a Roman military physician (medicus—I CE) describes the use of "lac" in a section of his *Compositiones* dedicated to mystery rite poisons/venoms. Greek and Roman medical references to "milks" include mammalian breast milk as well as the "milky" exudates of numerous botanicals. Classical medical "milks" are fluids produced by a plant or animal that can be used to "nourish" or "alter" living tissue. Scribonius lists "The Milk" as one of numerous transformative toxins.

[33]Callimachus, *Hymn to Zeus*. 1–55. As per the *ios*-rite, the priestess of the rite possesses the divine milk and is herself guarded by a dog (κύων). In the case of Zeus the dog/guardian is "golden," reflecting the Orphic mystery.

[34]Nonnos, *Dionysiaca* 13.171.

[35]Antoninus Liberalis, *Metamorphoses*, 30. Consider also the myth of the suckling of Romulus and Remus by a "wolf" named Lupa (lit. "prostitute" or in some cases "priestess').

[36]The Priapic dildo (alabastron/fascinum) was employed in magico-medical ritual as a pharmaceutical applicator. Drugs were rubbed onto the dildo before it was inserted into the rectum where the drug was absorbed. The phallus was used in Roman wedding ceremonies: Augustine, *City of God*. 6.9. The worship of Priapus was cultivated by ancient "sorceresses," or "workers in drugs." For descriptions of ancient dildos and evidence concerning ritual, forced sodomy, see: Richard Burton, *Priapeia*. pp. 137–68.

[37]Petronius satirizes the rite in his "Satyricon." The rites are Priapic and betray the use of medical vocabulary that implies association with contemporary medical/pharmaceutical practices. On the overlap of Priapic worship, medicine, and eroticism, see Sir Richard Burton's *Priapeia* (Wordsworth, 1995).

[38]Galen's multiple works on "antidotes" contain numerous descriptions of the process of balancing lethal drugs with their apparent antidotes. His formulae are sometimes based on famous "Theriaca" the priestly formulae for cult-drugs that schematically present the poisons and their antidotes. See Galen, K14.

venom-investigating mystics and sorceresses who pushed the medical frontier by pioneering the use of lethal venoms as potential curatives.[39]

After receiving both elements of the communion, namely the arrow toxin and its balancing "antidote," initiates of the *ios*-rite were manually masturbated by priestesses performing sexually explicit dances. During these ritual performances, priestesses and their female attendants reportedly urinated and ejaculated on psychotic, restrained initiates while forcefully bringing them to sexual climax.[40] Licking the genitalia of pubertal and prepubertal priestesses in order to receive medications was a prominent ritualistic practice.[41] The initiate's experience was considered visionary.

Ios-rite drugs applied to the anus of initiates were characterized by their purple or deep-scarlet color (*ios* can also be translated as "violet"), which would stain the anus, vagina, and mouth.[42] It should be noted that the vaginal excreta were themselves considered drugs and drug components. The Greek Magical Papyri (PGM) preserve the actual expressions for specific types of vaginal excreta. For example, ἰχῶρα παρθένου ἀώρου, was

---

[39]On the contribution of venom research to medicine in antiquity, see Philostratus' *Life of Apollonius*, 3.44: "And who," he said, "can deprive the art of divination of the credit of discovering simples which heal the bites of venomous creatures, and in particular of using the virus itself as a cure for many diseases? For I do not think that men without the forecasts of a prophetic wisdom would ever have ventured to mingle with medicines that save life these most deadly of poisons."

[40]The medical use of bodily fluids, including urine and saliva was not unusual. See Pliny, HN, 28.6, 22. The paelex (L) or πάλλας (G) was the assistant of the priestess, and was represented as approximately 9–12 years. See Seneca. *Hercules on Oeta*, 278 ff.

[41]Cunnilingus as a medico-religous act was common enough for Aeschylus to gingerly play with the theme in his *Agamemnon* 1478–80: ἐκ τοῦ γὰρ ἔρως αἱματολοιχὸς νείρᾳ τρέφεται, πρὶν καταλῆξαι τὸ παλαιὸν ἄχος, νέος ἰχώρ. ... "The love for licking seminal fluid is nourished by the groin ..." Of course, νείρᾳ for "nethers" or "groin" is properly the region of the body under the belly, or the genitalia.

[42]The purple color of the ios-rite concoctions was associated very early on with Medea and the preparation of her "Colchian Fire." The Latin word for "poison" itself is influenced by the Medic ios-rite traditions: "venenum" is also a "purple dye." For example: *alba neque Assyrio fucatur lana veneno*. Georgics. 2.465. Assyria is used by poets for Media and the Medes and the other ios-rite stars like Europa and the Tyrian purple. (See OLD): Assyrĭa, ae, f., = Ἀσσυρία, I.a country of Asia, between Media, Mesopotamia, and Babylonia, now Kurdistan, Plin. 5, 12, 13, § 66 al.—Hence, Assyrĭus, a, um, adj., = Ἀσσύριος, Assyrian, Verg. E. 4, 25; Luc. 6, 429; Stat. S. 3, 3, 212 al.; and Assyrĭi, ōrum, m., the Assyrians, Cic. Div. 1, 1, 1; Plin. 6, 13, 16, § 41; Vulg. Gen. 2, 14; ib. Isa. 7, 17 al.—Sometimes poetic for Median, Phrygian, Phœnician, Indian, etc.; so, "puella," i. e. the Phœnician Europa, Sen. Herc. Oet. 554: "venenum, i.e.," Tyrian purple, Sil. 11, 41: "stagnum," i. e. Lake Gennesareth, in Palestine, Just. 18, 3: "ebur," i. e. Indian, Ov. Am. 2, 5, 40: "malus, i. e. Medica," the citrontree, Plin. 15, 14, 14, § 48; cf. Voss ad Verg. G. 2, 126.

the serous exudate of nonmenstruating, prepubertal girls.[43] Ios-rite *Echidnae* priestesses were themselves known to self-apply the "burning purple" (the *ios* drugs) to their own vaginas before the performance of sacred sexual rites.[44]

Cunnilingus was sometimes referred to as "going Phoenician," an expression that plays on the perpetuation of the *ios*-rite and the use of one of its main ingredients—a famous purple dye derived from sea mollusks.[45] Echidnae priestesses administered their arrow poisons (snake venoms) to initiates via forced, ritual cunnilingus and by small stones passed into the rectum. Interestingly, these priestesses claimed to have complete immunity to the snake venoms used in their rites.

26.7 IDENTIFYING DRUGS USED IN THE IOS-RITE

Ancient texts written by physicians, priests, mystics, and naturalists preserve countless cult-related drug recipes, including those used by specific cult associations like that of the *Echidnae*. Despite an abundance of literary sources, it is notoriously difficult to accurately identify the

---

[43]PGM 4.2645.

[44]ἰχώρ (LSJ): 2. serous or sero-purulent discharge, Hp.VC19, Arist.HA630a6 (pl.), Gal.10.184, etc.; ἰχῶρες ὑδαρεῖς ὕπωχροι, from women in childbirth, Arist.HA586b32; of the putrefied blood of a viper, Id.Mir.845a8; of naphtha (prob.), regarded in legend as due to the putrefaction of giants' corpses, ib.838a29. This ichor, the discharge of prepubertal girls was associated with both Medea and her Naphtha. Both were said to burn, and to contain the breath of fire. Serum of a virgin, blood of a viper, and naphtha all have their source in the Medea.

[45]See *The Cunnilinges* in Richard Burton, *Priapeia.* pp. 165–168. The ios-rite is symbolized in the Orphic *Argonautica* (707–713), where the initiates are presented as the guardian dogs who are turned from calm to madness by the goddess/priestess. "At one of the gateposts there stood [a statue of] the far-seeing queen, scattering with her motion the radiance of fire, whom the Colchians propitiate as Artemis of the gate, resounding with the chase, terrible for men to see, and terrible to hear, unless one approaches the sacred rites and purification, the rites kept hidden by the priestess who was initiated, Medea, unfortunate in marriage, along with the girls of Cyta. No mortal, whether native or stranger, entered that way, crossing over the threshold, for the terrible Goddess kept them away by all means, breathing madness into her fire-eyed dogs." (Translated from the Latin by Jason Colavito, The Orphic Argonautica (2011). The Greek captures the use of the Orphic technical vocabulary—the vox Orphica in its references to the drugs: ει μή τις τελεταίς πελάσει καί θύσθλα καθαρμῶν, ὅσσα περ αρήτειρα καθάρματα μύστις ἔκευθε, δεινολεχής Μήδεια Κυτηιάσιν μίγα κούραις. Οὐδέ τις ενδοτέρω κείνην οδόν εισεπέρησεν ενδάπιος, ξείνός τε βροτῶν υπέρ ουδόν αμείψας: είργει γάρ πάντη δεινή θεός ηγεμόνεια, λύσσαν επιπνείουσα υπέρ γαληνοίς σκυλάκεσσιν. Galene, the antidote for the Medic ios-rite is even referenced: γαληνοίς.

specific species of plants and animals found in classical manuscripts. The process frequently degrades into something resembling educated guesswork rather than verifiable scholarship.

While there are thousands of pages of toxicology-related texts in Greek and Latin, particularly the drug books of Galen, Dioscorides, Nicander, and Scribonius Largus, the overwhelming majority of these texts have yet to be translated into English. Classical scholars, trained in mastering dead languages like Latin and Greek, are frequently discouraged by the highly technical subject matter and vocabulary contained in surviving toxicology-related classical sources.

Despite such research obstacles our texts abound with innumerable references to plant- and animal-derived substances used for centuries as both simple and compound mixtures in the maintenance of health and the performance of religious ritual. By compiling and examining references to specific drug formulae scattered in multiple genres (medical, philosophical, dramatic, religious, magical), it is possible to consider the bigger picture of widespread drug use within the context of ancient cult performance.

There are numerous drugs associated with the activities of Dragon priestesses and the colleges of *Echidnae*. Many of these compound drug mixtures are consistently linked with the historical figure of Medea, whom the Greeks and Romans (and arguably the Etruscans) considered the progenitor of the "Medes" and the founder of the magicoreligious order called the Magi.

26.8 MEDEA THE FIRST TOXIC PRIESTESS

Medea, princess of Colchis—a late Bronze Age (c. 1200−1000 BC) kingdom of the Black Sea—was personally associated with the cult rites of Diana, Hera, and Hecate; known for her physical beauty and potentially nefarious pharmaceutical knowledge, she was reportedly involved with the monarchies of several Greek city-states, including Athens and Corinth, and the foundation of several temples. In Corinth, she infamously murdered the reigning monarch and his daughter with an incendiary compound, called naphtha, typically reserved for ancient naval combat. She is said to have invented this ancient napalm.[46]

The name "Medea" carries the root-meanings "good counsel," "healing," "governance," and "genitals." The connections between Medea

[46]According to the *Suda* (nu,90), the military weapon Naphtha, was invented by Medea and and was named by her Medic contemporaries: *Naptha* has three genders: Found most commonly in its feminine form, a neuter is found as well, and its masculine form is used by Plutarch. What the Medes call *naptha* the Greeks call "Medea's Oil."

and words for healing and drugs is not confined to Greek.[47] In addition to the basic root for "healing" from Latin "medicina," a look at "venenum" the word for "poison" or "toxin" betrays linguistic influence from the Colchian (Medic) mystery rites and even the person of Medea herself.[48] In other words, the English word "medicine" is derived directly from the name of a Bronze Age queen known for her pharmaceutical expertise.

Medea, a "master-sorceress," had a longstanding reputation throughout the Mediterranean for possessing particular skill with the use of so-called drug-induced "fires," including the group of drugs that caused "oistros," the ecstatic state induced in ios-rite initiates who reportedly suffer from sensations of extreme "burning."[49] Like the *Echidnae* in Greece, Asia Minor, Italy, and North Africa—the priestesses who perpetuated her oracular rites—Medea prepared drugs in her vagina, and administered her own vaginal excreta as the communal "burning purple" of ios-rite mystery initiation.[50]

---

[47]Aphrodite is closely associated with all aspects of this root as the goddess born from the severed genitals of her father: ἠδὲ φιλομμηδέα, ὅτι μηδέων ἐξεφαάνθη. Hesiod, Theogony 200. See: (OLD): Venenum...2. Lit., a magical potion, charm: "item ut Medea Peliam concoxit senem: Quem medicamento et suis venenis dicitur Fecisse rursus ex sene adulescentulum," Plaut. Ps. 3, 2, 81: "dira Medeae," Hor. Epod. 5, 62: "Colcha," id. C. 2, 13, 8: "Colchica," id. Epod. 17, 35; Cic. Or. 37, 129; Hor. C. 1, 27, 22; id. Epod. 5, 22; 5, 87; id. S. 1, 8, 19; 2, 1, 48; Ov. M. 7, 209; 14, 55; 14, 403: "qui quodam quasi veneno perficiat, ut veros heredes moveat," Cic. Off. 3, 19, 76: "id quod amatorium appellatur, venenum est," Dig. 48, 8, 3.—. The name Medea, is literally synonymous with "medicine." This is due to the expert pharmaceutical experimentation and medico-magical practices of a single pre-Homeric priestess/princess.

[48]On the significance of the Med-root and its connections to Medea see: Karen (1951) JulPMC195119. In addition, PGM 4.2441–2621 contains a link between the performance of ios-rites and the Echidnae and the name Medixa, which appears to be a reference to Medea.

[49]The madness of "oistros" was considered a form of Bacchic inspiration; it was directly associated with the "entheotic" state (ἔνθεος, G.) at least as early as Euripides: *Trojan Women*, 365 ff. In the same play, Euripides incorporates all the elements of mystery initiation: ἔρως ἐτόξευσ᾽ αὐτὸν ἐνθέου κόρης (line 255). *Eros (erotic desire) is the source of the arrow poison that stings the entheogenic maiden (priestess). The ios-rite summarized in a single line.*

[50]Medea's use of snake venom in Bacchic mystery ritual was the subject of much of Seneca's *Medea*. See: *Medea* 705 ff. In the mythic narrative, Medea even plays with her own name: In Ap. Arg. 3.779 ff. Medea asks the question of how she will evade her parents' attention should she act the "Medea:" πῶς γάρ κεν ἐμοὺς λελάθοιμι τοκῆας φάρμακα **μησαμένη**; the name Medea is very much a title used to describe the pharmaceutically related actions of very specific gods and mortals.

The "burning purple," a mixture of Medea's vaginal excreta combined with ritual drugs, was also used as an ink. One ancient source tells us that the Golden Fleece of the Argonauts (the famous mythic expedition led by Medea and Jason) was a ram's skin, upon which was written the formula for the cult drug compound used in the mystery rite, otherwise known as the "burning purple (Carl Ruck, p. 381 ff.)." In this way, the quest for the fleece was very much an attempt to obtain the complex drug formula of the emerging *Echidna* priesthood, along with its chief priestess, who was herself the biological source of the drug.[51] Herodotus places Medea with Io, Europa, and Helen, as four abducted oracles of the Heroic Age; each young girl was associated with the manifestation of the maddening Bacchic "oistros," a psychotic, delusional state induced by the *ios*-rite.[52]

26.9 TOXIC PLANTS AND ANIMALS USED IN MEDEA'S IOS-RITE

Medea's ritual communion was the combination of one particular *ios* (arrow toxin) and its "antidote," aptly named *Galene*, or "Calm."[53] Medea's *ios* was the product of an ἔχις, a frustratingly general Greek term for "viper."[54] Ancient medical sources tend to group all vipers into a single family of snake-derived drug sources; envenomation by any

---

[51]Medea is typically presented as the source of the compound sacrament. For example, Scholia on Apollonius' Argonautica: ON 3.1013. προπρὸ δ᾽ ἀφειδήσασα θυώδεος ἔξελε μίτρηςφάρμακον. On Medea taking the drug from her mitra and giving to Jason. The scholiast says the language indicates an explication (epitasis) of the erotic and the acquisition/dosing of the drug. All relates back to the Titans and the pitch strain of the song used in the rite...stretching of the neura. The scholiast implies that the drug is derived from Medea during sexual congress. The neura traditionally represent the stretched erection of the mystic phallus seen in temples associated with mystery rites, such as in those found in Lucian's *De Dea Syria*.

[52]Io, Europa, Helen, and Medea were the four abducted priestesses discussed by Herodotus in the opening of his *Histories* (1.1 ff.).

[53]Pausanias 2.1.9, makes reference to a temple in Corinth with an image (ἄγαλμα) of Galene. Medea has extensive historical connections with the Corinthian royal line and the foundation of several temples. Cult operations involving the goddess Galene may have directly involved or indirectly invoked pharmaceutical associations of Galene/Eudion as a cult potion.

[54]Nicander discusses the viper and its ios (venom) in the context of the "Way." The priest discusses the sexuality of vipers and the unique nature of live-birth, reflecting the mystery rites of the Echidnae: *Theriaca*, 128 ff.

species of viper was treated simply as "viper-strike." To make matters more confusing, receiving the sacrament from an *Echidna* priestess was referred to as being "struck by the viper."

Medea's *ios* was called "Colchian Fire," and was said to contain as a principal ingredient, ἐφήμερον, or *meadow saffron* (*Colchicum autumnale*), called κολχικόν, or *colchicum* by ancient physicians.[55] The root contained a liquid consumed for the pleasant feelings it conveyed and was associated with the ritual sexual union of oracular priestesses with Delphic Apollo.[56] Meadow saffron was said to cause the same symptoms as toxic mushroom ingestion, and treatment for poisoning from *colchicum* was identical to that of certain mushrooms. According to one source, drinking Medea's "Colchian Fire" caused an intense itching of the lips and distressing intestinal discomfort.[57]

Colchian Fire, as a ritual sacrament, was mixed within and administered from a Medic priestess' vagina (κόλπος; νηδύς). The vagina served as a means of both mixing drugs and administering them to initiates. This ritual medico-religious act was common enough to be recognized by noted poets like Horace.[58]

Mushrooms are frequently associated with *ios*-ritual associated myth. According to one ancient priest connected with oracular Bacchic mysteries and their drugs, mania-inducing mushrooms are naturally infused with the venom of vipers (literally Echidnaic *ios*).[59] The combination of mushrooms and venom in ritual-associated myth is reinforced throughout texts related to Bacchic worship. Phsychotropic mushrooms like Amanita were a common element of compound sacramental compounds. In fact, the very name "Mycenaean," a name that came to

---

[55]Nicander, *Alexipharmaca*. 249 ff. Scribonius, *Compositiones*, #193.

[56]Euripides' Ion, 887 et seq., Ion is an example of the "child passed through the fires." His mother, Creusa, applied her colchicum (flowers of the meadow saffron) directly to her vagina before engaging in intercourse with the god. Ion, the product of the divine union, was dedicated to the temple of Apollo with "serpents" as guardians. These serpents are members of a *drakaina* priesthood.

[57]Scribonius Largus #193 includes a description of the initiate who has drunk Medea's ritual potion, known as Ephemeron, or Colchicum. Scribonius was a physician (medicus) attached to the Roman army. He includes numerous lengthy descriptions of the *mala medicamenta* employed by mystery rite participants. In addition, he lists the symptoms of these drugs and potential treatments.

[58]Cales venenis officina Colchicis, Horace, *Epode* 17.35. The vagina is the instrument used to mix and administer the drug. The medicated vagina becomes the Latin "officina" or "a workshop; manufactory." LSJ.

[59]Nicander, *Alexipharmaca*. 521–28.

represent Heroic Greek culture in general, was derived from Perseus' fondness for licking wild fungi.[60]

In addition to viper venom, psychotropic mushrooms, and meadow saffron, Medea's *ios* contained the famous "purple," or πορφύρα, a textile dye obtained from marine mollusks (murex). The dye was responsible for the vivid purple color of the *ios*.[61]

Ancient *Echidnae* were said to be identified by the dark purple drool mixed with foam that came from their mouths.[62] The product of this apparent excessive salivation, a dark spume, was harvested and considered a "lethal drug" to others but not harmful to the *Echidna* herself. The Echidna's skin was covered with hardened, dark patches, and her eyes were chronically infiltrated with blood and reportedly photosensitive. In some cases it appears that Medea's *ios* may have been introduced from the mouth of the Echidna directly into the eyes of the initiate. Medea's meadow saffron-containing compound was also called the "drug of Prometheus," and was closely associated with the color of the meadow saffron and the stunning purple dye of the murex.[63]

---

[60]Pausanias 2.16.3. Perseus once quenched his thirst by licking fluid flowing from a wild mushroom cap. He named the location where this mushrooms was found Mycene (Μυκήνη, from G. μύκης: "mushroom"). Mycene, in addition to being the name of an archaic Greek city, was also a nymph. Nymphs, the sexually promiscuous natural spirits of Greek myth, were frequently conceived of as growing alongside a host tree, and are often connected with fungal growth and a serous, ambrosial fluid released during sexual excitation. For example, to "nymph" (νυμφιάω) in Greek, is to rave in sexual mania, like a horse in season. (LSJ). The connection must not escape our attention, because the vaginal exudate of sexually receptive horses was used in ritual pharmacological preparations, and was considered a potent aphrodisiac associated with Orphic mystery rites, the magi, and Medea. In antiquity, amanita was intentionally cocultivated with the fig. See Nicander, frag. 78−79; Ruck (2018), 177 et seq; Carl A.P. Ruck, *The Son Conceived in Drunkenness*, 155 et seq.

[61]The last time the classic, Greco-Roman entheogenic "burning purple" of the mystery rites performed by the Dragon priestess is referenced appears to be the classicizing Latin Hymns of Michael Marullus (1458−1500 CE), who, like his mystic Greek and Roman predecessors, relied upon the Vox Orphica to detail drug-induced rites derived from blood and milk of the Bacchic orgy: Sancte pater, sive aetherio delapsus Olympo enthea divino praecordia concutis oestro..." (*Hymn to Jupiter Optimus Maximus*). Marullus fought with Vlad Dracula, who was himself made infamous for reported cultic banquets accompanied by the consumption of human blood and the impaling of infants to the breasts of their mothers.

[62]Ancient priestesses involved with drugs and necromancy (a.k.a. "witches") were often portrayed as foaming and drooling. See: Lucan, *Civil War*. 6.719 ff. In addition, their skin was characteristically marked by dark patches: Valerius Flaccus, *Arogonautica*. 2:105.

[63]Scholia on Apollonius: 5:859. "The Promethean drug is made in the κόχλος." Medea's drug was known as the Colchian Fire, or the Burning Purple.

26.10 MEDEA'S ANTIDOTE AS MEDICINE

The primary ingredients of Medea's Echidnaic *ios* were meadow saffron, marine mollusks, psychotropic mushrooms, and viper venom. Medea's "antidote," called *galene*, was administered concomitantly with the *ios* in order to ameliorate its apparent negative side effects. *Galene* was a nymph of the sea who brought "calm" to the waves.[64] Medea, like the nymph, was considered to be the source of the antidote (the "calm")—in the manner of the *Echidnae* priestesses—Medea breastfed initiates. However, Medea was also said to cut herself during the performance of Bacchic mystery rites, and Seneca tells us her flowing blood was itself the source of the cult's drug.[65]

Necromantic sources relate that cult "antidotes" like *galene* were actually mixtures of blood and breast milk. The playwright Aeschylus—nearly executed for revealing the secrets of the Eleusinian mysteries—portrayed the Drakaina as an inspired priestess who dispenses milk mixed with blood by means of the strike of a viper.[66]

According to Galen, the personal body physician of the emperor Marcus Aurelius (II CE), there existed multiple recipes for compound antidotes called *"galene,"* preserved by mystics and eccentric kings alike.[67] Galen claims that *galene* derived from *Echidnae* and held the reputation for being able to counteract the most deadly of drugs, including potions made from venoms, opium, henbane, and even deadly monkshood.

Medea's *galene* was also called *"castor"* and *"eudion."* Galen's formulae for the antidote *galene* include dozens of botanicals and seemingly numerous compound oils. Many of these same oils were universally ubiquitous in antiquity, and were used for the maintenance of sexual arousal, the prevention of sexually transmitted disease, and as abortifacients.[68] Mythic depictions portray Medea as constantly applying

---

[64]On *Galene* or *Galenaia*, see Callimachus, *Epigrams* 6. See also Galen's numerous works about pharmaceutical "antidotes." Galen K 14 p.32 ff. Galene is linked with the Orphic Eudion.

[65]Seneca, *Medea* 787 ff. Seneca's Medea includes numerous ios-rite drugs in her ritual performance, during which she performs ritual incisions during drug preparation and the acquisition of venom (via milking). Medea's necromantic mystery rites are called "sacra letifica" by Seneca (577), or "death-bringing mysteries."

[66]Aeschylus, *Libation Bearers* 514 ff.

[67]Galen, *On Antidotes*, K. 14 p.32 ff.

[68]Galen and Dioscorides are not shy about the use of abortifacients. Chemically induced abortions were common in antiquity and consistently associated with the activities of midwives, "witches," and mystery initiations—as child sacrifice. Lucan provides a sensationalized portrayal of cult consumption of aborted fetuses: *Civil War*, 6.554 ff.

compound oils to her skin and genitalia; in one case, she instructed her twelve drug-skilled disciples to administer her bathwater as a healing drug.[69] It is likely that the *galene* found in Galen's drug recipes was based on a simpler recipe for ointments and "anointing oils" used to cover the breasts of nursing *Echidnae* like Medea.

26.11 CHRISTIAN IOS RITES

The widespread use of ios-based pharmaceuticals by mystery religion initiates of Greco-Roman antiquity created and sustained significant social conflict. Priestesses involved in the preparation and administration of drugs—derived from their own body fluids—became valued commodities, to be traded and/or influenced for religious and political advantage. With time, these priestesses established oracular institutions, built temples, promoted educational associations, and guided the foundation of democratically inclined city-states; it is fair to say that ios-driven Bacchic (ecstatic) mystery religions influenced much of the intellectual and cultural development of Western civilization.

Christianity employed the very same specific technical vocabulary used by ios-rite priestesses in both the New Testament and Apocrypha. Like its pagan predecessors, Christian communion became an exclusionary pharmaceutical movement. Followers of "the Way" claimed the "blood and body" of their chief priest was the exclusive "salvation" of the "kingdom of god." St. John, in his visionary Apocalypse, foretells the overthrow of the infamous, "Mother of Prostitution," who "intoxicates by means of sex acts."[70] John specifically identified his so-called

---

[69]For a current, in-depth treatment of Medea's personal application of drugs (dermal and vaginal), see Carl Ruck, *The Great Gods of Samothrace and the Cult of the Little People* (Regent, 2017).

[70]St. John, Apocalypse 17:3–5. It is notable that this Mother of ritual prostitution offered up a cup full of her sexual acts: "ἔχουσα ποτήριον χρυσοῦν ἐν τῇ χειρὶ αὐτῆς γέμον βδελυγμάτων καὶ τὰ ἀκάθαρτα τῆς πορνείας αὐτῆς," The language of John's text matches the Orphic Hymns and Orphic literature that preserves the ios-mystery rite. For example, the rite of the mystery is an ablution of that which is "impure" (akatharta). The same term was used in magico-medical literature for mixed solutions used in ritual drug administration. The word can also be found in association with the drawing down of the sorceress-priestess' menses and appears to be used as a drug—in addition to the bathwater of the Echidna priestess (Medea is recorded bathing and providing her used bathwater as medicine.

"whore" as the purple-clothed Dragon Priestess who "sits on a venomous scarlet beast" and bears the name "Mystery."[71]

The Baptists (Baptae L.), mystery cult practitioners contemporary with Jesus and his students, were best known for the ritual ablutions they performed. Pagan and Christian sources tell us the Batptists practiced restricted diets, wore women's lingerie, drank ritual substances from penis-shaped drink ware, and promoted a "purification" of the soul through the practices of the Magi.[72] During the rise of "the Way," a Baptist leader named John was famously decapitated at the request of a young female child named Salome, who had earned the right to ask for the Baptist's head after performing a presumably sexually explicit dance for the reigning monarch and his guests.[73]

John the Baptist's premier student, Simon Magus, was ostracized by both the Baptists themselves and the leaders of "the Way," for his reliance upon a priestess-prostitute aptly named Luna/Helen.[74] Simon Peter's resistance to Simon Magus was mirrored in St. Paul's admonition and re-baptism of Christians whom he claimed were engaged in rites with "drug-using prostitutes."[75] Despite Paul's admonition to avoid these priestesses, and despite the fact that one of these young priestesses, specifically a Pythia, persistently followed Paul while

---

[71]St. John, Apocalypse 17:5. καὶ ἐπὶ τὸ μέτωπον αὐτῆς ὄνομα γεγραμμένον, μυστήριον, ΒΑΒΥΛΩΝ Η ΜΕΓΑΛΗ, Η ΜΗΗΡ ΩΝ ΠΟΡΝΩΝ ΚΑΙ ΩΝ βΔΕΛΥΓΜΑΩΝ ΗΣ ΓΗΣ. "And upon her forehead was a name written, MYSTERY, BABYLON THE GREAT, THE MOTHER OF HARLOTS AND ABOMINATIONS OF THE EARTH." King James version (KJV).

[72]Juvenal, Satires 82–116. Descriptions of the baptists in feminine garb like the girdle (ζώνη) worn on the night of a girl's first sexual union with her husband. The religious order of the Magi was founded by Medea, the first "Maga." The word is the same as that used for those who followed a "star" to find Jesus.

[73]On Salome: See Matthew 14:6–11; Mark 6:21–29. Salome was only a little girl (κοράσιον), and the nature of her dancing is obscure. Whether or not the dance was sexual in nature, it was performed in front of prominent community leaders and generals, and was so successfully arousing as to provoke Herod to promise her considerable wealth. Salome was herself a member of a priestly line so it may be the case that she was performing a specific Hellenized "maiden-dance," similar to those seen in the Greek world. Like St. Mary, Salome may have been dedicated to the Temple in Jerusalem and nurtured by the vestals there.

[74]Acts 8:9–24. The Apocrypha and early Church fathers contain a myriad of stories about Simon Magus and his relationship with the oracular witch-priestess Helen (aka. Luna—following Ios-ritual titles).

[75]Paul, Galatians 5:16. The term translated as "prostitution" is often tainted by modern English and Western mores. It does not have an explicitly negative connotation. Acts of "porneia" are strictly sexual, and are sometimes linked with divinities, heroes, and priestesses. Christianity, under the influence of scholars the likes of St. Jerome promoted a negative view of pagan sexual standards and practices.

publicly humiliating him, the followers of "the Way" were known to have the ability to drink snake venom without suffering any negative symptoms.[76] In fact, like the Echidnae under Medea, the earliest Christians were known to have pioneered the use of venom in the treatment of disease: "they will pick up snakes with their hands; and when they drink deadly poison, it will not hurt them at all; they will place their hands on sick people, and they will get well."[77]

Peter and Paul, prominent leaders of "the Way," directed born-again mystery initiates to abstain from the sexual rites of contemporary Echidnaic mysteries.[78] This command to abstain from "sexual idolatry" by the followers of Jesus and "the Way" would have necessitated the use of an alternate source of antidote for any ios-rite participant. The only source for such an antidote was the semen or "milk" of a prepubertal boy; if Jesus and the apostles were following the Echidnaic Ios-rite, Jesus, as "savior" would have had to derive his antidote by the performance of a sexual act on a 9–12-year-old boy. The gospel of Mark records that Jesus was arrested in a public cemetery after midnight in the company of an anonymous naked male child.[79]

Christian dislike for rituals involving "prostitutes," the likes of Mary Magdalene and Salome, (Mary used a dildo-shaped drug applicator on Jesus called an *alabastron*) would have necessitated a switch from

[76]On St. Paul's confrontation with a Pythia (an oracular priestess): Acts 16:16 ff. Paul was bitten by a viper (*echidna*) while traveling but suffered no observable ill affects, which surprised observers: Acts 28:3–5.

[77]Gospel of Mark, 16:18. (NIV translation in text). The Greek of this New Testament passage employs specific medical ios-cult terminology (in bold here): ὄφεις ἀροῦσιν κἂν **θανάσιμόν τι πίωσιν** οὐ μὴ αὐτοὺς βλάψῃ, ἐπὶ ἀρρώστους χεῖρας ἐπιθήσουσιν καὶ καλῶς ἕξουσιν. A θανάσιμόν is a specific category of drug found in surviving medical texts. In other words, the first generation of Christians following Jesus openly acknowledged drinking drug concoctions derived from snakes. See also Scholia on Nicander. It should also be noted that in the first century CE, Roman physician Scribonius Largus describes a series of symptoms suffered by the use of drugs (mala medicamenta) associated with the ios-rite mysteries that is similar to those suffered by Jesus in the Garden of Gethsemane after he "drank of the cup," and experienced a vision of an angel. See Scribonius Largus (#188 ff. esp. #193). Scholia of Nicander indicate the "ephemeron"—the possible drug of the cup of Jesus—was the personal discovery of Medea. The word used to describe the physical substance that exudes from Jesus in the Garden is also used to describe the exudate of the anus of Dionysus in Aristophanes' *Frogs*.

[78]Paul, *Galatians* 5:16: Paul, like others, joins πορνεία and ἀκαθαρσία with ἀσέλγεια, εἰδωλολατρεία and φαρμακεία as prohibited acts, all of which are ritual cult terms found in the Vox Orphica. Paul prohibited the drug-induced magic of the mysteries and even re-baptized followers of "the Way" in Ephesus who had been baptized by John the Baptist (*Acts* 19:1–5).

[79]Mark 14:51–52.

girl-derived to boy-derived antidote. Jesus, whose name in Greek is "Jason" (Jason was the guardian of the Medea—the cult drug formula inscribed on the fleece, and his name contains the root of the word "ios" and the Greek verb for "healing"), as a classical mystery "savior" involved in the perpetuation of an ios-rite, would have needed the semen of a young boy to balance the symptoms brought on by imbibing the necromantic (resurrection) "cup" of the mystery.

26.12 TOXICOLOGICAL QUESTIONS

The ιóς-rite employed numerous plant- and animal-derived toxins in an attempt to achieve and maintain an artificially induced state of psychosis in initiates while limiting any dangerous side-effects. This multivalent pharmacological approach to medicine and religion—the creation of a pharmaceutical communion—presents a myriad of important toxicological research questions. For example:

1. *What molecular mechanisms contributed to the inhibition of potentially lethal symptoms, including hemorrhage and blood-pressure collapse, in subjects ritually exposed to viperid proteases?*
2. *How do viperid venoms, toxins derived from the glandular secretions of marine mollusks, colchicine, and fungal GABA receptor agonists react with each other and with mammalian muscle and nervous tissue?*
3. *How did chronic viper venom exposure enable prepubertal females to lactate, while conferring upon them an immunity that could be passed on to other subjects? Does the developing mammary gland produce antibodies following chronic exposure to viperid venom?*
4. *What were the chemical consituents of the ejaculate collected from human subjects chronically exposed to viperid venom? Are these constituents biologically active?*
5. *What are the potential medical uses of the poly-pharmaceutical combinations of toxins employed in ancient mystery-religions? Why are psychoactive mushrooms associated with the use of viper venom?*

26.13 CONCLUSION

Ancient Mediterranean religious microcultures developed a unique and pioneering method of pharmacy that focused on the interaction of viperid toxins with the human body and other pharmacologically active substances. Drug use within the context of ritual communion was far more complex than the simple and much later theoretical models of medical pharmacology offered by classical Hippocratic physicians, which had come to rely too heavily on the exercise of dogmatic philosophical theories rather than rigorous experimentation on human subjects—as did *ios*-rite priestesses within their protected ritual settings. The ancient Colchian priestess Medea is very much a forgotten pioneer of Western pharmacy and perhaps the least recognized scientific genius of her day. Her experimentation and discoveries drove all of Western medicine, and her practices influenced the very formation of Western medical vocabulary.

The practices of Medic snake-venom-driven cults influenced prominent ancient mystery religions, both pagan and Christian alike. There are indelible linguistic marks upon the language of the Christian holy writings that place the activities of Jesus and the early Church squarely within the realm of communal drug users who employed sexuality as a means of compounding and administering drugs. For example, based on the technical vocabulary employed by the Christian texts themselves, it is likely that Jesus was performing fellatio on the naked boy that was with him in the Garden of Gethsemane in order to secure the antidote to the sacramental cup from which he drank before he had his vision of an angel. If Jesus had been unable to access the antidote after partaking of the ritual ios-cup, he would have died from its contents within a few hours.[80]

While there is academic value in elucidating the history of ancient Mediterranean cultures, a knowledge of the complex pharmaceutical practices of ancient priestesses may also yield medicinally relevant information for modern toxicologists talented enough to disentangle the complex biochemical pathways involved in ancient communal polyvalent toxicology.

---

[80]According to our written evidence, Jesus was crucified, but died prematurely, as noted by the Roman authorities involved in his execution. Crucifixion was not his cause of death (Mark 15:44–45): "*Pilate was surprised to hear that he was already dead. Summoning the centurion, he asked him if Jesus had already died. When he learned from the centurion that it was so, he gave the body to Joseph.*" (NIV). In addition, Jesus' last words, according to the gospel of Mark were: Ἐλωΐ Ἐλωΐ λαμὰ σαβαχθανεί; while the language is still debated, it is interesting that λαμά is the title of the Echidnaic priestess "Lamia" (G. λάμια) and σαβαχθανεί appears to be a derivative of the name of Chthonian Sabazeus, the divine father of the Bacchic mystery initiation. Lamia was herself linked directly with Medea through the Orphic Melinoe, who was herself called Medea, and the oracle of Medea found in Pindar (Pythian 4).

References

Karen, Thelma, 1951. The etymology of medicine. Bull. Med. Libr. Assoc. 39 (3), 216–221.

John Lydus, *De Ostentis*. 2.6.B.

Carl Ruck, The Great Gods of Samothrace and The Cult of the Little People. p. 381 ff.

Ruck, Carl A.P., 2018. The Son Conceived in Drunkenness: Magical Plants in the World of the Greek Hero. Regent Press, Berkeley, CA.

Harmful Botanicals

Alain Touwaide[1,2]

[1]Brody Botanical Center, The Huntington Library, Art Collections, and Botanical Gardens, San Marino, CA, United States [2]Institute for the Preservation of Medical Traditions, Washington, DC, United States

27.1 CLASSICAL TOXICOLOGY

Medicinals of classical antiquity are generally considered the basis of the relatively recent development of Western drugs. In pharmacotherapy, a wide range of medicinals was drawn from the vast biodiversity of the

Toxicology in Antiquity
DOI: https://doi.org/10.1016/B978-0-12-815339-0.00027-5

Mediterranean environment. Significantly, the many plants used as *materia medica* for the preparation of remedies were also consumed daily as preventative formulations, of which the so-called Mediterranean diet—now redefined as the "Greek diet"—is the modern.

Unsurprisingly, the knowledge and judicious use of natural resources for the management of health, be it curative or preventative, went together in antiquity with awareness of the potential dangers (from mind alteration to death) of some botanicals. This resulted in the development of methods aimed at counteracting their effects.

27.2 SOURCES AND DATA

Knowledge of potentially harmful botanicals is evident in the written legacy of ancient Greece and mythological references are ubiquitous in this literature. The magician Circe, daughter of the Sun and of the Oceanid Perse, i.e., the elements that shaped the cosmos, is said to have used botanicals to convert the companions of Odysseus into swine (*Odyssey*, 10.475–541).

In this chapter, knowledge of harmful botanicals in classical antiquity is drawn from the ancient medical and scientific literature. Although information on natural toxic substances can be found in the most ancient body of Greek medical literature currently preserved (namely, the series of over 60 treatises ascribed to Hippocrates [between 375 and 350 BCE] and forming what is called the so-called *Hippocratic Collection*), it is not specific and does not communicate an exact understanding of the ancient knowledge about toxic plants. Nevertheless, it is useful for chronological purposes and for a reconstruction of the development of what can be called *classical toxicology*—in other words, a body of data on a determined object, organized according to specific parameters, aiming at a relevant objective, and forming, if not a discipline, at least a specialty that is explicitly recognized as such. The best source about classical toxicology defined in this way is a pair of treatises ascribed to Dioscorides (himself dating from the 1st century CE), the author of *De Materia Medica*, the most famous encyclopedia of *materia medica* of classical antiquity. One of these two treatises deals with poisons of all kinds (*De Venenis*), and the other covers animal venoms (*De Venenosis Animalibus*).

Though certainly not written by Dioscorides himself, these two treatises reflect the development in the knowledge of poisons and venoms during the 1st century CE. At any rate, they had a diffusion that was almost as exceptional as that of *De Materia Medica*, as they provided the basis of virtually all the subsequent written works on toxicological matters—from Galen (129–c. 216 CE) himself, however eager he was to always reformulate the legacy of previous generations and to reinterpret and integrate it into his own thinking; to such typical Byzantine medical encyclopedists as Oribasius in the 4th century; Aetius and Alexander of Tralles in the 6th century; and Paul of Egina, the last physician of Alexandria, in the 7th century; and Theophanes Chrysobalantes in the 10th century. Arabic and medieval physicians also used the pseudo-Dioscoridean treatises as a source of the treatment of toxicology in their medical works, be they all-encompassing encyclopedias or shorter treatises specifically devoted to toxicology. The two treatises ascribed to Dioscorides thus appear to have formed the backbone of most premodern knowledge of natural harmful substances, be they botanicals, minerals, or animals, the latter of which included venoms injected by snakes, scorpions, fishes, and insects, and the bites of rabid dogs. The continuity in the transmission and use of the treatises ascribed to Dioscorides does not preclude the fact that original works developed over time, including theoretical reflections on the concept of poison itself. The fact is that the body of data contained in these works provided the basic information for almost all material on the topic for approximately 15th centuries, even though this set of data may have been reformulated, expanded, or revised.

Of the two treatises, the most relevant here is the one entitled *On Deleterious Substances and Their Prevention* (=*De Venenis*). It contains 34 chapters totaling over 600 lines and 5000 words. After a first long chapter dealing with general considerations, each of the subsequent 32 chapters is devoted to one harmful natural substance: 10 animals (plus the honey produced by bees that have taken the pollen of a toxic plant), 16 plants, and five minerals. The common denominator of all these substances (regardless of the natural kingdom to which they pertain) is that they are taken orally, as opposed to those in the twin treatise *De Venenosis Animalibus* incorrectly ascribed to Dioscorides, in which all substances (exclusively animal in nature) are injected into the body or come into contact with the skin.

In the treatise *De Venenis*, all the chapters (whatever the nature of the harmful substance) are built from a template made of three parts: (1) the organoleptic description of the substance, (2) the description of the clinical signs following its absorption, and (3) the therapies aimed at counteracting its effects. The parts are not dealt with in the same way or with the same degree of detail in each chapter. As an example of this template, here is the discussion of hemlock (Chapter 11: Mithridates of Pontus and His Universal Antidote):

> **On hemlock**. Hemlock taken in a draught may cause scotomy (dizziness and dimness of sight), possibly leading to blindness, as well as hiccups, impaired intellectual capacity, and cold extremities. It may ultimately result in convulsions and suffocation due to choking.
>
> At the beginning, we will eliminate it [= hemlock] as in any other case, that is, by [having the patients] vomiting. Further on, using evacuation, we will eliminate the portion [of hemlock] that has arrived into the intestines. Then, we will use wine as the best remedy, giving it with intervals during which we will give donkey milk or absinth

with pepper and wine; castorium and rue with wine; cardamon or storax or pepper with nettle seeds and wine; or laurel leaves, silphium and its juice with oil and sweet wine; and also sweet wine itself abundantly administered will help pretty much.

The plants the harmful effects of which are analyzed in *De Venenis* are listed in the following table in alphabetical order according to their current English names when available. The table provides the following information for each such plant:

1. The current English name;
2. Number of the chapter to the plant in the treatise *De Venenis*;
3. The Linnaean binomial designation of the plants according to current taxonomical literature;
4. The reference to the chapter in Dioscorides, *De Materia Medica*, where the plant is analyzed according to the standard critical edition;
5. References to the literature cited at the end of this chapter, where current scientific information about each plant can be found.

English name	Chapter number	Scientific identification	Dioscorides	Notes
Buttercup	14	*Ranunculus sardous* Crantz	2.175	Barceloux (2008, pp. 690–2)
Coriander	9	*Coriandrum sativum* L.	3.63	Frohne and Pfänder (2004, pp. 39–41). The toxicity attributed to coriander in antiquity (that is, mind-altering) probably results from contamination or the ingestion of high doses of the plant Leclerc (1954)
Doruknion	6	Possibly *Convolvulus oleaefolius* Desr.	4.74	van Wyk and Wink (2004, p. 406)
Efemeron	5	Most probably *Colchicum autumnale* L.	4.83	Barceloux (2008, pp. 693–702)
Farikon	19	N/A	N/A	Not identified; probably a Solanacea (Fig. 27.3).
Fleawort	10	*Plantago psyllium* L.	4.69	van Wyk and Wink (2004, p. 245)
Hemlock	11	*Conium maculatum* L.	4.78	Barceloux (2008, pp. 796–9)
Henbane	15	*Hyoscyamus* spp.	4.68	Barceloux, (2008, pp. 776–83)
Honey from Heraklea	8	Honey from bees having absorbed the pollen of *Rhododendrum* spp. (Fig. 27.1).	2.82	Barceloux (2008, pp. 870–2)
Horned poppy	18	*Glaucium flavum* L.	4.65	van Wyk and Wink (2004, pp. 275–6)

Mandrake	16	*Mandragora autumnalis* L. or *M. officinalis* L. (Fig. 27.2).	4.75	Barceloux (2008, pp. 779–80)
Mushrooms	23	Multiple species	4.82	For gastroenteritis-producing species Barceloux (2008, pp. 290–3)
Pine thistle	21	*Atractylis gummifera* L.	3.8	Barceloux (2008, pp. 514–16)
Poppy juice	17	*Papaver somniferum* L.	4.64	van Wyk and Wink (2004, p. 225)
Toxikon	20	N/A	N/A	Not identified; supposed to be the plant whose juice was smeared on arrows (hence its name, as *toxon* means arrow)
White hellebore	13	*Veratrum album* L.	4.148	Barceloux (2008, pp. 815–18)
Wolfsbane	7	*Aconitum napellus* L.	4.77	Barceloux (2008, pp. 736–42)
Yew	12	*Taxus baccata* L.	4.79	Barceloux (2008, pp. 899–901)

27.3 ANALYSIS

27.3.1 Aim and Scope of Toxicology

In classical antiquity, the aim of the branch of medicine dealing with harmful substances (including botanicals) was to cure the victim of poisoning, be it the result of a mistake or a criminal action. The therapeutic strategy of the time was guided by the principle that each toxic substance has a specific, typical action. Hence, the therapy needed to counteract such action in an equally specific way, with appropriate therapeutic means according to a principle explicitly mentioned in the treatise ascribed to Dioscorides (see 8.12–13). This general principle, which clearly results from a long-term examination of the differentiated effects of the range of botanicals above (and probably of many others, but not necessarily those leading to life-threatening conditions), defined the therapeutic strategy of physicians in their treatment of cases. It also defined the content and structure of the chapters devoted to each substance and the general structure—that is, the sequence of the chapters—of the treatise ascribed to Dioscorides and those which followed it.

27.3.2 Structure and Purpose of Information

With regard to the chapters devoted to each substance, each is divided into three parts in most cases, as I have mentioned. The first two parts

FIGURE 27.1 Rhododendron, the source of the so-called *mad honey*.

FIGURE 27.2 The root of mandrake and its representation in a Latin medieval manuscripts (Tacuinum sanitatis in medicina, manuscript Österreichischen Nationalbibliothek, Series nova 2644, Wien, Austria).

(that is, the organoleptic description of the poisonous substances and the description of their effects, possibly including some pathological mechanism, supposed or real) aim to identify the poison in order to apply the appropriate therapy. The first one relies on the assumption that the draft through which the noxious substance has been administered is still available. The examination of the organoleptic qualities of a substance—which

FIGURE 27.3   Deadly Nightshade (*Solanum nigrum* L.).

includes everything but tasting it, which would harm the physician—should lead to the substance's identification, after which a specific therapy can be directly administered. If the causal agent is no longer available (which may be the case, particularly with criminal poisoning), it should be eliminated from the body of the patient via vomiting. If this can be done soon after the absorption of the lethal drink, the causal agent can be identified again by its organoleptic properties.

If the poisoning cup is not available or vomiting cannot be quickly provoked, identification of the harmful substance should be made through the disturbances that it causes to the victims. This is the rationale of the clinical description of the symptoms following the absorption of the poisons.

27.3.3 Creation of the System

These symptomatologies are relatively short, as they aim to stress the major characteristics of the toxic action of each substance in order to reach a rapid and precise diagnosis. They clearly rely on the repeated observation of multiple clinical cases, the recording of all the signs following the absorption of harmful substances by humans, and the isolation of the most common signs on the basis of these multple cases, further expressed in an essential way, as an abstract concept. Such abstraction is certainly not the result of the work of a determined individual (such as, for example, the author of the treatise ascribed to Dioscorides whoever he might have been). Actually, it is the product of years, if not centuries, of observations, possibly first transmitted in an unorganized way by means of oral tradition and popular knowledge, and then transferred to the world of physicians and written down, codified, and standardized into proper technical language. During the last phase of its codification, such knowledge may have continued to be refined, clarified, and made more accurate and complete (and, consequently, also more efficient), finally taking the form that it has in the treatise attributed to Dioscorides.

The role of a practitioner suspecting a case of poisoning consisted of interpreting the symptoms of the patient he was treating, and to repeat in a certain sense the exercise of abstraction of his predecessors (which resulted in the data conveyed in writing by medical literature). This was done for the purpose of extracting from an actual case its most salient pathological signs, to compare them with the clinical descriptions in the literature (that is, a treatise like the one attributed to Dioscorides), and to ascertain that the case he was treating corresponded to one in the literature. On this basis, he had to apply the therapy specifically prescribed for this case.

27.3.4 Application of the System

In this diagnostic process, physicians probably proceeded in a gradual way, step by step. In the treatise *De Venenis*, we note that the several harmful substances are grouped not by natural kingdom (as one might expect), but by major categories of effects. Using the descriptions from the ancient text—that is, the symptoms attributed to each poison, with no attempt to identify their causal mechanisms—there are six major groups of actions, which are listed below. For each one, the substances that provoke it are mentioned between parentheses, and, for each substance, the number of the chapter in *De Venenis* is given in square brackets (substances are listed according to the alphabetical order of their English name when available numbers between square brackets are the chapter numbers in the original Greek text of *De venenis*):

- Lesions of the digestive system [*doruknion* [6], (Touwaide, 2008), *efemeron* [5], (d'Abano, 1949), honey from Heraclea [8], wolfsbane [7]];
- Dramatic reduction of body temperature (fleawort [10], hemlock [11], white hellebore [13], yew [12]);
- Troubles of the mind (buttercup [14], coriander [9], henbane [15], mandrake [16]);
- Deep sleep and loss of consciousness (horned poppy [18], poppy [17]);
- Impaired mobility and paralysis (*farikon* [19]);
- Harsh irritation of the mouth; swelling and, consequently, obstruction of the alimentary tract (mushrooms [23], pine thistle [21], *toxikon* [20]).

The first phase in the diagnostic process consisted of identifying one of these major actions. As is shown by the chapter numbers, all substances with a similar or identical action were grouped together.

Once a major action had been identified, the physician needed to distinguish more precisely the agent within the group. The symptoms attributed to each substance are listed with increased differentiation, proceeding almost by binary choices. This process can be summarized as follows, assuming that we have a group of three substances and that symptom A is the one that characterizes the whole group:

- If only symptom A is present, then the substance is labeled 1.
- If symptom A is accompanied by symptom B, then the substance is labeled 2.
- If symptom B is absent, if it was replaced by symptom C, or if it is present and accompanied by symptom C, then the substance is labeled 3.

This means that the method for the proper identification of a specific poison was differential, consisting of grouping together all the substances with a similar or identical action, comparing their symptoms, and finally identifying exclusive symptoms, leading to a precise identification of the poison. Therapy was administered on that basis.

27.3.5 Therapeutic Principle

As a consequence of the lack of understanding of the pathological mechanisms, therapies were limited to treating the symptoms rather than the causes, something that drastically reduced their efficacy in spite of their sheer number and great variety. Nevertheless, in the case of subacute intoxication, it is highly probable that therapies may have compensated for the actions of the toxic agents during the crisis phase, allowing the patients to survive.

27.3.6 Period of Creation of the System

Regarding the period when this system was created, we noticed that it cannot be traced in any form to the vast collections of treatises ascribed to Hippocrates, which span between the 5th century BCE and the 2nd century CE. It can be recognized, though only partially, in the *De Materia Medica* of Dioscorides (1st century CE), since the work, dealing with *materia medica* specifically, does not include considerations of the pathologies that may have been generated by botanicals. However, the grouping of *materia medica* by action (and, in our specific case, by pathological action) is present in the work. In the 2nd and early 3rd centuries, this system was no longer present as such in the treatise *De Venenis*, as the substances were grouped by natural kingdoms and each one was listed in alphabetical order according to the Greek name. All the information linked with the grouping, therefore, is lost. Nevertheless, the general principle of the system that we have reconstructed through the pseudo-Dioscoridean *De Venenis* (that is, the necessity of a specific treatment based on the identification of relevant symptoms), can be found in a scattered form in Galen's many works. It thus seems that the development of a toxicological method with a theoretical system accounting for the written works that have come to us dates back to the 1st century CE, and probably somewhat earlier, because the achievements of the 1st century CE may have caused the disappearances of their predecessors.

Extant documentation includes fragments or traces of works on toxicological matters by physicians of the Alexandrian school during the 3rd century BCE. Nevertheless, judging from the scanty remains, such works seem to have been limited to anatomopathological analysis, rather than to a theorization about a body of facts observed by experience. Furthermore, during the 2nd century BCE, the work of Nicander, though it demonstrates good knowledge of the activity of poisons, neither includes a classification nor, on this basis, elements of differential diagnosis as in the pseudo-Dioscoridean treatise *De Venenis*. It is reasonable to speculate that the system of toxicology presented above and defined here as classical toxicology dates back to the period between the fading of the medical school of Alexandria and the zenith of Greek medicine in the 1st century of the Roman Empire—that is to say, between the 1st century BCE and the end of the 1st century CE.

27.4 HISTORICAL IMPORTANCE OF ANCIENT TOXICOLOGY

The treatment of toxicology in the treatise ascribed to Dioscorides experienced exceptional fortune through the centuries. Not only was it abundantly reproduced in Greek in Byzantium (over 40 manuscripts have been preserved in spite of the devastation of time), but it also provided the narrative for the section on toxicology in the encyclopedias by Byzantine physicians such as Oreibasios (4th century), Aetios of Amida (6th century), and Paul of Aegina (7th century). Indeed, its impact went beyond the Byzantine world, as its impact was felt in the West in the 13th/14th century through the labors of the physician, natural scientist, and philosopher Pietro d'Abano (c. 1250–c. 1316).

A native from the town of Abano in the vicinity of Padua, Pietro compiled a small book entitled, in Latin, *De Venenis* (*On poisons*, and *On*

Venoms) since the Latin term designates both venom and poison as Latin does not have a specific term to distinguish these two categories of toxic agents). The treatise is divided into two major parts: the first is a general analysis of the concept of venom and poison (*venenum* in Latin), and the second contains a clinical description of the symptoms following dermal or oral exposure to venoms and poisons, together with methods for eliminating them from the organism or counteracting their effects. The substances under consideration (76 in total) come from the three natural kingdoms: mineral (13 substances), vegetable (38 plants or parts of plants), and animal, including humans (25 venoms, animal products, human physiological liquids, and bites by humans).

Science and medicine flourished during Pietro d'Abano's time. Greek scientific medical treatises that had been translated into Arabic from the 9th century on, were known in the West through subsequent translation into Latin. Pietro d'Abano, however, went further and returned to the ancient scientific texts in Greek. At some point, he traveled to Constantinople, the capital of the Byzantine Empire, in order to learn Greek. Whether through good fortune or good connections, he was able to gain access to the most important scientific center in Constantinople in the 13th and 14th century, the *Hospital of King Milutin* (*Xenodochion tou Krali*, in Greek). There, Pietro d'Abano had access to a multitude of Greek texts, precisely those which had been translated into Arabic in the 9th century AD and were the basis of scholarly study and elaboration by Arabic scientists.

A close examination of Pietro d'Abano's chapters on poisonous plants reveals that these chapters are in large part a translation and adaptation of the small treatises ascribed to Dioscorides.

In Pietro d'Abano's chapter on the toxic effects of the ingestion of coriander (*Coriander sativum* L.), the symptoms of the intoxication are the following in the two treatises (the original text in Latin of Pietro d'Abano and in Greek of the Pseudo-Dioscorides, respectively, is followed in each case by an English translation which is mine):

Pietro d'Abano, *De suco coriandri*
   Ille cui sucus coriandri datus fuerit patietur quasi destructionem intellectus, ac sicut ebrius videatur et tandem moritur stupide.
On coriander juice
   He who has been given coriander juice, suffers from an almost destruction of his intellect, seems as though drunk, and finally dies struck senseless.
Pseudo-Dioscorides, Alexipharmaka, **9**
   τὸ δὲ κόριον . . . μανίαν ἐπιφέρει τοῖς διὰ μέθην ὁμοίαν . . . .
   Coriander . . . . provokes a madness similar to that of drunkenness
   . . .

Similarly, the chapter devoted to cursed crowfoot (*Ranunculus sceleratus* L.), describes the so-called sardonic laughter (which is in fact a labiofacial paralysis that might be compared to a smile) in terms very similar to those of the Pseudo-Dioscoridean treatise:

Pietro d'Abano, *De apio risus*
Ille cui datum fuerit apium risus in potu, facit hominem extra mentem et continue ridet, propter hoc vocatur apium risus.
On the sardonic laugh
He who will be given a draught of cursed crowfoot, goes out of his mind and laughs continuously, and this is why it is called the sardonic laughter.
Pseudo-Dioscorides, *Alexipharmaka*, 14
ἡ δὲ σαρδόνιον λεγομένη πόα ... παραφορὰν διανοίας ... ἐπιφέρει καὶ σπασμὸς μετὰ συνολκὸς χειλέων ὥστε γέλωτος φαντασίαν παρέχειν ἀφ' ἧς διαθέσεως καὶ ἡ σαρδώνιος γέλως οὐκ εὐφήμως καθωμίληται ἐν τῷ βίῳ ...
The plant called Sardonion ... provokes a transport of the mind and spasms with a contraction of the lips such as to give the impression of laughing, a disposition which has rightly been called sardonic laughter in daily life ...

As can be seen from these few examples, Pietro d'Abano's Latin text adheres closely to the Greek, to such an extent that we can almost consider his text a translation of the Greek, although he did not render the Greek original in a word-for-word translation, but expanded on the original text in what could almost be considered a paraphrase.

27.5 CONCLUSION

The system described here is not explicitly presented in the pseudo-Dioscoridean treatise *De Venenis* or in any of its subsequent heirs. Nevertheless, both the grouping of the substances and the presentation of the symptoms of each substance within the groups are clear enough to allow such a reconstruction. Although the pathological mechanisms generating the effects taken into consideration in order to identify the causal agents were not determined, the system was efficient—at least for the identification of harmful botanicals, if not for the treatment of the effects following their absorption by humans. This probably accounts for the exceptional fortune of the body of knowledge created in antiquity.

References

Barceloux, D.G., 2008. Medical toxicology of natural substances. Foods, Fungi, Medicinal Herbs, Plants and Venomous Animals. Wiley, Hoboken, NJ.

Benedicenti A., Pietro d'Abano (1250-1316). Il Trattato "De Venenis" (Biblioteca della "Rivista delle scienze mediche e naturali" 2). Florence: Leo S. Olschki, 1949, the first part of which contains the Latin text of Pietro d'Abano.

Frohne, D., Pfänder, H.J., 2004. Poisonous plants, A Handbook for Doctors, Pharmacists, Toxicologists, Biologists and Veterinarians, second ed Timber Press, Portland, OR.

Leclerc, H., 1954. Précis de phytothérapie. Essais de thérapeutique par les plantes françaises[4]. Masson, Paris.

Touwaide, A., 2008. Pietro d'Abano sui veleni. Tradizione medieval e fonti greche. Medicina nei Secoli 20, 591–605. which demonstrates that it is a translation of Greek texts.

van Wyk, B.-E., Wink, M., 2004. Medicinal plants of the world. An Illustrated Scientific Guide to Important Medicinal Plants and Their Uses. Timber Press, Portland, OR.

28

Pearl, An Ancient Antidote of Eastern Origin

Maria D.S. Barroso

Director of the Department of History of Medicine Portuguese Medical Association, Av. Almirante Gago Coutinho, Lisbon, Portugal

28.1 EASTERN CRADLE

Pearls, perfected by Nature, not requiring the art of man, were the earliest naturally occurring gems discovered along the coasts of countries in the East. Fish-eating tribes, perhaps off the coast of India or bordering an Asiatic river, were attracted by their luster when they found them while opening mollusks in search of food. Pearls have been long beloved in the East, enhancing the charms of Asian beauty, adding splendor to the royal courts of Persia and India where they were widespread. Asiatic peoples have always found beauty and value in these jewels. They were likely among the first to collect them in large quantities. Pearls have long been suggestive of Oriental luxury and magnificence, while simultaneously praised as symbols of spirituality and power. No ancient symbol of Oriental divinity or object of veneration has been without this adornment. No Oriental poem has also lacked this symbol of purity and chastity (Kunz and Stevenson, 1993, p. 3).

The mythical origin of pearls in Hindu literature is associated with Krishna, the eighth avatar or incarnation of Vishnu, the most important Hindu god. Krishna drew pearls from the depths of the sea to adorn his daughter Pandaia on her wedding day. In another version of the legend, the pearl was a trophy of the monster Pankagna's defeat by Krishna, who subsequently used it to adorn his bride (Kunz and Stevenson, 1993, p. 4).

In ancient China, it was believed that pearls originated in the brain of the fabled dragon. They can be found shining in the center of images of gods and were frequently offered as tributes from foreign princes to emperors (Kunz and Stevenson, 1993, p. 5). They were also used in imperial funerary rites. In the Kan period (1206–1368 AD), the emperors were embalmed, dressed with garments garnished with pearls, and enclosed in cases of jade (De Mely, 1896, p. 178).

28.2 ORGANIC ORIGIN AND CHEMICAL COMPOSITION

Pearls are produced by marine bivalve mollusks. They are considered gems, although they are not minerals. Instead, they are organic concretions like coral, sea shells, and mother-of-pearl, an iridescent material produced by some mollusks as an inner shell layer or outer layer (Schlüter and Rätsch, 1999, p. 26). Pearl-producing oysters Pinctata from the family of Pteriidae, are not related to edible oysters which belong to the family of *Ostrea edulis*. All species of the family produce pearls. River pearls or freshwater pearls are produced by mollusks of the family of the Unionidae and Margaritiferidae. Pearls are made from the same material as the mollusk shell.

Numerous mollusks and whelks can produce pearls. Their quality is not ideal. They look like porcelain, and do not have the shimmer of mother-of-pearl. Yet they produce beautiful pearls of different sizes and colors, appreciated by collectors and sellers.

The production of Oriental pearls is documented in the Persian Gulf, the Red Sea, and in the Pearl Bank of the Gulf of Mannar, located between India and Sri Lanka. Few oysters produce pearls. In the Sri Lankan coast, a few dozen were found among millions of oysters in the past. Oyster pearl fishing in the Gulf of Mannar has been recorded for at least three thousand years. Pearl fishers diving for pearl oysters were known as early as 300 BC. Pinctata oyster families are also found in the sea off the coast of Venezuela, in the Pacific coast of California, in the Seychelles Islands, and in the coasts of the Indian and Pacific Ocean. Oyster culture currently provides the largest quantity and biggest size of natural pearls (Schlüter and Rätsch, 1999, pp. 4−7).

Pearls are composed of 82%−86% calcium carbonate (as aragonite $CaCO_{29}$) 10%−14% conchiolin ($C_{32}H_{43}N_9O_{11}$) and 2%−4% water (Mohsen, 2000, p. 103). The main component of pearls, calcium carbonate ($CaCO_3$), is one of the most common earth minerals. It takes the form of calcite ($CaCO_3$), of which marble is composed, and aragonite which forms stalactites at low temperatures.

X-rays reveal that pearls and mother-of-pearl share the same structure and composition. Both pearls and mother-of-pearl are composed of a mosaic of aragonite crystals bound together with conchiolin, a protein composed of a matrix of organic macromolecules and polysaccharides, secreted by the outer layer, the mollusk epithelium (mantle) (Schlüter and Rätsch, 1999, pp. 22−23). The production of pearls by oysters appears to be a reaction to foreign material introduced in the oyster shells by worms and small crabs (Schlüter and Rätsch, 1999, pp. 30−31).

While this theory is true to a large extent, microscopic examination of some pearls suggests that a foreign substance is not always essential to their formation. They may originate in calcareous concretions of minute size, termed "calcopherules" (Kunz and Stevenson, 1993, p. 45).

Pearls can be formed in three ways. "Ampullar pearls" are formed in the pockets or ampullae of the epidermis. "Muscle pearls" are formed around calcospherules at the insertion of the muscles. "Cyst pearls" are formed when parasitic worms build cysts, located in the connective tissue of the mantle (outer layer) and within the soft tissues of the body, around which concentric layers of nacre are deposited (Herdmann, 1903, pp. 9−10) (Fig. 28.1).

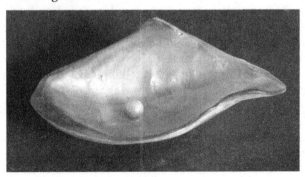

FIGURE 28.1   Encysted pearl in the shell from a Dutch Kunstkammer, 17th century (Távora Sequeira Pinto Collection, Oporto).

28.3 PEARLS IN INDIAN AND CHINESE MEDICINE

In the East, the pearl was more than a jewel, a symbol of power, or a valuable amulet. According to the gemologist, George Frederick Kunz (1856—1932), pearls were used in medicine since ancient times. Their use is mentioned in the oldest Sanskrit medicinal work, the "Charaka Samhita," composed early in the Christian era. Therapeutic properties have been particularly credited to pearls in Oriental countries. Their powder has been rated very highly, and is still currently used. It was considered beneficial in cases of malaria, indigestion, hemorrhage, and was used as a stimulant. Medical literature of the Orient contains many accounts of the uses of pearls and their assorted formulations into pills and ointments, for example (Kunz and Stevenson, 1993, p. 308).

Pearls, together with mother-of-pearl, are a component of ancient Chinese traditional medicine. Perforated pearls were not considered for medicinal use. Pearls were credited as good for eye diseases, as aphrodisiacs, and to promote fertility. Mollusk shells were credited with specific therapies. The shell of the giant abalone, *Halitotis gigantea*, was said to soothe the liver and be an effective tranquilizer, reducing irritability, dizziness, delirium, and cramps (Schlüter and Rätsch, 1999, pp. 133—134).

In India and the Himalayas, pearls were believed to originate in moon dew, in full moon nights during which the gods drank the soma or amrita, the elixir of life. Pearls represented the source of eternity of the healing gods. They were also seen as universal antidotes. Further, they were credited with curing hemorrhage, jaundice, mania, mental disturbances, eye diseases, and lung tuberculosis. In traditional Ayurvedic medicine, even as practiced today, pearl powder is considered a fortifier and aphrodisiac. In this tradition, different mythical origins and properties were attributed to pearls and mother-of-pearl shells, according to their colors (Schlüter and Rätsch, 1999, pp. 137—138).

The therapeutic indications of pearls did not change much over time. According to a treatise written by Narahari, a physician of Kashmir (about 1240 AD), pearls reduced to powder and dissolved in liquids were advised in topical applications to relieve eye disorders. Taken orally, they should treat poisonings, lung tuberculosis, and increase strength and wellbeing, helping to fight morbid conditions and stabilize any imbalance of the humors (phlegm, bile) (Garbe, 1882, p. 74).

Statements attesting to the curative properties of pearls also come from modern India and Japan. The Indian Prince Sourindo Mohun Tagore (1840—1914) added further therapeutic indications of pearls. In the treatment of hematemesis, he notes "The burnt powder of this gem, if taken with water as *Sherbet*, cures vomiting of blood of all kinds." Pearl powder would be otherwise helpful: "It prevents evil spirits working mischief in the minds of men, takes off bad smell from the mouth

FIGURE 28.2 Inrō with God of Longevity (Jurōjin) and a stag (obverse) and long-tailed turtle (Minogame) under bamboo (reverse) depicted on four cases; lacquered wood with gold hiramaki-e, togidashimaki-e, nashiji ("pear-skin ground"), gold foil cutouts, and mother-of-pearl inlay on black lacquer ground Netsuke: manju type, basket with chrysanthemums; lacquered wood with mother-of-pearl inlay Ojime: carved red lacquer bead by a Japanese Somada School Artist from the Edo period (1615–1868), Accession Number: 29.100.729. *Courtesy of the Metropolitan Art Museum, New York.*

(...) Burnt pearls mixed with water and taken into the nostrils, as a powder, takes away head sickness, cures cataract, lachrymal, and swelling of the eyes, and the painful sensation such as is caused by the entry of sand into them, and ulcers. (...) Whether taken internally or externally, it is a sure antidote to poison. It drives way all imaginary fears and removes all body pain" (Tagore, 1881, p. 871).

In Japan, the curative properties of pearls were also credited in soothing the heart, lessening phlegm, curing fever, smallpox, and bleary-eyedness, and serving as antidotes to poison (Kunz and Stevenson, 1993, p. 309). As in China, pearls were associated with longevity. A God of Longevity (Jurōjin) is depicted on an inrō of lacquered wood garnished with pearls and mother-of-pearl inlay in Fig. 28.2.

28.4 PEARLS IN WESTERN MEDICINE

Pearls were adornments of Olympian goddesses, associated with love and seduction. Juno wore pearl earrings in the Iliad (XVI, 183) and in the Odyssey (XVIII, 298). An Eros pendant hangs from a Greek pearl

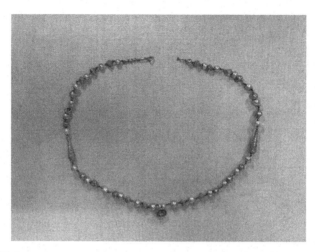

FIGURE 28.3   Pearl and gold necklace with pendant of Eros (330–300 BC), said to be from Mytilene, Accession Number 99.23. *Courtesy of the Metropolitan Art Museum.*

and gold necklace from the Hellenistic period in Fig. 28.3. Theophrastus (371–287 BC) refers to it as a special stone and valuable jewel, produced by oysters in India and in the Red Sea (Caley and Richards, 1956, p. 52). He doesn't mention it as a medicine.

Pearls were known to the Romans as margaritae and union. A famous story of magnificence was told by Pliny the Elder (23–79 AD) about Cleopatra VII Philopater, the last ruler of Ptolemaic Egypt (69–30 BC). Pliny recounts that she enriched a drink with a large and valuable pearl that she was wearing as an earring as a tribute to the Roman general, her ally and lover, Marc Anthony (83–30 BC). Pliny refers to the pearl as one of "those choicest, most rare and unique productions of Nature" (Pliny N.H. Book 9, Chapter 53 in Bostock and Riley, 1855, Volume II, p. 439). Pliny does not refer to its use in medicine.

In the Middle Ages, Renaissance and Baroque eras, European courts and monarchs highly praised the queen of gems. Christianity devoted pearls to the Virgin Mary.

Pearls were equally esteemed medicinally by the Arabs who adopted some of the Indian and Chinese learning about them. A rare Syriac manuscript copied and translated by Sir Walis Budge (1857–1934) transcribes the recipes of three pearl antidotes among prescriptions for heart diseases. Although the recipes are called antidotes they are not antidotes against poison, just medicines for heart conditions.

The first antidote was for diseases of the heart; the second for pain of the heart and all diseases that produce black bile; the third for fear and palpitation of the heart. A special Pearl Antidote figures among royal medicines bearing the names of eminent physicians and kings, such as the Antidote of Esclepiades, Antidote of Galen, Antidote of Caeser, and recipes made of valuable and expensive materia medica such as the Gold Antidote and the Musk Antidote (Budge, 2009, pp. 301−302).

Pearls were usually the main ingredient of European cordials. Coral, also composed of calcium carbonate, was also a typical component, along with other gems, and vegetable and aromatic ingredients, mixed in water (frequently rosewater), wine, and honey (On the medicinal use of coral, see Barroso, 2017, pp. 267−282).

In the middle of the 13th century, a work titled *Grabadin* by the author known as "Mesue Junior," "Mesue Junior," or "Johannes Mesue Damascenus" was very popular. Owing to the mystery surrounding his writings, this author is known as Pseudo Mesue. There is great doubt as to the origin of the work mentioned as the Arabic originals have never been found. The very existence of Mesue Junior is doubtful. It is considered that a Latin compiler assumed the name. The *Grabadin*, composed in twelve parts, dealing with the *materia medica*, was very important since it familiarized Latin Europe with Arabian pharmacopeia and therapeutics. Many hundreds of editions of this book were issued during the Middle Ages. It was among the first medical works to be printed at Venice in 1471 (Campbell, 2006, pp. 76−77). This work was particularly relevant in the introduction of Eastern mineral *materia medica* to the European pharmacopeia.

One of its popular recipes was a combination of gems: the Gem Electuary. It combined white pearls, blue sapphire, pieces of emerald, garnet, red coral, amber, gold, and silver reduced to powder, ivory filings, vegetable ingredients such as zedoary, doronicum (a kind of daisy belonging to the botanical family Asteraceae), saffron, cardamom, cinnamon, bound together with musk, honey, and rose sugar in a dilution of water and wine (Damascenus, 1581, 95). It was intended to strengthen the brain, the heart, the liver, and the uterus. It was prescribed to improve melancholy, sadness, shyness, and solitude, and to soothe heart tremors, and alleviate syncope. The British geologist, Christopher Duffin, has dedicated a very enlightening article to the Gem Electuary (Duffin, 2013, pp. 81−111).

Many cordials followed the Gem Electuary in which pearls continued to be the first ingredient. Pearls were less often used as simple medicines. The English naturalist, Robert Lovell (c. 1630−1690), summarized the properties and therapeutic uses of pearls in Europe during his day: "Pearls cherished the spirits and principal parts of the body, cleansed the eyes, dried up water thereof, helped their filth, strengthened the nerves by which moisture flowed into them, when put into collieries; were good against melancholic grief, cardiac passions, defended against pestilent diseases, and were mixed with cordials. Pearls were also good against diarrhea, fainting, and helped against trembling of the heart, and dizziness of the head. They were put into antidotes or corroborating powders. They helped the flux of blood, stopped the terms and cleansed the teeth, helped in fevers, The *Oil of Pearles* helped in the resolution of the nerves, convulsions, decay of old age, frenzy, kept the body sound and recovered it when out of order, rectified women's milk and corrected the natural parts and seed." It cured abscesses, stomach ulcers, cancer, and hemorrhoids.

He concluded, stating: "The best are an excellent cordial, by which the oppressed balsam of life and decayed strength are recreated and strengthened, therefore they resist poison, the plague, and putrefaction, and exhilarate, and therefore they are used as the last remedy for the sick persons" (Lovell, 1661, pp, 77–78). The pearl appears to have been among the most respected stabilizers of imbalances of all kinds and an analeptic.

Another very expensive European recipe included pearls: the The Goa Stone or Cordial Stone, an artificial bezoar (gastric concretions of indigestible material) created by the Jesuitic friar Gaspar António in middle 17th century. Bezoars, produced by ruminants' second stomach, were very appreciated by Arabic authors as antidotes. They were introduced in European medicine during the Middle Ages. In the Renaissance, porcupine bezoars were introduced in Europe by the Garcia de Orta (1490–1568), an outstanding Portuguese Jewish doctor who sailed to India and described Oriental *materia medica*. He described the porcupine bezoar, one of the most expensive Renaissance medicines, and referred to its medicinal use (Orta, 1563, pp. 226–227).

Seed pearls (small pearls, not suitable for jewelry), were the first ingredient of cordial stones. Others ingredients were: musk, ambergris, red coral, white coral, emerald, fossilized shark's teeth, topaz, terra sigillata, ruby, Cananor stone, jacinth, deer horn tips, sapphire, and Oriental bezoar (A.A V.V, 1766, p. 262).

Cordial stones were cordials and antidotes for poison. They were intended to provide a product that could be less expensive than bezoar stones (On this subject, see Barroso, 2014, pp. 77–98). However, the ingredients of the Goa Stone were also very expensive. Like bezoar stones, they were carefully kept in richly decorated containers of gold or silver filigree worked in patterns of Indo-Portuguese jewelry (Fig. 28.4).

28.5 PEARLS AND MODERN CHEMISTRY

Calcium carbonate is a known antacid. It is the primary structural mineral in the body. It helps to regulate cell permeability, maintains

FIGURE 28.4   Goa Stone and gold and silver container. Indo-Portuguese filigree, end of the 17th century, Távora Sequeira Pinto Collection (Oporto).

acid—base balance, is involved in nerve transmission and muscle relaxation, and is required for phosphorus metabolism and energy production in the Krebs cycle.

Calcium is also important as an antagonist to some toxic metals, especially lead and cadmium, by opposing and blocking the absorption and utilization of these toxic metals. It is involved in blood clotting, and in the fight-or-flight response (Wilson, 2016).

These features highlight the cited beneficial use of pearls, especially in treatment of heart conditions, and as an antidote.

28.6 CONCLUSION

From ancient times, naturally occurring pearls have served as symbols of beauty and magnificence, and as an exponent of health. Besides being esteemed as jewels related to gods and emperors, pearls have been widely used as medicines in Eastern civilizations. The three main therapeutic indications are as cardiac soothers, fortifiers, and antidotes. European countries adopted their use as medicines from the Middle Ages until the 18th century.

To some extent, calcium carbonate, their main chemical component, accounts for some effectiveness, especially in the treatment of heart conditions, and as an antidote against poisons, at a time when the science of chemistry had not yet developed and therapeutic resources were scarce and difficult to assess.

Acknowledgments

I wish to thank Dr. Philip Wexler for kind reading of the manuscript, Dr. Álvaro Sequeira Pinto for kind permission to reproduce the images of the Távora Sequeira Pinto (Oporto) collection, and the Metropolitan Art Museum of New York for kind permission to reproduce the Museum pieces.

References

A.A V.V. Collecção de Varias Receitas [...] (Manuscript from the Archive of the Jesuitic Company), Rome, 1766.

Barroso, M.D.S., 2014. The bezoar stone: a princely antidote, the Távora Sequeira Pinto Colletion:Oporto. Acta Med. Hist. Adriat. 12 (1), 77–98.

Barroso, M.D.S., 2017. Coral in petrus hispanus 'treasure of the poor'Special Publications In: Duffin, C.J., Gardner-Thorpe, C., Moody, R.T.J. (Eds.), Geology and Medicine: Historical connections, 452. Geological Society, London.

Bostock, J., Riley, H. (Eds.), 1855. The Natural History, Pliny the Elder. Taylor and Francis, London.

Budge, W.E.A., 2009. Book of Medicines, Ancient Syrian Anatomy, Pathology and Therapeutics. Routledge, London and New York.

Caley, E.R., Richards, J.F.C., 1956. Theophrastus. On Stones. The Ohio State University, Columbus, OH.

Campbell, D., 2006. Arabian Medicine and its Influence on the Middle Ages. Martino Publishing, Hertford.

Damascenus, I.M., Opera De medicamentorum purgantium delectu [...] (s. ed.): Venetiis, 1581.

De Mely, F., 1896. Les Lapidaires de l'Antiquité et du Moyen Age. Tome I Les Lapidaires Chinois. Ernest Lecoux Éditeur, Paris.

Duffin, C.J., 2013. The gem electuary. In: Duffin, C.J., Moody, R.T.J., Gradner-Thorpe, C. (Eds.), Geology and Medicine, 375. Geological Society, Special Publications, London, pp. 81–111.

Garbe, R. (hrg.), Die Indischen Mineralen und ihre Namen und die ihnen zugescriebene Kräfte. Naraharabi's Râĝanighantu Varga XIII, Sanskrit und Deutsch, Verlag von S Hirzel, 1882.

Herdmann, W.A., 1903. Reposit to the Government of Ceylon of the Pearl Oyster Fisheries of the Gulf of Manaan. Colonial Government of the Royal Society, London.

Kunz, G.F., Stevenson, C.H., 1993. The book of the pearl. The History, Art, Science and Industry of the Queen of Gems. Dover Publications Inc, New York, NY.

Lovell, R., 1661. Panmineralogicon. An Universal History of Mineralls [...]. Joseph Godwer, Oxford.

Mohsen, M.D., 2000. Dicionary of Gems and Gemology. Springer Verlag, Berlin, Heidelberg.

Orta, A.G., 1563. Coloquio dos Simples, e drogas he cousas medicinais da India [...]. Ioannes de Endem, Goa.

Schlüter, J., Rätsch, C., 1999. Perlen und Perlmutt. Ellert & Richter Verlag GmbH, Hamburg.

Tagore, R.S.W., 1881. Mani-mála: or a Treatise on Gems, Vol. II. Stanhope Press, Calcutta.

Wilson, L., July 2016. Calcium. L. D. Wilson Consultants, Inc, http://drlwilson.com/Articles/calcium.htm accessed on the 23th August 2017.

29

Rhetoric, Demons, and the Poisoner's Tongue in Judaism and Early Christianity

John F. DeFelice

Department of History, University of Maine at Presque Isle,
Presque Isle, ME, United States

OUTLINE

Toxicology in Antiquity
DOI: https://doi.org/10.1016/B978-0-12-815339-0.00029-9

411

29.1 THE HEBREW BIBLE

Unlike many accounts of ancient times, there is no account of poisoning in the Hebrew Bible. Only in the Apocrypha is there a single case of suicide by an unspecified poison[1] (Trestrail, 1993). In most cases, poisoning is mentioned metaphorically or figuratively in terms of words suggestive of toxicity and actions of people. Nonetheless, there appears to be a knowledge of the dangers of venomous creatures that denizens of Israel and the surrounding area had to deal with on a daily basis.

29.2 KNOWLEDGE OF VENOMOUS DANGERS

29.2.1 Spiders and Insects

The writers of the Hebrew Bible were quite familiar with several kinds of venomous pests. The area corresponding to modern Palestine is home of two species of the Black Widow (*Latrodectus tredecimguttatus* and *Latrodectus pallidus*) and the Mediterranean Recluse (*Loxosceles rufescens*) (http://www.hadassah-med.com/about/tips/poisonous-bites-and-stings). Spiders (עַכָּבִישׁ, *akkabish*), venomous or harmless, have scant mention in the Bible and are better known for their webs, used as a metaphor for fragility or entrapment than any poison. These are not mentioned in the New Testament[2,3,4].

Hornets make an appearance in the Bible, usually in reference to the conquest of Canaan. The text describes God sending hornets (Hebrew צִרְעָה *tsirah*) to drive out the Canaanites and two Amorite kings.[5,6,7] While the idea of using stinging insects in warfare may seem far-fetched, there are a number of other examples from antiquity. The Maya in central America, the Romans, German states during the 30 Years War, and Ethiopians fighting Mussolini all used beehives and wasp nests against enemies. In one key campaign, the Mesopotamian city of Hatra repulsed the siege of Septimus Severus twice, in part by hurling clay pots full of stinging insects down on the Roman attackers from their walls (Neufield, 1980; Mayor, 2004).

[1]II Maccabees.
[2]Job 8:14.
[3]Proverbs 30:28.
[4]Isaiah 59:5.
[5]Joshua 24:12.
[6]Deuteronomy 7:20.
[7]Exodus 23:28.

The specific wasp was most probably *Vespa orientalis*, which is still found in Palestine today. Such wasps will give a painful sting when agitated and unlike bees can sting multiple times. Toxicity from wasp stings varies (Goddard, 2007; Edery et al., 1972). Pliny speculated that 27 stings from a wasp could kill a man (Pliny 11.73). But if one has allergies to wasp venom, one sting still may be fatal. More damaging would be the sheer panic and disorder a mass of stinging insects could cause. Other biblical texts refer to some attackers metaphorically as bees.[8,9]

29.2.1.1 Scorpions

Scorpions are mentioned, in a generic sense, several times in both the Hebrew Bible (עֲקְרָב *aqrab*) and the New Testament (σκορπίος skorpios). The name is possibly derived from σκορπίζειν τον ιον: "scattering the poison" (Liddell and Scott, 1935). The sparsity of references is surprising as there are at least seven species of scorpions in modern Palestine today. All these can cause painful stings, as every camping archeologist who put on a boot or shoe without looking first usually finds out. All scorpion stings are extremely painful. This is due to the nature of their venom, a powerful mix of neurotoxins that varies from species to species. The scorpions in this area feed primarily on insects. Perhaps the most dangerous is the Deathstalker, also known as the Israeli or Palestine Yellow Scorpion, the Omdurman Scorpion, or the Naqab Desert Scorpion (*Leiurus quinquestriatus*). Those stung may experience anaphylaxis. But usually only children, the elderly, or infirm risk fatalities (Qumsiyeh et al., 2013; Simard and Watt, 1990). Scorpions are mentioned as dangerous pests and one boundary of ancient Israel was known as "Scorpion Pass."[10,11,12]

In the Hebrew Bible, scorpions symbolize both rebellious and arrogant people. Ezekiel was told to speak boldly as a prophet. "Do not be afraid, though briers and thorns are all around you and you live among scorpions. Do not be afraid of what they say or be terrified by them, though they are a rebellious people."[13] And Reheboam, the son of Solomon, decided to tell his people, after consulting with his young drinking companions, how tough a ruler he would be upon his coronation by proclaiming, "My father made your yoke heavy; I will make it even heavier. My father scourged you with whips; I will scourge you

[8]Psalm 118:12.

[9]Isaiah 7:18.

[10]Deuteronomy 8:15.

[11]Numbers 34:4.

[12]Judges 1:36.

[13]Ezekiel 2:6.

with scorpions." It is more likely that he is describing a whip that is set with sharp objects at its tip than actually whipping a victim with a scorpion. The result was a permanent breach in the kingdom into two competing kingdoms.[14,15] The scorpion, along with briars and thorns, is also used as a type to describe Jews who are apostate from their faith.[16]

29.2.1.2 Snakes

Snakes are yet another groups of venomous creatures mentioned in both the Hebrew Bible and the New Testament. As the children of Israel traveled through the wilderness, they encountered "fiery serpents." There are several Hebrew words describing different kinds of snakes, but trying to match them with particular species is, for the most part, problematic. Currently Israel and Jordan have around 40 species of snakes, of which 10 species are venomous.

These include the Palestine (or Israeli) Saw Scaled Viper, also known as the Burton's Carpet Viper (*Echis coloratus*), the Palestine (or Israeli) Viper (*Vipera paleastinae*), The Desert Black Cobra (*Walterinnesia aegyptia*), the Palestinian Mole Viper (*Atractaspis engaddensisi*), the Desert sidewinding horned viper (*Cerastes cerastes*), and the Persian Horned Viper (*Pseudocerastes persicus* or *fieldi*). The most dangerous is the Palestine Saw Scaled Viper, which can inject twice the fatal dose for humans in a single strike (Kochba, 1998). The powerful procoagulant effect of this venom results in a 20% rate of mortality even with modern treatment. Though slightly less toxic, snake bites are more common with the Palestine Viper. Its venom also has a procoagulant effect and death from cardiovascular collapse usually comes between 12 and 20 hours after a bite. The Desert Black Cobra's venom contains a powerful neurotoxin, while the Palestine Mole Viper is a smaller burrowing viper that prefers moist areas around oases. It is not an aggressive snake, but its venom contains isotoxins (sarafotoxins) that tend to target cardiac activity. All these snakes are nocturnal (Kochva et al., 1993).

Oddly enough, despite what were most likely daily encounters with venomous creatures, in both the Hebrew Bible and New Testament, there is no record of a historical figure being stung by a scorpion and only two references to people bitten by snakes.[17,18]

In general, snakes are viewed negatively in both the Hebrew Bible and New Testament. This starts in Genesis with the role of the serpent

---

[14]1 Kings 12:11, 14.

[15]2 Chronicles 10:11, 14.

[16]Ezekiel 2:6.

[17]Numbers 21:6.

[18]Acts 28:3.

in the fall of man. This is also where, according to rabbinic traditions, the angel Samael starts his career as "The Poisoner" as he either partners with or becomes the tempting serpent in the Garden of Eden (Dulkin, 2014).

Snakes were used by God to judge the children of Israel when they complained about Moses as a leader as wood and water ran out in the wilderness in their flight from Egypt. Numbers 21:6 states that "Then the Lord sent venomous snakes among them; they bit the people and many Israelites died." Many translations use the phrase "fiery serpents" (שָׂרָף *saraph*) to describe the snakes. The word literally means "fiery," and is the base Hebrew word used to describe fiery angelic seraphim. But here and in other places is it used to describe the burning a victim feels with a snake bite. The cure is also interesting. Moses was told by the Lord to fix a bronze serpent on a pole, and when those who were bit looked upon it, they were healed.[19] Hezekiah, King of Judah, later destroyed this bronze serpent (נְחֻשְׁתָּן *Nechushtan*), most probably because it was associated with the local deities of Canaan and people offered sacrifices to it[20] (Handy, 1992; Jones, 1968; Murison, 1905).

29.2.1.3 Venom

The early Hebrews were aware that some snakes and scorpions injected venom. But the words translated as "venom" vary, often within the same text. For example, one of Job's comforters, Zophar, states that the wicked man may find evil sweet in his mouth, but the food in his stomach will be turned into the venom of cobras (or perhaps asps). The word for venom here is *merorah* (מְרֹרַת), often translated as gall or bitterness. Later in the same text Zophar notes that he will suck down the poison of cobras (or asps).[21] This time the word is *rosh* (רוֹשׁ), which can also mean gall or bitterness. In one psalm the writer laments that the wicked constantly lie and that their poison is like the venom of a serpent.[22] Here both times the word for poison is *hemah* (חֵמָה), which is translated as poison, venom, heat, anger, or rage, often joined poetically to other words for venom or poison (Harris, 1980a,b,c; Harrelson, 1962; Harrison, 1986). This serves as a reminder that there is little precision in these ancient texts. They are poetic not clinical. This observation holds true both with venoms and types of snakes.

Even over consumption of alcohol is described as a snake bite. Proverbs 23:31−33 warns, "Do not gaze at wine when it is red, when it

[19]Numbers 21:7−9.

[20]2 Kings 18:4.

[21]Job 20:12−16.

[22]Psalm 58:4−5.

sparkles in the cup, when it goes down smoothly! In the end it bites like a snake and poisons like a viper. Your eyes will see strange sights, and your mind will imagine confusing things."[23] This saying incorporates two Hebrew words for snake: *nachash* (נָחָשׁ), usually translated as serpent, and *tsepha'* (צֶפַע), which is translated as viper, adder, or the odd word cockatrice. This often is described as a mythical serpentine beast, a two-legged flying dragon, and appears as a far more threatening creature elsewhere in the Hebrew Bible (Murison, 1905).[24] Jeremiah, the last prophet of Judah before the Babylonian conquest in 586 BCE, warned that the Lord was "sending serpents upon you, adders, for which there is no charm and they will bite you, declares the Lord." Like the previous verse in Proverbs 23:32, he couples the words *nachash* (serpent) and *tsepha'* (perhaps a mythical dragon) to indicate the worst kind of judgment. Most commentators now agree that the word *tsepha'* cannot be translated as "viper" as it is said to lay eggs and vipers are ovoviviparous.[25] The writers probably identified this beast with the mythical basilisk of Egyptian and Greek lore.

Though there are a number of Hebrew words for snakes, attempts to identify them with specific species has proven problematic. One that seems correct can be found in Genesis, describing one of the 12 tribes of Israel: "Dan will be a snake by the roadside, a viper along the path, that bites the horse's heels so that its rider tumbles backward." The Hebrew word used is *shephiphon* (שְׁפִיפֹן) and may indicate the Persian horned viper (*P. persicus* or *fieldi*), which tends to wait for prey partially under the sand or under low bushes (Murison, 1905; Mendelssohn, 1965).[26]

29.3 POISONOUS PLANTS AND POISONOUS WATER

For the most part, poisonous plants are described in very general terms in the Hebrew Bible. Plants or water with a bitter taste are often described as poisonous by biblical writers.

The best known of poison plants was the toxic gourd. In this story, Elisha the prophet commanded a group referred to as the "sons of the prophets" to prepare a meal for guests even though it was a time of famine. One found some wild gourds of an unknown type, cut them

[23]Proverb 23:31–33.

[24]Isaiah 11:8,14:29; 59:5.

[25]Isaiah 59:5.

[26]Genesis 49:17.

up, and put them in the stew. As the guests eat, they became ill, and cried out to Elisha, "O thou man of God, there is death in the pot!" Elisha then cast barley meal into the pot, which rendered it (miraculously) edible and harmless.[27] This gourd has long been identified as the bitter or wild gourd, *Citrullus colocynthis*. It grows as a vine along the ground and produces a fruit the size of an orange. Its leaves resemble grape leaves (Macht, 1919–1920; Helser, 1979). In small doses, it has use as a purgative, but in high doses will cause hemorrhagic colitis accompanied by vomiting, convulsions, and in some cases, death. Its bitterness makes accidental ingestion unlikely, though there are some medicinal uses as a laxative and it is used as part of a mixed concoction in Egypt applied to a pessary as an abortifacient in the Eber Papyrus (1550–1500 BCE). *C. colocynthis* may also be the "vine of Sodom." The vine is described as being poisonous and bitter[28] (Javadzadeh et al., 2013; Riddle, 1997b).

Generally, in the Hebrew Bible, water and foods that are "bitter" (מָרָה *marah*) are considered poisonous. For example, during the flight from Egypt, the writer of Exodus reported, "When they came to Marah, they could not drink the waters of Marah, for they were bitter; therefore, it was named Marah." Moses purified the water by throwing in a piece of wood.[29] More often bitter water is considered a sign of divine judgment. For example, in the waning days of Judah, Jeremiah warned, "Why are we sitting still? Assemble yourselves, and let us go to fortified cities. And let us perish there. Because the Lord has doomed us and given us poisoned (bitter) water to drink. For we have sinned against the Lord."[30] The first miracle of Elisha also involved purifying water. The people of the city of Jericho came to Elisha complaining of the local water. They stated that "the water was bad and the land causes miscarriages. Elisha called for a new bowl, filled it with salt, and threw it into the spring. The Lord then declared the water pure."[31] This is similar to the purification of water by Moses. The prophet Hosea compared faithless oaths and corrupt judgments with poisonous weeds in a field.[32] In this case the word translated as "poisonous weed" is translated elsewhere as "gall." It is the Hebrew word רוֹשׁ *rosh*, and it is sometimes translated as hemlock, venom, or simply as poison. Two other Hebrew words, *ra'al* and *tar'elah* may also mean

[27]2 Kings 4:38–41. Macht (1919–1920)

[28]Deuteronomy 32:32.

[29]Exodus 15:22–25.

[30]Jeremiah 8:14. See also 9:15; 23:15.

[31]2 Kings 2:19–22.

[32]Hosea 10:4.

poison, only because they appear in parallel poetic verses with *hemah* and *mererah*.

29.4 POISON AND WORMWOOD

Arrows dipped in poison were obviously known to the Hebrews, though there is no record of them being used. However, lies and slander are compared to arrows, poisoned or otherwise, several times in the Hebrew Bible.[33,34,35,36,37] The potential for a sovereign to be poisoned was known as well. The word *mashqeh* (מַשְׁקֶה), often translated as butler, is better translated as cup bearer. He poured wine as well as tested the wine for his sovereign. The word occurs 12 times (in several grammatical forms) in the Hebrew Bible, nine times describing the court of Pharaoh in Egypt, twice in the context of the Queen of Sheba visiting Solomon's court, and once describing the role of Nehemiah at the court of the king of Persia.[38,39,40,41] Solomon is the only king of Israel recorded as having a cup bearer, but the office and role were obviously known. No king of Israel or Judah is recorded as being poisoned (Yamauchi, 1980; Younger, 1997; Poison, http://www.jewishvirtuallibrary.org/poison).

The Hebrew Bible mentions wormwood (לַעֲנָה *laanah*) as a poison and a bitter substance. Identifying the specific plant mentioned is difficult, but most scholars identify it with *Artemisia judaica* also known as *Artemisia herba-alba*. In Deuteronomy, the writer compares those who fall away from the Lord of Israel as a "root bearing poisonous fruit and wormwood."[42] The prophet Jeremiah proclaimed that the Lord feeds both fallen people and apostate prophets with poison (bitter) water and wormwood.[43] The prophet Amos complained that the people "turned justice into poison (or venom: רוֹשׁ *rosh*) and the fruit of righteousness

---

[33]Job 6:4.

[34]Psalm 64:3.

[35]Proverbs 25:18.

[36]Proverbs 26:18.

[37]Jeremiah 9:7.

[38]Genesis 40–41.

[39]1 Kings 10:5.

[40]2 Chronicles 9:4.

[41]Nehemiah 1:11.

[42]Deuteronomy 29:18.

[43]Jeremiah 9:15; 23:15.

into wormwood."[44] Wormwood has a strong, bitter taste. It is often used as fodder for sheep and camels, resulting in bitter milk. When burned as fuel, it has have a pleasant, aromatic smell. But, in general, it is used as a symbol of poison bitterness in the Bible. This idea appears only once in the New Testament. In Revelations, the star named Wormwood falls to earth as a sign of God's judgment, rendering one third of the planet's water bitter and poisonous. Appropriately, the Greek word for Wormwood is ἄψινθος *apsinthos*, which literally means, "not drinkable." The Greek version of the Hebrew Bible, the Septuagint, uses the same word for wormwood. While it is certainly used as a metaphor for bitterness, it also had several medicinal purposes. Every species of *Artemisia* has anthelmintic qualities and was used to remove intestinal parasites in antiquity. It may also have been rendered into an alcoholic drink (as absinthe is today). It has antiinflammatory properties and was occasionally used in antiquity as an abortifacient[45,46] (Riddle, 1997a).

Wormwood usually is not considered a lethal poison, but beyond the reference to fatalities associated with the star of Revelations, it is associated with the demonic. Samael, the fallen angel some traditions associated with the devil, has the reputation of being known as "The Poisoner." In some Jewish traditions, Samael kills his victims with a drop of gall of wormwood in a drink. In another, he killed the first born of Egypt in the time of Moses as the Angel of Death by poisoning. His very name means "The Venom of God" (Godbey, 1930; Samael, http://www.jewishvirtuallibrary.org/samael; Demons and Demonology, http://www.jewishvirtuallibrary.org/demons-and-demonology; Angel of Death, http://jewishencyclopedia.com/articles/5018-death-angel-of).

29.5 THE NEW TESTAMENT

29.5.1 Common Territory, Common Problems

The early church, born in Palestine, inherited both familiarity with local venomous creatures, the rhetoric of its holy book, and the cultural instincts of its people. Again, there is no historical account of a poisoning, or a scorpion bite and one instance of a snake bite. But apart from some practical parables and references, the rhetorical use of scorpions, serpents, and even wormwood is lifted to the level of a cosmic struggle. Later, as the church develops beyond its Jewish roots into the Greco-

---

[44]Amos 6:12.

[45]Revelations 8:10–11.

[46]Lamentations 3:15,19.

Roman gentile world, part of the signs believed to follow the message of the Gospel (and orthodox doctrine) were signs of toxicological immunity. Finally, as the church developed its own system of apologetics and morality, it targeted both contraceptives and abortifacients as elements of sorcery and murder.

29.5.2 Jesus, Snakes, and Scorpions

In the Gospels, Jesus followed the pattern of associating serpents with human enemies and apostate religious leaders. Three times he referred to his enemies as "a brood of vipers."[47] And his cousin, John the Baptist, labeled sinners with the same designation.[48] In fact, serpents are elevated to being a symbol of the devil himself.[49]

Jesus, while teaching on prayer, referred twice to scorpions, again in a generic sense. In Luke's Gospel he asked the crowds following him, "Which of you fathers, if your son asks for a fish, will give him a snake instead? Or if he asks for an egg, will give him a scorpion?"[50] This has puzzled numerous Bible commentators, who focus more on how a snake can look like bread or a scorpion can look like an egg rather than how it is an inappropriate thing to give a child!.

In Luke 10, Jesus sent out 72 disciples to preach in all the towns that he intended to visit. They were specially charged to heal the sick and live off what their hosts offered. They returned reporting their success, even noting that demons fled, perhaps a reference to exorcisms. Jesus then said to them "I saw Satan fall like lightning from heaven. I have given you authority to trample on snakes and scorpions and to overcome all the power of the enemy; nothing will harm you. However, do not rejoice that the spirits submit to you, but rejoice that your names are written in heaven." From the context, "snake and scorpions" clearly represented evil spiritual entities rather than literal venomous animals.[51]

More problematic is Mark 16:17−18: "And these signs will accompany those who believe: In my name they will drive out demons; they will speak in new tongues; they will pick up snakes with their hands; and when they drink deadly poison, it will not hurt them at all; they will place their hands on sick people, and they will get well." In this case there is clearly a call for disciples to do specific actions are "signs" (σημεῖον, sémeion) to validate their ministries. This word appears 77

---

[47]Matthew 3:7; 12:34; 23:33.

[48]Luke 3:7

[49]Rev 12: 1−17, 20:1−10.

[50]Luke 11:11−12.

[51]Luke 10 18−20.

times in the New Testament. Not all references are positive. For example, in Matthew 24:4, Mark 13:22, and Revelation 13:13 demonic forces also try to validate their message with signs. Jesus himself often accused the faithless of demanding signs.[52] But here there is no stepping on snakes and scorpions by accident. Snakes are picked up and poison is drunk purposely as signs.

29.5.3 Mark 16, Toxicological Immunity, and Early Church Hagiography

Interpretation of Mark 16:17−18 is not without controversy. Even in the present day, there are snake handling cult in parts of the United States (Covington, 1996). But these controversial passages are missing in the earliest surviving texts of this Gospel and appear to be a much later addition. The earliest extant copies of the Gospel of Mark end at 16:8 and the transition between verses 8 and 9 is stylistically awkward. In all, there are four different endings in various manuscripts. Eusebius, the 4th-century church historian and St. Jerome, in the 5th century, considered the long version as inauthentic. It was probably written because the short ending of Mark ends so abruptly (without a resurrected Christ) or the original ending was lost (Swete, 1952; Carson, 2010; Cranfield, 1959; Metzger, 2005).

It is possible that signs associated with the apostles in the *Book of Acts* were incorporated into the long ending of the Gospel of Mark. There is no shortage of healings, resurrections, exorcisms, speaking in tongues, and even a snake bite in its text (for examples, see Acts 2:3−4; 3:1−11; 5:16; 16:16−19; 20:7−12, 28:1−6) (Fig. 29.1).

29.5.3.1 The Missing Snake

The sole snake bite incident in the New Testament is itself of great interest. In Acts 28:3−6, the Apostle Paul was on his way to stand trial in Rome. On this trip he was shipwrecked on Malta where he encountered a venomous snake while tending a fire. According to the text, "Paul gathered a pile of brushwood and, as he put it on the fire, a viper, driven out by the heat, fastened itself on his hand. When the islanders saw the snake hanging from his hand, they said to each other, 'This man must be a murderer; for though he escaped from the sea, the goddess Justice has not allowed him to live.' But Paul shook the snake off into the fire and suffered no ill effects. The people expected him to swell up or suddenly fall dead; but after waiting a long time and seeing nothing unusual happen to him, they changed their minds and said he was

---

[52]Matthew 16:4.

FIGURE 29.1   American Pentecostal snake handler.

a god." The main problem with this text is that there are no poisonous snakes on Malta that are dangerous to humans (Fig. 29.2)!

This has resulted in several interesting traditions. One is that St. Paul drove poisonous snakes out of Malta, as St. Patrick did in Ireland. Another is that he cursed the vipers and now any poisonous snakes arriving on the island instantly die. A later tradition claims that St. Paul lived in a cave at Rabat for several months and that limestone from this cave ground into powder cured snake bites (St. Paul's Earth). Sometimes this ground material was incorporated into metal or ceramic vessels to provide medicinal properties for its contents to cure snake bites and diseases. Other cures from Malta included "St Paul's Tongue," which upon examination is fossilized shark teeth (Ventura, 1990).

The fact that there is no snake on Malta dangerous to man either currently, in antiquity, or in the fossil record has caused some issues concerning the authenticity of the text. There is only one mildly poisonous snake on Malta, the Cat Snake (*Telescopus fallax*), possessed of a rather weak venom. It is a rear-fanged *colubrid* and presents no danger to humans. Furthermore, it may not be indigenous, only being noted in the late 19th century. Perhaps Paul's bite was from a nonpoisonous snake mistaken for a venomous one, or the island itself has been misidentified by translators. The Greek word in the text is *Melite* (Μελίτη), which most interpreters assume is the modern island of Malta, known in antiquity as *Sicula Melita*. However, in the Adriatic is another island of the same name in antiquity, which is also known as *Melita Illyrica*, modern day Mljet. While this is not the opinion of the majority of commentators, it does solve the snake problem. Mljet is home to a deadly viper, *Vipera ammodytes*, also known as the Horn Nosed Viper (Shuker, 2014, 1976; Meinardus, 1979; http://reptiledatabase.reptarium.cz/species?genus = Vipera&species = ammodytes).

FIGURE 29.2 St. Paul being bitten by a snake in Malta after his famous ship wreck.

29.5.3.2 Poison Stories

While Paul certainly accounts for a number of the signs listed in Mark 16:17−18, there is no report where he ingested poison. There are examples in apocryphal and pseudepigraphical works. For example, in the *Acts of John*, the apostle is challenged to consume deadly poison by Aristodemus, priest of the "idols." It was first tested on human subjects, who died. John makes the sign of the cross over the cup and drank it with no harm. After being challenged again by Aristodemus, he raised the other victims from the dead by means of his cloak[53] (http://www. earlychristianwritings.com/text/actsjohn.html). In a later work, John was challenged directly by the Emperor Domitian to drink a cup of the "strongest poison." After a fervent prayer, he drank it with no harm (The Acts of the Holy Apostle; http://gnosis.org/library/actjnthe.htm). Eusebius, the church historian of the 4th century mentions that Joseph,

---

[53] Acts of John XX−XXI.

also known as Barsabbas in the biblical Book of Acts, drank deadly poison and suffered no harm.[54,55]

These types of stories continue in the 6th century. Monks rebelling against the austerity of St. Benedict of Nursia tried to poison him with a glass of poisoned wine. Knowing their plans, Benedict made the sign of the cross and the glass exploded[56] (http://www.tertullian.org/fathers/gregory_02_dialogues_book2.htm#C3). Scattered throughout the lives and miracle stories of early saints are occasional tales of toxicological immunity.

29.5.4 The Early Church: Contraceptives, Abortifacients, and the Rhetoric of Poisoning

A final subject of interest is the long history of invective against contraceptives and abortion in the early church. The subjects do not appear in the New Testament or for that matter in the Hebrew Bible.

29.5.4.1 Rabbinic Judaism

This issue was not raised to a significant degree in rabbinic Judaism. There is no mention of a deliberate abortion in the Hebrew Bible. For all practical purposes, the law was focused on protecting the child after birth. An injury against a pregnant woman resulting in miscarriage results in a fine paid to the father (not the mother). Further punishment was required if the woman herself was injured or died.[57] If the fetus threatened the life of the mother during the pregnancy, it could be aborted up until the time the head appears, which was considered the point of child birth. For all intents and purposes, the fetus was the property of the father (Riddle, 1997c; Riddle, 2010). There were specific times when contraceptives and abortion were allowed, but the general rule was that marriage was for procreation as part of the mandate to "fill the earth"[58] (Yamauchi, 2014a,b).

The chemicals used to promote sterility, birth control, or abortion are a matter of speculation. Rabbinic sources refer to women and men taking "a cup of roots" (*ikarin*). What these roots were is unknown. Rabbinic sources are divided as to whether this is permitted or not and if permitted, under what circumstances. Commenting on the Jewish

[54]Acts 1: 23.

[55]Eusebius, Ecclesiastical History, 3.39.9–10.

[56]Gregory the Great, Dialogues, 2.3.

[57]Exodus 21: 22–25.

[58]Ge 1:27–28.

traditions in this period, John Riddle noted, "No one can read the Hebrew accounts and believe that contraception and abortion were encouraged, but neither were they banned altogether. There were circumstances in which each was appropriate" (Riddle, 1997c).

29.5.4.2 Pharmakeia and Poisoning

The New Testament also does not mention birth control and abortion, but in Galatians 5:20 Paul discusses a sin of the flesh being *pharmakeia* (φαρμακεία). It is a word that can mean anything from sorcery to administering drugs for treatment. It occurs in only one other place in the New Testament, Revelations 18:23, where it is clearly meant to mean sorcery. A related term, *pharmakos* (φάρμακος), occurs three times in the New Testament, only in the Book of Revelations. This refers to one who practices sorcery.[59]

As the church continued to grow and expand after the 1st century, it was increasingly composed of gentiles as opposed to Jews. It was influenced by the Greco-Roman world as much as it impacted it. Thus to understand what was available to early Christians as birth control or abortifacients, one must study the pharmacology of their Greco-Roman host. There is rarely anything specific about birth control and abortion in early Christian literature other than multiple condemnations. These writings, while revealing the negative attitudes and doctrines of the early church on contraceptives and abortions (by any means), do not expand our knowledge of medicinal methods used. The writers did not wish to preserve and disseminate information of this kind to their readers. In fact, their goal was quite the opposite. A rare exception is found in Tertullian, a church father of the late 2nd century, who was well aware of current procedures for surgical abortions (Tertullian; http://www.newadvent.org/fathers/0310.htm).

Some early Christian literature clearly applied *pharmakeia* to medicines used for birth control and as abortifacients. In the *Didache*, a 1st- or 2nd-century work also known as *Teaching of the Twelve Apostles*, *pharmakeia* is usually translated as witchcraft as part of the Way of Death (θανάτου ὁδός)[60] (http://www.newadvent.org/cathen/04779a.htm). The *Epistle of Barnabas* condemns both abortion and infanticide[61] (http://www.earlychristianwritings.com/barnabas.html). While some of the antiabortion writings in early church literature do not differentiate between surgical abortions and those induced by drugs, others do discuss the use of chemical abortifacients. Many of the public writings

[59]Revelations 9:21, 21:8, 22:15.

[60]Didache 5: 1–2.

[61]Epistle of Barnabas 19:5.

of Christian leaders from this period tend to be in the genre of apologetics. In other words, they explain Christian doctrines and behavior to a (usually imaginary) audience and defend the church against popular accusations. For example, the apologist Athenagoras of Athens, towards the end of the 2nd century, wrote a treatise addressed to Marcus Aurelius. "And when we say that those women who use drugs to bring on abortion commit murder, and will have to give an account to God for the abortion, on what principle should we commit murder? For it does not belong to the same person to regard the very fetus in the womb as a created being, and therefore an object of God's care, and when it has passed into life, to kill it; and not to expose an infant, because those who expose them are chargeable with child-murder, and on the other hand, when it has been reared to destroy it"[62] (http://www.newadvent.org/fathers/0205.htm). In this case, Christians had been accused of infanticide and using infants in human sacrifices[63] (http://www.newadvent.org/fathers/0205.htm; Wahemake, 2010). The apologetic defense was to claim innocence of this crime by going even further and noting that even contraceptives and abortions were forbidden among them.

Beyond this, as the religion spread throughout the Roman Empire, defining and regulating every aspect of sexual activity was a means of both social control and setting the social boundaries of the new religion. This is clear from a comment of Hippolytus of Rome in the late 2nd century, where birth control, fornication, greed, and abortion are condemned in two short sentences: "Whence women, reputed believers, began to resort to drugs for producing sterility, and to gird themselves round, so to expel what was being conceived on account of their not wishing to have a child either by a slave or by any paltry fellow, for the sake of their family or excessive wealth. Behold into how great impiety that lawless one has proceeded by inculcating adultery and murder at the same time!"[64] (http://www.newadvent.org/fathers/050109.htm). Equating birth control and abortion with poisoning and murder continued as standard rhetoric to the end of our period of study. The 6th-century Council of Trullo, also known as the Quinisext Council, noted: "Those who give drugs for procuring abortions and those who receive poisons to kill the fetus are subject to the penalty of murder"[65] (http://www.newadvent.org/fathers/3814.htm).

---

[62]Athenagoras of Athens, A Plea for the Christians, 35:6.

[63]Athenagoras of Athens, A Plea for the Christians, 15.2,34.

[64]Hippolytus of Rome, Refutation of All Heresies, 9:7.

[65]Quinisext Council. XCI.

References

Angel of Death, ⟨http://jewishencyclopedia.com/articles/5018-death-angel-of⟩ (accessed 17.12.17).

Carson, D.A., 2010. In: Rev. Ed. Longman III, Tremper, Garland, David E. (Eds.), The Expositor's Bible Commentary, Matthew ～Mark, vol. 9. Zondervan, Grand Rapids, MI, pp. 986–989.

Covington, Dennis, 1996. Salvation on Sand Mountain: Snake Handling and Redemption in Southern Appalachia. Penguin, New York, NY.

Cranfield, C.E.B., 1959. The Gospel According to Mark. Cambridge University Press, London, pp. 471–476.

Demons and Demonology, ⟨http://www.jewishvirtuallibrary.org/demons-and-demonology⟩ (accessed 17.12.17).

Dulkin, Ryan S., 2014. The Devil Within: a rabbinic traditions-history of the Samael story in Pirkei de-Rabbi Eliezer. Jewish Stud. Q. 21 (2), 153–175.

Edery, H., Eshay, J., Lass, I., Glitter, H., 1972. Pharmacological activity of oriental hornet (Vespa orientalis) venom. Toxicon 10 (1), 13–23.

Godbey, Allen H., 1930. Incense and poison ordeals in the ancient orient. Am. J. Semitic Lang. Lit. 46 (4), 229–230.

Goddard, Jerome, 2007. Physicians Guide to Arthropods of Medical Importance. CRC Press, Boca Raton, FL, pp. 405–407.

Handy, Lowell K., 1992. Serpent, bronze. In: Freedman, David Noel (Ed.), The Anchor Bible Dictionary, vol. 5. Doubleday, New York, NY, p. 1117.

Harrelson, W.J., 1962. Poison. The Interpreter's Dictionary of the Bible. Abingdon, New York, NY, pp. 838–839.

Harris, R. Laird, 1980a. Theological Wordbook of the Old Testament, vol. 1. Moody, Chicago, IL, pp. 374–375.

Harris, R. Laird, 1980b. Theological Wordbook of the Old Testament, vol. 2. Moody, Chicago, IL, p. 826.

Harris, R. Laird, 1980c. Theological Wordbook of the Old Testament, vol. 4. Moody, Chicago, IL, pp. 462–464.

Harrison, Roland K., 1986. Poison; Venom. In: Bromley, Geoffrey W. (Ed.), The International Standard Bible Encyclopedia, vol. 3. Eerdmans, Grand Rapids, MI, pp. 899–900.

Helser, Charles, 1979. The Gourd Book. University of Oklahoma Press, Norman, OK, pp. 11–13.

⟨http://gnosis.org/library/actjnthe.htm⟩ (accessed 17.12.30).

⟨http://reptiledatabase.reptarium.cz/species?genus = Vipera&species = ammodytes⟩ (accessed 17.12.23).

⟨http://www.earlychristianwritings.com/barnabas.html⟩ (accessed 17.12.29).

⟨http://www.earlychristianwritings.com/text/actsjohn.html⟩ (accessed 17.12.30).

⟨http://www.hadassah-med.com/about/tips/poisonous-bites-and-stings⟩ (accessed 17.15.12).

⟨http://www.newadvent.org/cathen/04779a.htm⟩ (accessed 17.12.28).

⟨http://www.newadvent.org/fathers/0205.htm⟩ (accessed 17.12.19).

⟨http://www.newadvent.org/fathers/0310.htm⟩ (accessed 17.12.30).

⟨http://www.newadvent.org/fathers/050109.htm⟩ (accessed 17.12.31).

⟨http://www.newadvent.org/fathers/3814.htm⟩ (accessed 17.12.30).

⟨http://www.tertullian.org/fathers/gregory_02_dialogues_book2.htm#C3⟩ (accessed 17.12.30).

Javadzadeh, Hamid Reza, Davoudi, Amir, Davoudi, Farnoush, et al., 2013. Citrullus colocynthis as the cause of acute rectorrhagia. Case Rep. Emerge. Med. vol. 2013. Available from: https://doi.org/10.1155/2013/652192. Article ID 652192, 5 pages.

Jones, K.R., 1968. The bronze serpent in the Israelite cult. J. Biblical Lit. 87, 245–256.

Kochba, E., 1998. Venomous snakes of Israel: ecology and snakebite. Public Health Rev. 26 (3), 209–232.

Kochva, E., Bdolah, A., Wollberg, Z., 1993. Sarafotoxins and endothelins: evolution, structure and function. Toxicon 31 (3), 541–568.

Liddell, H.G., Scott, Robert, 1935. Greek-English Lexicon, first ed. Oxford University Press, Oxford, p. 1615.

Macht, David I., 1919. A pharmacological study of biblical gourds. Jewish Q. Rev. N.S.10 (2/3), 185–197.

Mayor, Adrienne, 2004. Greek Fire, Poison Arrows and Scorpion Bombs. Overlook Ducksworth, Woodstock, pp. 177–181.

Mendelssohn, H., 1965. On the biology of the venomous snakes of Israel, Part II. Isr. J Zool. 14, 185–212.

Metzger, Bruce, 2005. A Textual Commentary on the Greek New Testament, second ed. Hendrickson, Peabody, MA, p. 123.

Murison, Ross G., 1905. The serpent in the old testament. Am. J. Semitic Lang. Lit. 21, 115–130. esp. 123–126.

Neufield, Edward, 1980. Insects as warfare agents in the Ancient Near East. Orientalia 49, 30–57.

Pliny, Natural History, 11.73.

Poison, ⟨http://www.jewishvirtuallibrary.org/poison⟩ (accessed 17.12.17).

Qumsiyeh, Mazin B., Salman, Ibrahim N.A., Salsaa, Michael, Amr, Zuhair S., 2013. Records of scorpions from the Palestinian territories, with the first chromosomal data (Arachnida: Scorpiones). Zool. Middle East 59 (1), 70–76.

Riddle, John, 1997a. Eve's Herbs. Harvard University Press, Cambridge, pp. 47–48, 52.

Riddle, John M., 1997b. Eve's Herbs. Harvard University Press, Cambridge, pp. 35–37.

Riddle, John M., 1997c. Eve's Herbs. Harvard University Press, Cambridge, pp. 72–75.

Riddle, John M., 2010. Goddesses, Elixirs, and Witches. Palgrave Macmillan, New York, NY, pp. 49–50.

Samael, ⟨http://www.jewishvirtuallibrary.org/samael⟩ (accessed 17.12.17).

Shuker, Karl, 1976. St. Paul shipwrecked in Dalmatia. Biblical Archaeol 39 (4), 145–147.

Meinardus, Otto F.A., 1979. St Paul's Last Journey. Caratzas Brothers, New Rochelle, NY, pp. 79–85.

Shuker, Karl, 2014. On the trail of Mediterranean mystery snakes. Edge Sci. 20, 16–19.

Simard, J. Marc, Watt, Dean D., 1990. Venoms and toxins. In: Polis, Gary A. (Ed.), The Biology of Scorpions. Stanford University Press, Stanford, CA, pp. 435–441.

Swete, Henry Barclay, 1952. The Gospel According to St. Mark. Eerdmans, Grand Rapid, MI, pp. 400–408.

Tertullian, Treatise on the Soul, XXV. 1675–1678.

The Acts of the Holy Apostle and Evangelist John the Theologian 2439–2440.

Trestrail, John H., 1993. Poisons in the Bible. Mithridata 3 (1), 5–6.

Ventura, Charles S., 1990. Maltese medical folklore: man and the herpetofauna in Malta: a review. Maltese Med. J. 41 (2.1), 1–3.

Wahemake, Bart, 2010. Incest, infanticide, and cannibalism: anti-Christian imputations in the Roman Empire. Greece Rome 57 (2), 337–354.

Yamauchi, Edwin M., 1980. Was Nehemiah the cupbearer a eunuch? In: Fohrer, Georg (Ed.), Zeitschrift für die Alttestamentliche Wissenschaft, 92. Walter de Gruyter, New York, NY, pp. 132–142.

Yamauchi, Edwin, 2014a. Contraception and control of births. In: Yamauchi, Edwin M., Wilson, Marvin R. (Eds.), Dictionary of Daily Life in Biblical and Post Biblical Antiquity, vol. I. Hendrickson, Peabody, MA, pp. 359–366.

Yamauchi, Edwin, 2014b. Abortion. In: Yamauchi, Edwin M., Wilson, Marvin R. (Eds.), Dictionary of Daily Life in Biblical and Post Biblical Antiquity, vol. I. Hendrickson, Peabody, MA, pp. 5–10.

Younger Jr., K. Lawson, 1997. In: VanGemeren, Willem A. (Ed.), New International Dictionary of Old Testament Theology and Exegesis, vol. 2. Zondervan, Grand Rapids, pp. 242–243.

Further Reading

Chadwick, Henry, 1993. rev. ed. The Penguin History of the Church: The Early Church, vol. 1. Penguin, New York, NY.

Polis, Gary A. (Ed.), 1990. The Biology of Scorpions. Stanford University Press, Stanford, CA (Chapters 10, 12).

Riddle, John M., 1997. Eve's Herbs. Harvard University Press, Cambridge (Chapters 2, 3).

30

Poisonous Medicine in Ancient China

Yan Liu

Department of History, University at Buffalo, Buffalo, NY, United States

For all things under the heaven, nothing is more vicious than the poison of aconite. Yet a good doctor packs and stores it, because it is useful. **Masters of Huainan (*second century BCE*)**

The standard Chinese word for poison is *du* 毒. Modern readers often frown upon the word, because it invites associations with danger, harm, and intrigue. But this translation is misleading, as *du* in the past had diverse, even opposite meanings. At the core of them lay the notion of potency, the ability not just to harm as a poison but also to cure as a medicine. Accordingly, instead of avoiding poisons entirely, classical Chinese medicine strategically harnessed them for therapy. This chapter probes the roots of this vital pharmacological practice in ancient China. The history of Chinese medicine cannot ignore the history of poison.

30.1 ETYMOLOGY OF *DU*

Let us start with the first dictionary of Chinese history, *Explaining Simple and Analyzing Compound Characters* (*Shuowen jiezi*), compiled in 100 CE. The dictionary explains the basic meaning of *du* as thickness, which refers to the physical shape of mountains. Thickness implies heaviness, abundance, and potency; the word does not carry a negative sense.

Further study of the scripts of *du* reveals conflicting implications. Two Han dynasty (206 BCE–220 CE) variants of *du* are shown in Fig. 30.1 (Shi, 2004). The two scripts share the upper part of the character *tu* 土, which means soil, and relatedly, growth. The lower parts of the two differ. In the first variant, it is written as *wu* 毋, which means stop. In the second, we see a different character, *mu* 母, which means mother, and implicitly, nurture. Therefore, depending on how the lower part is written, *du* signifies either prohibiting or promoting growth. Thickness could elicit opposite outcomes.

30.2 *DU* IN CHINESE PHARMACOLOGY

How do these two meanings manifest themselves in ancient Chinese pharmacology? Let us examine the first drug treatise in China, *Divine Farmer's Classic of Materia Medica* (*Shennong bencao jing*), compiled by the Han officials who possessed drug knowledge. It names 365 drugs, and parses them into three groups. Drugs in the top group are mostly nontoxic (*wudu*), which promise to lighten the body, supplement *qi*, avert aging, and prolong life; drugs in the middle group, defined as either nontoxic or toxic (*youdu*), can prevent maladies and replenish depletions; drugs in the bottom group, most of which are toxic, effectively cure illnesses (Unschuld, 1986, pp. 23–25).

This classification reveals two key features of Chinese pharmacology. First, drugs are categorized and defined by their toxicity or *du*. This *du*-

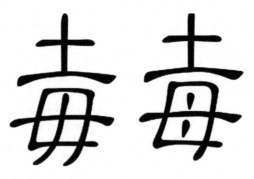

FIGURE 30.1   Two variants of *du* in the Han.

centered grouping of drugs would remain fundamental to Chinese medicine throughout the premodern period. Second, toxicity figures not as something to be avoided at all costs, but as potency valuable in the cure of illnesses. At the same time, the text also notes the danger of toxic drugs, warning that one should only use them temporarily and cautiously. A toxic substance is a double-edged sword; it cures if handled properly, and harms if it is not.

What, then, is the proper use of a toxic drug? The *Divine Farmer's Classic* specifies two methods.

The first is dosage control. The text declares that when ingesting a toxic drug, one should start with a dose as small as a millet seed. If this proves insufficient, one should increase the dosage gradually until the patient is cured. The amount of the drug, in short, has to be carefully calibrated to the patient's state of recovery.

The second is drug combination. Chinese pharmacy frequently combines drugs to maximize their remedial power. The *Divine Farmer's Classic* defines six ways of combination, ranging from mutual facilitation to mutual annihilation. Relevant to the use of toxic materials is the strategy of "mutual inhibition" (*xiangwei*), which adds a nontoxic drug to a toxic one to curb its potency, yet still preserve its therapeutic efficacy.

Although toxic substances are powerful medicines, the *Divine Farmer's Classic* places most of them in the bottom group, regarding them as inferior to nontoxic ones in the middle and top groups. Rather than treating illnesses, drugs in the middle group aim to strengthen the body to keep one healthy. This fits well with a medical philosophy prominent at the time, manifested by the Han dynasty medical classic, *Yellow Emperor's Inner Classic* (*Huangdi neijing*): It is better to prevent illnesses from occurring than to cure them after they arise (Huang Di nei jing su wen, 2011, p. 57).

Furthermore, drugs in the top group, mostly nontoxic, target a higher goal of longevity, which resonates with the ideal of "nourishing life" (*yangsheng*) in China. This ancient set of practices proposed regular ingestion of tonics (mushrooms, resin, mica, etc.) combined with bodily movements, breathing, and meditation techniques to enhance life (Kohn, 1989). All these methods aimed to purify the body and eliminate toxic substances that triggered its decay. Nontoxic drugs for "nourishing life," therefore, induce an opposite bodily effect to that of toxic ones: The former promote purification to "lighten the body" by releasing its toxic burden, whereas the latter, with their characteristic "thickness," invigorate the body to combat maladies. Understanding *du* in this context allows us to appreciate its paradoxical meaning in Chinese medicine: A toxic drug effectively cures illnesses, but impedes one from achieving a higher goal—the cultivation of the body to attain a healthy, long life.

30.3 ACONITE, THE POWER TO CURE

We have explored thus far the principle underlying the use of toxic drugs in ancient China. Let us now consider a specific example, aconite, one of the most commonly used drugs in Chinese medicine. Aconite encompasses a group of herbs of the *Aconitum* genus that Chinese sources refer to by a variety of names (*fuzi, wutou, tianxiong, cezi, wuhui, jin,* etc.). In general, these names correspond to different parts of the herb, or parts collected in different seasons. For example, *wutou*, which literally means "black head," refers to the parental tuber of the herb harvested in the spring, whereas *fuzi*, which literally means "attached offspring," depicts the daughter tuber that grows in the summer (Fig. 30.2). Chinese pharmacology attributes distinct medicinal functions to each of these varieties of aconite (Obringer, 1997, pp. 91–143).

More than 50 species of aconite are utilized in Chinese medicine today, with *Aconitum carmichaeli* from northern Sichuan (southwest China) being the major produce. According to modern pharmacology, the main toxic component in aconite is aconitine-type alkaloid: 0.2 mg of it, taken orally, suffices to poison a person, while 3–5 mg may cause death by cardiovascular and neurological failure. When administered in

FIGURE 30.2   Aconite (*fuzi*) depicted in the 11th-century *Classified Materia Medica* (*Jing shi zhenglei beiji bencao*, the edition of the 18th-century *Siku quanshu*).

smaller doses, however, aconitine and other alkaloids in the herb can relieve pain, reduce inflammation, and strengthen the heart (Bisset, 1981). The use of aconite, therefore, becomes the art of moderating the herb's toxicity while preserving its therapeutic power.

Early evidence for the medicinal use of aconite in China comes from a collection of medical formulas excavated from Mawangdui in modern Hunan (southern China), which date from the 2nd-century BCE in the early years of the Western Han dynasty (206 BCE−9 CE). Among the over 400 substances in these formulas, aconite was the second most-used drug (21 times), only surpassed by cinnamon. In most cases, aconite was applied externally, often mixed with other drugs, to treat wounds, abscess, scabies, and itch. When taken internally, it acted as a tonic to replenish *qi*, boost sexual energy, and prolong life. Intriguingly, several formulas employed aconite to achieve speed of travel, evincing its magical power (Harper, 1998, p. 105).

Aconite remained popular in the following centuries, but its functions altered. During the Eastern Han dynasty (25−220 CE), the herb was harnessed mainly as an internal medicine to treat cold illnesses, whereas its tonic and magical activities faded. The influential medical work *Treatise on Cold Damage Disorders* (*Shanghan lun*), compiled by Zhang Zhongjing in the early 3rd century, prescribed aconite frequently; we find the drug in 20% of all formulas (23 in total) (Zhang, 1999). Conspicuously, these formulas always mixed aconite with other herbs, most often licorice and ginger, which echoed the method of drug combination to mitigate toxicity mentioned earlier. Moreover, most formulas employed processed aconite, the product upon broiling. This method effectively reduces toxicity, as modern tests show that they trigger the physical loss of alkaloids in the herb. Crude aconite was still used in some formulas but mainly for medical emergencies.

What is the therapeutic rationale for using this toxic drug in Chinese medicine? The *Divine Farmer's Classic* defines aconite as a warming drug, which, according to the principle of opposites, treats cold disorders. The symptoms include wind-induced coldness, coughing, pain in the joints, and blood clotting. Aconite, with its heating power, can dissipate cold and break up stagnation in the body. This process often generates strong, if not violent, bodily sensations. Zhang Zhongjing, for instance, noted that after patients took one dose of an aconite decoction, they fell into a state of impediment; after three doses, they became dizzy. Zhang, however, did not regard these reactions as pathologies, but rather as positive signs of the drug's efficacy. This interpretation expresses the core philosophy of deploying toxic drugs in ancient China, aptly summarized by an aphorism—If a drug does not deliver a spell of dizziness, it cannot cure severe illnesses.

30.4 ACONITE, THE POWER TO KILL

Now that we have reviewed the medicinal use of aconite, we will turn to its sinister functions. In fact, before becoming a medicine, aconite had long been used to kill. The earliest evidence comes from a 4th-century BCE text, *Discourses of the States* (*Guoyu*), where the herb appeared in a court murder. In 656 BCE, Li Ji, a concubine of Duke Xian of the Jin state, planned a conspiracy to remove the heir apparent, Shen Sheng, so her own son could succeed to the throne. She asked Shen Sheng to offer food to his father, but beforehand secretly added aconite (*jin*) into the meat. When the Duke was about to eat the food, Li Ji asked him to first test it on a dog, which died instantly. Seeing himself in trouble, Shen Sheng fled, and later committed suicide.

Six centuries later, during the Han dynasty, aconite figured in another court murder. According to a 2nd-century CE source, *History of the Western Han* (*Hanshu*), in 71 BCE, Huo Xian, the wife of a powerful general, planned to murder the Empress Xu to seize the position for her daughter. She hired a court physician named Chunyu Yan to present medicine to the empress, who had just given birth to a child. Chunyu Yan mixed aconite (*fuzi*) in the medicine, and persuaded the empress to take it to recover from childbirth. The empress died soon after. The murder was ingenious, covered by the reality that women often died after delivery at the time. Intriguingly, after the empress ingested some of the "medicine," she complained that her head felt dizzy, wondering whether she was poisoned. Chunyu Yan claimed that the effect was normal for healing, and convinced her to take more, leading to her tragic demise. The line between medicine and poison is thin; so is that between life and death (Li, 2011).

In addition to being favored for murder, aconite was also employed in hunting and warfare, mainly as an arrow poison. The dictionary *Shuowen jiezi* includes a pre-Han script of *du* with a radical that signifies knife. This ancient form of *du* implies its association with weapons, possibly coated with poison for hunting or military purposes. This may be the original function of poison in China. Aconite as arrow poison is suggested by one of its many names—*shewang*, which means "shoot and ensnare." We find evidence for this function in a medical manuscript from Mawangdui (2nd-century BCE), which includes seven formulas to treat aconite (*wuhui*) poisoning. One of them prescribed applying a few herbs onto the wound, which was likely caused by arrow poison (Harper, 1998, p. 238). The example indicates that by the 2nd-century BCE, the use of aconite-coated arrows was so prevalent that an antidote had been developed.

Finding antidotes went hand in hand with making poisons. Among its six drug combinations, the *Divine Farmer's Classic* specifies one as "mutual annihilation" (*xiangsha*), in which one drug is harnessed as antidote to counter the other as poison. Some oft-used antidotes in Chinese medicine were licorice, ginseng, ginger, honey, and salt. To counteract aconite poisoning, soybean soup was the primary choice in antiquity. Modern studies indicate that the rich proteins in soybean can coat and precipitate toxic materials in the body, explaining its antidotal efficacy. Not surprisingly, many of these antidotes, such as licorice and ginger, overlap with those that mitigate toxic drugs by "mutual inhibition," discussed earlier. Dosage is probably the determining factor that allows the same substance to act in these two different manners.

30.5 FROM *DU* TO *PHARMAKON*

Thus far we have examined how *du* in Chinese pharmacology ambiguously straddled medicine and poison. This duality, of course, was not unique to China. The English word "pharmacy" derives from the Greek word *pharmakon*, which means both remedy and poison. The use of toxic materials in Greek medicine had already appeared in the Hippocratic Corpus (5th- or 4th-century BCE), which frequently prescribed a herb called hellebore. The herb served as a strong purgative to treat humoral imbalance, gynecological problems, and lung disorders, almost reaching the status of a panacea. On the other hand, the Hippocratic writers observed the harmful, even lethal effects of hellebore when administered carelessly (Girard, 1990).

We encounter more toxic drugs in Dioscorides's *De Materia Medica*, a monumental work in Western pharmacology compiled in the 1st-century CE (Dioscorides, 2011). Dioscorides discussed many toxic substances in his treatise, which can be divided into two groups. The first group contained about 50 drugs that produced unpleasant but mild side effects such as headache, dim-sightedness, and stomach discomfort. In the second group, we find a dozen drugs with stronger toxicity, including opium poppy, henbane, thorn apple, mandrake, and hemlock, the last of which was presumably the deadly poison that Socrates drank. Dosage was the key to administering these drugs; Dioscorides repeatedly warned that consuming them in excess could harm and even kill. Evidently, the therapeutic use of toxic materials was prominent in ancient Greek pharmacology.

This continuity between medicine and poison, evinced by *pharmakon*, parallels the Chinese concept of *du*. If we look further, however, we detect a subtle yet visible change in Greek pharmacology that began to distinguish poison from medicine. As early as the 3rd-century BCE, Apollodorus wrote a treatise on poison, which has been lost. Some of his toxicological knowledge was preserved a century later in Nicander's two poems, *Theriaca* and *Alexipharmaca*. The poems, arguably the earliest toxicological writing in Greek antiquity, offered detailed accounts of poisonous animals and plants as well as poison-induced symptoms (Scarborough, 1977 and 1979). Nicander's poems were highly influential because later authors, including the famous Roman physician Galen, extensively cited him as a source on toxicology.

The separation of poison from medicine became more patent in *De Materia Medica*. Although Dioscorides utilized many toxic materials for therapy, he also identified some poisons in which he discerned no medicinal value, such as wolfsbane, yew, and meadow saffron. He listed these poisons simply to warn against them. Despite the blurred boundary between medicine and poison in Greek antiquity, then, a group of toxic substances began to gradually move out of the *pharmakon* continuum. This separation became more pronounced in medieval Europe. In China, however, we do not see this phenomenon. All drugs in Chinese pharmacy, including highly toxic ones, were perceived to have medicinal value; no absolute poisons existed. This does not mean that toxicological knowledge was absent in ancient China. Quite the contrary, we find abundant discussions about poisons, poisoning, and antidotes, as shown above. Yet in classical Chinese medicine, toxicology was always part of pharmacology.

No example better illustrates this divergence than the distinct fates of aconite in Greek and Chinese medicine. In *De Materia Medica*, the herb, there called wolfsbane, was only used as a poison to kill wolves, without any perceived curative functions. Correspondingly, the 1st-century CE Roman naturalist Pliny the Elder called the herb "plant arsenic," evoking its harmful nature. But in China, as we have seen, aconite was highly valued for its therapeutic power, hailed as the "chief of the hundred drugs." This marked difference points to diverging philosophies of utilizing toxic substances between Greek and Chinese pharmacology. If we compare Dioscorides's comments on the side effects of toxic drugs to the aforementioned Chinese aphorism that if a drug does not cause dizziness, it cannot cure severe illnesses, this much seems clear: Greek medicine prescribed drugs in spite of their toxicity; Chinese medicine, because of it.

The prominent use of poisons as curative agents in Chinese history urges us to rethink the principles and practices of Chinese medicine today. Contemporary views of classical Chinese medicine often contrast the benign naturalness of Chinese herbal remedies with the dangerous side effects of Western synthetic drugs. This dichotomy is misconceived.

It not only overlooks the abundant poisons in Chinese pharmacy but also, at a fundamental level, fails to understand the paradoxical nature of drug therapy. For any drug we consume, be it aconite or aspirin, ginseng or vitamin, we introduce a foreign agent—hence, in a broad sense, poison—to our body. A drug's potentials for curing and harming are always intertwined. It is ultimately not the substance itself, but *how* we use it that matters. We cannot avoid poisons, for the art of medicine is the art of poison.

References

Bisset, N.G., 1981. Arrow poisons in China. Part II. Aconitum—botany, chemistry, and pharmacology. J. Ethnopharmacol 4, 247–336.

Dioscorides, P., 2011. De materia medica [Beck L.Y., Trans.]. Olms-Weidmann, Hildesheim/New York, NY.

Girard, M.C., 1990. L'hellébore: panacée ou placébo? [Hellebore: Panacea or placebo?]. In: Potter, P., Maloney, G., Désautels, J. (Eds.), La malade et les maladies dans la Collection hippocratique [The patient and the illnesses in the Hippocratic collection]. Editions du Sphinx, Québec, pp. 393–405. [in French].

Harper, D.J., 1998. Early Chinese Medical Literature: The Mawangdui Medical Manuscripts. Kegan Paul International Press, London/New York, NY.

Huang Di nei jing su wen: an annotated translation of Huang Di's inner classic—basic questions, 2011. [Unschuld P.U. and Tessenow H. in collaboration with Zheng J.S., Trans.]. University of California Press, Berkeley, CA.

Kohn L. (Ed), in cooperation with Sakade Y., 1989. Taoist Meditation and Longevity Techniques. University of Michigan Center for Chinese Studies, Ann Arbor, MI.

Li, J.M., 2011. Nüyi sha'ren: Xihan xupingjun huanghou mousha'an xinkao [Killing by a female physician: a new study on the murder of Empress Xu in the Western Han]. In: Li, J.M. (Ed.), Lüxingzhe de shixue [Out of place: travels throughout Chinese medical history]. Yunchen wenhua shiye gufen youxian gongsi, Taipei, pp. 285–324. [in Chinese].

Obringer, F., 1997. L'aconit et l'orpiment: drogues et poisons en Chine ancienne et médié- vale [Aconite and orpiment: drugs and poisons in ancient and medieval China]. Fayard, Paris. [in French].

Scarborough, J., 1977 and 1979. Nicander's toxicology I: snakes. Pharm. Hist 19, 3–23. Nicander's toxicology II: spiders, scorpions, insects, and myriapods. Pharm. Hist 21, 3–34.

Shi, Z.C., 2004. Zhongguo gudai duzi jiqi xiangguan cihui kao [A study of the character *du* and related words in ancient China]. Dulixue shi yanjiu wenji [Research papers on the history of toxicology] 3, 1–9. [in Chinese].

Unschuld, P.U., 1986. Medicine in China: A History of Pharmaceutics. University of California Press, Berkeley, CA.

Zhang, Z.J., 1999. Shang Han Lun: On Cold Damage [Mitchell C., Feng Y., Wiseman N., Trans.]. Paradigm Publications, Brookline, MA.

31

Toxicity of Ayurvedic Medicines and Safety Concerns: Ancient and Modern Perspectives

P. Rammanohar

Amrita Centre for Advanced Research in Ayurveda, Amrita Vishwa Vidyapeetham, Kollam, Kerala, India

OUTLINE

31.1 INTRODUCTION

Ayurveda, which literally means the knowledge of life, is the traditional system of medicine indigenous to the Indian subcontinent, with an impressive history spanning more than two thousand years. Ayurvedic medicines are derived from plant, animal, and mineral resources that make up the biodiversity of the earth. Ayurveda has adopted a very inclusive approach in building up its pharmacopoeia. One of the primary guiding principles that underpins the evolution of Ayurvedic Pharmacopoeia is the axiomatic statement that any substance under the sun can be used as a medicine (Vaidya, 2002c). There are explicit references to this effect in the works of *Caraka, Susruta,* and *Vagbhata,* the most authoritative medical writers in the tradition of Ayurveda. It is pertinent to note that Ayurveda also includes synthetic or manmade substances within the purview of its pharmacopoeia. There is a Sanskrit term for synthetic substances—*Krtrima* (Trikamji, 1980d). Any natural or synthetic substance of plant, animal, or mineral origin can in principle become a potential drug source in Ayurveda.

The *Susrutasamhita,* the celebrated textbook of Ayurveda on surgery, beckons us to search the globe at every nook and corner, emphasizing that the earth is bountiful and that there are potential medicinal substances waiting to be discovered everywhere (Trikamji, 1980a). Although anything in the world can be used as a medicine, we find that a very limited number of substances actually got codified and listed formally in the Ayurvedic Pharmacopoeia. This is because a very stringent process of evaluation, testing, and selection was employed to build and authenticate the Pharmacopoeia. In other words, the Pharmacopoeia of Ayurveda endorsed the most effective and tested remedies only after assessing safety and potential harm that can accrue from their use in humans (Trikamji, 1992h).

Although Ayurveda has been the mainstream healthcare system in India for centuries, it was relegated to the status of traditional medicine (WHO, 2013) as an outcome of colonial rule and in more recent times is considered to be an alternative or complementary medicine in the developed world (https://nccih.nih.gov/health/ayurveda, 2018). There has been renewed interest in Ayurveda in the wake of the global search for safer alternatives to modern medicine. A large number of people in India look up to Ayurveda as a safer option for many chronic illnesses. Ironically however, Ayurvedic medicines and treatments have also been under the scanner for safety issues. Especially the herbomineral formulations containing substances like mercury, arsenic, and other minerals have been red-flagged and are being scrutinized for toxicity. Nevertheless, these medications continue to be manufactured and prescribed to millions of people in India. There are published papers reporting adverse events and toxicity from Ayurvedic medicines (Dunbabin et al., 1992). The paradox is that, there are also publications vouching for the safety of Ayurvedic medicines (Kumar et al., 2012, Mar 6). The Ayurvedic Pharmacopoeia also includes medicinal plants that are recognized as highly poisonous by the tradition of Ayurveda itself under Schedule E1. The Indian Ministry for Traditional Medicine (Ministry of AYUSH) has issued regulatory warnings and guidelines for use of formulations containing such substances (http://ayush.gov.in/ sites/default/files/Public%20Notice%20in%20English.pdf, 2018) (see Table 31.1).

TABLE 31.1 Schedule E1 Issued by Government of India Listing the Poisonous Substances Used in Ayurveda

Poisons of plant origin	Poisons of animal origin	Poisons of mineral origin
Ahipena (*Papaver somniferum*)	Sarpavisha (snake venom)	Gauripashana (arsenic)
Arka (*Calotropis gigantea*)		Haritala (arsenic disulfide)
Bhallataka (*Semecarpus anacardium*)		Manahshila (arsenic trisulfide)
Bhanga (*Cannabis sativa*)		Parada (mercury)
Danti (*Baliospermum montanum*)		Rasakarpura (chloride salt of mercury)
Dhattura (*Datura metal*)		Tuttha (copper sulfate)
Gunja (*Abrus precatorium*)		Hingula (cinnabar)
Jaipala (*Croton tiglium*)		Sindura (red oxide of lead)
Karaveera (*Nerium indicum*)		Girisindura (red oxide of mercury)
Langali (*Gloriosa superba*)		
Parasika Yavani (*Hyoscyamus niger*)		
Snuhi (*Euphorbia neriifolia*)		
Vatsanabha (*Acontium chasmanthum*)		
Vishamushti (*Strychnox nuxvomica*)		
Shringivisha (*Acontium chasmanthum*)		

Even as Ayurveda is gaining popularity globally and increasing numbers of the population are looking for safer and alternative solutions for many illnesses, an objective assessment and review of the potential toxicity and safety concerns of Ayurvedic treatments and medications becomes extremely relevant.

This chapter is an attempt to review the concept of drug safety and toxicity as described in the classical textbooks of Ayurveda. The chapter also aims to have a close look at the issue of safety in the context of Ayurvedic practice today. Knowledge gaps will be identified and pointers to directions for future research on the topic will be provided.

31.2 CRITERIA FOR INCORPORATION OF A NOVEL AND UNKNOWN SUBSTANCE INTO THE AYURVEDIC PHARMACOPEIA

We will start our discussion with the guidelines laid down by the ancient physicians for formally incorporating a substance into the pharmacopoeia for medicinal use. The purpose of this discussion is to understand whether enough care was taken to rigorously assess the efficacy and safety of a substance before accepting it as a drug source.

Ayurvedic texts warn against the use of novel or inadequately known substances for medicinal purposes. One of the earliest textbooks on Ayurveda, the *Carakasamhita*, states that the use of any substance, which has not been adequately understood in terms of nomenclature, identity, properties, and clinical application, can result in undesired consequences (Trikamji, 1992c). The Ayurvedic Pharmacopoeia was constantly evolving, and new substances were being introduced after proper evaluation of their medicinal properties and safety profile. A substance would get incorporated into the Pharmacopoeia as a medicine only when satisfactory knowledge about its safety and efficacy was generated. Using unknown or inadequately known substances for medicinal purposes was considered to be as dangerous as fiddling with poison, a sharp weapon, fire, or lightning (Trikamji, 1992c). Assessment of safety was a cardinal principle emphasized in developing and expanding the Ayurvedic Pharmacopoeia.

Ayurveda developed elaborate protocols for evaluation of new substances, so much so that only a thousand plants got introduced in Ayurvedic Pharmacopeia over a span of more than 2000 years. On the other hand, more than 8000 plants were being used in the folk medical practices [All India Coordinated Project on Ethnobiology (AICPE)].

There are four important guiding principles on the basis of which Ayurveda developed its Pharmacopeia: (1) the first principle is that any substance of natural or synthetic origin is a potential medicine (Vaidya,

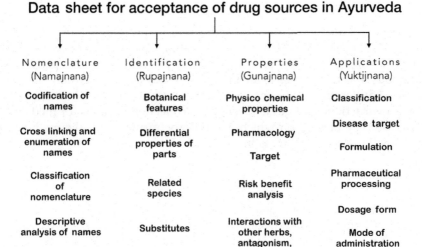

FIGURE 31.1   Data sheet for acceptance of drug sources in Ayurveda.

2002c); (2) the second principle is that any substance that is to be used as medicine must be thoroughly studied with reference to its nomenclature, identity, properties, and application (Trikamji, 1992b); (3) the third principle is that there is a high risk of harm in the use of a substance that has not been adequately understood (Trikamji, 1992c); and (4) the fourth principle is that a well-understood substance should not be abused or misused (Trikamji, 1992c).

In other words, even a known medicine can be hazardous if used improperly. Hence the ancient texts are replete with warnings about the misuse or abuse of established drugs.

Information pertaining to the medicinal properties of substances used as drugs in Ayurveda was classified under four major headings (Sharma, 1982) (see Fig. 31.1).

31.2.1 Namajnanam

A polynomial system of nomenclature was developed and systematized in the tradition of Ayurveda. One substance would be known by several names and one name would be common to several substances. These names were also suggestive of morphological, anatomical, phytochemical, and pharmacological characteristics of the drug source and helped in fixing the identity. The polynomial system of nomenclature also integrated the diverse nomenclature of drugs that were in vogue in a country like India with its astounding diversity with respect to language, culture, geography, and climate.

31.2.2 Rupajnana

Nomenclature gives only an indication of the identity of the plant. The identity had to be fixed by a thorough analysis of morphological, anatomical, as well as pharmacological properties of the plant and methods and techniques available in those days were used for this purpose. Adulteration and contamination were also discussed. There were clear instructions and guidelines for the collection of herbs for medicinal use, the type of land on which it should be grown, with the method as well as the season of harvesting.

31.2.3 Gunajnana

The next step is Gunajnana, which includes in modern parlance the study of both pharmacodynamics and pharmacokinetics. Ayurveda looks at the properties of a drug in isolation as well as the outcome of the transformation that it undergoes in the human body. In a nut shell, Ayurvedic pharmacology is complete understanding of what the drug does to the body and what the body does to the drug. Vipaka is the metabolized state of the medicine in which it exerts its pharmacological action.

31.2.4 Yuktijnana

The emphasis in Ayurveda is on discovering safe use of a medicinal substance. Even poison can become medicine if properly used (Trikamji, 1992c). The focus is on the protocol for safe use rather than inherent safety of a substance (Vaidya, 2002d). *Yuktijnana* deals with protocols and algorithms for the clinical application of various drugs and formulations (see Fig. 31.2). It deals with the methods for

FIGURE 31.2   Workers brewing herbal decoctions at an Ayurvedic Pharmacy.

combining different herbs into formulations, processing the formulations into various dosage forms, protocols for administration of medicines, indications and contraindications, and safety warnings as well (see Figs. 31.3 and 31.4). *Yukti* means rational use and therefore, *Yuktijnana* means the knowledge that helps the physician to use medicines rationally.

FIGURE 31.3   The Ayurvedic herbal decoctions being filled in bottles.

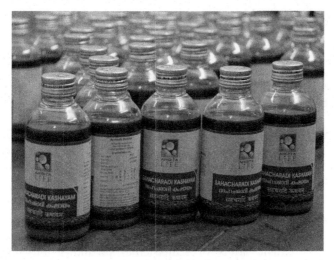

FIGURE 31.4   Packaged and labelled Ayurvedic herbal decoctions.

The patient has been warned not to consume a medicine prescribed by a physician without consideration of the above factors. This is perhaps one of the earliest public warnings issued to safeguard people from irrational prescription of medicines. It is also a very early reference to medical error (Trikamji, 1992c).

31.2.5 Absolute and Relative Safety of Medicines

The *Carakasamhita* explicitly states that there is no perfect medicine that is absolutely safe. Ayurveda does not claim that a medicine becomes safe just because it is of herbal origin. A foundational principle in Ayurvedic pharmacology is the dictum that there is no substance that does not produce desirable and undesirable effects at the same time (Trikamji, 1992m). The relative safety of a substance is determined by assessing the risk–benefit ratio. One can never presume that any drug is intrinsically safe. The text or tradition of Ayurveda does not endorse the widespread myth that anything and everything that bears the label "Ayurveda" is safe because it uses herbs as drug sources.

Only the drug sources which were studied thoroughly in terms of nomenclature, identity, properties, and applications were officially recognized in the Ayurvedic Pharmacopoeia. Therefore, out of approximately 8000 species of medicinal plants in the Indian ethnobotanical database, only about 1500 plants were recognized and listed in the classical textual traditions of Ayurveda as medicinal sources.

31.2.6 Substance Abuse

There are warnings about substance abuse in the traditional texts. The texts warn that even nectar can become toxic if used improperly (Trikamji, 1992c). Elaborate guidelines are laid out in the text on how to use a substance as medicine in the most balanced and harmless manner, taking into consideration various factors like the nature of the disease, the stage of the disease, the constitution of the patient, and so on and so forth. Ayurveda emphasized that the safety of a substance depends on the protocol for its proper use (Vaidya, 2002d). Sources of medicines are classified on the basis of the degree of toxicity both in the short term and long term.

31.2.7 Evaluating Safety Using Animal Models

Both the intrinsic safety of a medicinal substance (*ausadhadravya*) and the risk in its clinical application (*prayoga*) were given due consideration in the Ayurvedic texts. In modern parlance, both hazard and exposure were considered to assess the safety of Ayurvedic medicines.

Multiple factors influencing the safety of medicinal substances find mention in classical Ayurvedic literature. Even testing substances by administration to animals for determination of toxicity has been endorsed by the tradition of Ayurveda. There are textual references very clearly recommending the testing of new food substances by feeding animals to rule out toxicity (Vaidya, 2002b). Ayurveda also resorted to zoopharmacognosy to obtain clues about the medicinal properties of herbs and other natural substances. The *Atharva Veda* refers to medicinal herbs that are known to the mongoose, the eagle, and other animals and birds (Anonymous, 2003).

It seems that long-term evaluation of the safety of substances was also attempted by the ancient Ayurvedic physicians considering the fact that only a few herbs were added to the pharmacopoeia in a hundred years. While innumerable formulations were developed by combining well-known herbs in different ways, new herbs were incorporated into Ayurvedic Pharmacopoeia at a much slower pace. It appears that a potential source of medicine was studied for many decades before using it in clinical practice.

31.2.8 Classification of Poisonous Substances from a Point of View of Safety

Ayurveda classifies substances into three broad categories: medicine, poison, and food. All substances are generally classified into: those that normalize the body functions (*shamana*); those that disturb the body functions (*kopana*); and those that maintain the body functions (*svasthahitam*). Apparently, these categories refer to medicine, poison, and food. There are the intermediate substances like those that work as both food and medicine, and those that are poisons and at the same time medicines. In the backdrop of this broad categorization, several classifications were envisaged that looked at the safety of the drug. For example, we have substances that are recommended for regular use versus those that have to be used very sparingly (Trikamji, 1992a). When it comes to toxic substances, they are broadly classified into the fatal (*Visha*) and toxic (*Upavisha*) categories and those that cause immediate (*badhanam*) as well as long-term (*sanubadhanam*) harm (Trikamji, 1992g). Based on the source, substances are divided into vegetable poisons (*Sthavara Visha*), animal poisons (*Jangama Visha*), and synthetic poisons (*Kritrima Visha*). The 10 substrata for vegetable poisons are root, leaf, fruit, flower, bark, latex, heartwood, resin, plant tissues, and tubers Trikamji (1980b).

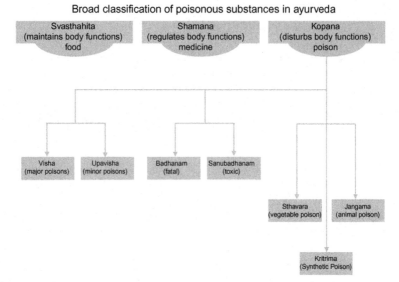

FIGURE 31.5   Broad classification of poisonous substances in Ayurveda.

The 16 substrata for animal poisons are eyes, breath, teeth, nails, urine, feces, semen, saliva, menstrual blood, face, bite, flatus, bone bits, bile, hair, and corpse (Trikamji, 1980c). Semitoxic substances, irritants, and even food substances that may cause mild imbalances and undesirable outcomes in the long run have been recognized in Ayurveda (see Fig. 31.5).

31.2.8.1 Food Safety

Food safety has also been discussed in classical Ayurvedic texts. Elaborate protocols have been formulated to ensure the safety of food. Toxicity of food was identified by inspection, testing in fire, and by administration to animals. The signs of skin contact with contaminated food and oral ingestion have been described in the texts. The effects of food poisoning in the alimentary tract, as well as systemic manifestations, have been discussed. Remedial measures include removal of the toxic food by flushing the alimentary tract and administration of medicines to neutralize the toxic effects. Special attention has been given to the protection of the vital organs (Vaidya, 2002a).

31.2.9 Developing a Protocol for Safe Use of a Drug

The safety and efficacy of any medicinal substance is dependent on its insufficient, excessive, inappropriate, or appropriate use. Using the medicine in an effective manner was referred to as *Samyak Yoga* (Trikamji, 1992c). The physician was expected to be aware of potential adverse events and had to be trained to handle such unexpected effects of medicines. The purity and quality of the medicine had to be assured before use. Use of substitutes was also regulated and the texts provide a list of approved substitutes for use in formulations.

31.2.9.1 Ensuring Microbiological Safety of the Raw Materials

Elaborate procedures have been laid out for harvesting and storage of substances to ensure probably microbiological safety and preventive procedures were used for this purpose. Fumigation of herbs, storage in hygienic locations, collection procedures and pretreatment of herbs were mandatory to ensure that the medicines were free from microbial contamination.

31.2.9.2 Checking for Mutagenic Effects of Drugs

Ayurvedic texts discuss the possibility of damage to genetic material that can result in anomalies and disturbances at both the structural and functional levels. The *Carakasamhita* talks about the blueprint of the whole body that is preserved in seed form (*bija*) and which can be broken down into two further sublevels, i.e., part of the seed (*bijabhaga*) and fraction of the part of the seed (*bijabhaga avayava*). If the seed, its part or the fraction of the part is damaged, then the part or organ of the body it represents would be affected (Trikamji, 1992f).

Extremely potent medicines (*tikshna*) are to be used with care. And, especially for children, such medicines should be avoided. Children need medicines of mild potency (Athavale, 1980c). For infants, the medicines are to be administered to the mother so that it becomes available through her breast milk. Ayurveda was thus aware of the fact that medicines are excreted in the breast milk. This knowledge was used for therapeutic purposes as well as to ensure safety by avoiding administration of certain drugs to lactating mothers.

There is a separate protocol for treatment of pregnant women. Highly potent medicines are to be avoided during pregnancy (Athavale, 1980c). The classical Ayurvedic texts have discussed in great detail genetic and congenital diseases that can arise from exposure to risk factors.

Ayurveda also advocated specific protocols for the aged. Many treatments and medications were contraindicated in old age and the principle of management of old age diseases was very conservative in outlook (Athavale, 1980c).

31.2.9.3 Checking for Long-Term Toxicity of Drugs

Ayurveda gives clues regarding the long-term effects of substances. There is mention in the text of substances that can remain in the body for a long period of time causing chronic inflammation without being metabolized or eliminated (Trikamji, 1992k). The long-term toxic effects of substances were understood as the cause of many chronic diseases like cancer. According to Ayurveda, cancer is one of the outcomes of chronic inflammation (Trikamji, 1992j).

31.2.9.4 Purification of Drugs

Ayurveda describes three approaches for safe use of toxic substances: (1) optimizing the dose to maximize therapeutic efficacy and minimize toxicity (Athavale, 1980a); (2) by coadministering another substance, which will mitigate the side effects of the toxic ingredient; and (3) by purification of the toxic substance, which is done before use by specialized methods of processing.

Below is a review of typical substances profiled for their safety and potential toxicity in classical Ayurvedic literature. These substances range from common food items to extremely toxic drugs and poisons.

Pepper: Ayurveda recommends purification of pepper by frying in ghee before use. It is considered as an irritant and can damage the mucosa on prolonged or excessive use. It is also not to be used in fever and such other conditions where pitta is aggravated.

Honey: Honey is also to be purified before use. It is interesting to note that honey is the only natural food that has been associated with the incidence of botulism in infants. We need to examine whether the purification of honey makes it safe in this regard.

Calamus root: *Acorus calamus* has been found to contain an alkaloid that may be carcinogenic. There has been much hue and cry about the toxicity of *A. calamus* in recent times. It is indeed interesting to see that Ayurvedic texts like the *Cakradatta* have recommended the purification of calamus root before use (Tripathi, 1991). This is a good example of the minute observations of the ancient proponents of Ayurveda.

Ghee: Ghee is not a herbal substance, but it is interesting to note that Ayurveda has described ghee as harmful or beneficial depending on how it is used (Vaidya, 2002e). Ghee from milk of grass-fed cows is superior and safe for use. Ghee is beneficial and safe only when there is no digestive or metabolic stress in the body.

Marking nut: Marking nut is a well-known irritant and can produce nasty blisters on the skin. Ayurveda uses marking nut seeds in many recipes after an elaborate process of purification (Sastri, 1980a). This is an example of safe use of a toxic substance after purification.

Nux vomica: *Nux vomica* seeds are known for their toxic effects on the central nervous system. Ayurveda advises purification of *Nux vomica* seeds before they can be used for therapeutic purposes (Sastri, 1980b). However, improper purification can lead to undesirable outcomes, which have been reported in several research journals.

Aconite: Aconite is a highly toxic substance that is used as a potent medicine in Ayurveda after purification. Sharadini Dahanukar has done extensive research on aconite and experimentally proved that Ayurvedic methods of purification render aconite safe for human therapeutic use. In these studies, aconite was purified using cow's urine and cow's milk independently as well as in combination. The aconitine content was found to be absent in aconite processed by cow's urine and it was reduced drastically in aconite processed by cow's milk. When complete purification is done, even the highest dose of aconite is nontoxic.

31.2.9.5 *Herbomineral Preparations*

Between the 8th and 12th centuries CE, a specialized branch of Ayurveda known as Rasasastra emerged with a focus on use of minerals and metals as medicine. Herbomineral preparations contain minerals and metals, which are not present in elemental form in the formulations. These medicines are prepared by subjecting the ingredients to very complex methods of processing, which converts them into compounds and sometimes nanoparticles (Mukhi et al., 2017). These methods of processing are expected to render these minerals and metals safe for human use. XANES and EXAFS-based analysis of the Ayurvedic Hg-based herbomineral preparation *Rasasindura* revealed that it is composed of single-phase α-HgS nanoparticles (size ~ 24 nm), free of Hg0 or organic molecules. The nonexistence of Hg0 appears to imply the absence of Hg-based toxicity (Ramanan et al., 2015).

Elaborate procedures are described in the classical Ayurvedic texts for the purification of mercury. These are called the *ashta samskaras* (eightfold processing of mercury) or *ashtadasa samskaras* (18-step purification of mercury). One of the primary steps in the detoxification of mercury is the grinding of mercury with sulfur to produce a compound called Kajjali. In this way, mercury is converted into its sulfide in an inorganic form. Inorganic mercury as a sulfide is the least toxic form of mercury, which has low bioavailability and also cannot cross the blood–brain barrier or the placental barrier. Studies in Chinese medicine have also pointed out that cinnabar (mercury sulfide)-containing traditional medicines are generally relatively nontoxic at therapeutic doses. Cinnabar toxicity is attributed to incorrect preparation methods, inappropriate doses, or wrong drug combinations (Liu et al., 2008).

Nevertheless, the use of herbomineral formulations in Ayurveda has raised eyebrows and Saper et al. (2004) pointed out the presence of heavy metals in Ayurvedic medicines sold in general stores in the USA. The toxicity of herbomineral formulations has raised many debates. Ayurvedic physicians attribute the reported toxicity of herbomineral formulations to improper pharmaceutical methods. On the other hand, scientists consider the presence of heavy metals like mercury as the source of toxicity of these formulations. Further research is needed to resolve such debates on the basis of scientific evidence while herbomineral formulations continue to be widely used in Ayurvedic practice in India even today.

31.2.9.6 Adverse Events

Even if all the prescribed precautions are taken, Ayurveda has recognized the possibility of adverse events in the course of therapy and the word *Vyapat* has been coined to denote the same (Trikamji, 1992l). *Vyapat* is a dangerous or hazardous outcome of a therapeutic intervention. The Ayurvedic texts emphasize that the success of a physician depends very much on his ability to anticipate and be prepared to handle *vyapats* or adverse events. Iatrogenic diseases were also recognized and mentioned as *mithyopacarajanya vyadhis* (Trikamji, 1992i). The celebrated *Carakasamhita*, the premier textbook on general medicine in Ayurveda concludes with the discussion on risk factor and safety measure (Trikamji, 1992n). Risk factor is known as *Vyapat* and safety measure as *Vyapat Siddhi*. These references testify to the fact that the safety of medicines and treatments were rigorously scrutinized in the tradition of Ayurveda. Elaborate guidelines have been provided to minimize the incidence of adverse events as well as iatrogenic diseases.

31.2.10 Guidelines for Safe Use of Medicines

The Ayurvedic approach to the safe use of substances lies in correct usage of the substance. The most important factors to be considered are the dosage and the timing (Trikamji, 1992d). Several other factors are also to be considered to ensure safe outcomes. For example, people living in marshy lands should avoid curds and those living in the tropics should avoid alcohol. Ayurveda has recognized physiological changes in the body in response to the seasons and therefore guidelines have been developed to fine tune the administration of drugs in accordance with the seasons to minimize adverse events (Trikamji, 1992e).

Ayurvedic treatment is customized according to the constitution of the individual, e.g., *Kantakari* (*Solanum xanthocarpum*) should be avoided for persons with a pitta constitution or suitable mitigating herbs need to be coadministered. Similarly, when administering the bark of *Shigru* (*Moringa oleifera*) in persons with pitta constitution, the addition of asparagus roots is recommended to prevent bleeding.

Ayurveda has given an emphasis on identifying unwholesome interactions between drug, food, and constitution (Athavale, 1980b). Therefore, the safety regimens consider drug–drug interactions, drug–constitution interactions, drug–food interactions, and food–food interactions.

The safety of a medicine ultimately depends on adherence to the guidelines for safe use.

31.2.11 Discussion

It is clear from the above discussion that the text and tradition of Ayurveda recognized the potential toxicity of medicines and advocated measures to ensure safety of Ayurvedic medicines and treatments. Even as substances were classified into food, medicine, and poison, it was understood that in principle no substance was absolutely safe or toxic. Ultimately, the safety of a medicine depended on its appropriate and rational use.

The tradition of Ayurveda has experimented with highly toxic substances for clinical applications, such as mercury, arsenic, lead, and many other such metals and minerals. The ancient physicians concluded that such toxic substances could be put to medical use by pharmaceutical processing and adjustment of dosage. While such formulations are widely used by Ayurvedic physicians, especially in India, their safety has not yet been accepted by mainstream science. There are reports of toxic side effects of Ayurvedic medicines even as there are studies that indicate the safety of these formulations. The Ayurvedic community asserts that toxicity of Ayurvedic medicines containing metals and minerals can be attributed to inadequacy of the manufacturing practices and that these medicines are safe if prepared meticulously in accordance with the traditional guidelines. It remains to be ascertained with certainty whether Ayurvedic methods of pharmaceutical processing can render highly toxic substances like lead safe for human use.

31.3 CONCLUSION

A careful study of classical Ayurvedic literature in the course of its evolutionary history reveals that the safety of medicines used in treatment was emphasized and detailed protocols and guidelines have been formulated to ensure the safety of medications and treatments.

Right from the assessment of a novel substance for medicinal use, its pharmaceutical processing and up to its application in specific clinical conditions, Ayurveda has emphasized the need for safety checks in the use of herbal formulations.

In the case of herbomineral formulations, debates have surfaced regarding their intrinsic safety, calling for rigorous research. The Ayurvedic viewpoint that toxic substances like mercury can be rendered safe for human use through pharmaceutical processing is not accepted by mainstream scientists. There are some scattered studies that have investigated the toxicity of Ayurvedic medicines containing toxic metals and minerals. However, further studies are needed to arrive at definite conclusions.

Ayurveda continues to be practiced in the country of its origin and is also gaining popularity around the globe. Contamination of raw drugs with pesticides, metals, and other harmful effluents as well as the use of toxic metals and minerals in certain Ayurvedic formulations and inadequate quality control in the preparation of Ayurvedic medicines have raised red flags that challenge the safety and credibility of Ayurvedic medicines and interventions. It is hoped that careful blending of traditional knowledge with modern advancements in science will provide answers for these challenging questions in due course of time.

References

All India Coordinated Project on Ethnobiology (AICPE). Final Technical Report 1992–1998. Ministry of Environment & Forests, Government of India.

Anonymous, 2003. Atharvaveda Samhita. Nag Publishers, Delhi, p. 188.

Athavale, A.D. (Ed.), 1980a. Ashtanga Sangraha. Anand Athavale Publishers, Pune, p. 63.

Athavale, A.D. (Ed.), 1980b. Ashtanga Sangraha. Anand Athavale Publishers, Pune, p. 115.

Athavale, A.D. (Ed.), 1980c. Ashtanga Sangraha. Anand Athavale Publishers, Pune, p. 116.

Dunbabin, D.W., Tallis, G.A., Popplewell, P.Y., Lee, R.A., 1992. Lead poisoning from Indian herbal medicine (Ayurveda). Med. J. Aust. 157, 835–836.

⟨http://ayush.gov.in/sites/default/files/Public%20Notice%20in%20English.pdf⟩ (accessed on 15 June 2018).

⟨https://nccih.nih.gov/health/ayurveda⟩ (accessed on 15 June 2018).

Kumar, G., Srivastava, A., Sharma, S.K., Gupta, Y.K., 2012. Safety evaluation of an Ayurvedic medicine, Arogyavardhini vati on brain, liver and kidney in rats. J. Ethnopharmacol. 140 (1), 151–160.

Liu, J., Shi, J.Z., Yu, L.M., Goyer, R.A., Waalkes, M.P., 2008. Mercury in traditional medicines: Is cinnabar toxicologically similar to common mercurials? Exp. Biol. Med. (Maywood). 233 (7), 810–817. Available from: https://doi.org/10.3181/0712-MR-336.

Mukhi, P., Mohapatra, S.S., Bhattacharjee, M., Ray, K.K., Muraleedharan, T.S., Arun, A., et al., 2017. Mercury based drug in ancient India: The red sulfide of mercury in nanoscale. J. Ayurveda Integr. Med. 8 (2), 93–98.

Ramanan, N., Lahiri, D., Rajput, P., Varma, R.C., Arun, A., Muraleedharan, T.S., et al., 2015. Investigating structural aspects to understand the putative/claimed non-toxicity of the Hg-based Ayurvedic drug Rasasindura using XAFS. J. Synchrotron Radiat. 22 (5), 1233–1241.

Saper, R.B., Kales, S.N., Paquin, J., Burns, M.J., Eisenberg, D.M., Davis, R.B., et al., 2004. Heavy metal content of ayurvedic herbal medicine products. JAMA 292 (23), 2868–2873.

Sastri, Ambikadatta (Ed.), 1980a. Bhaishajya Ratnavali. Chaukhambha Sanskrit Sansthan, Varanasi, p. 25.

Sastri, Lakshmipathy (Ed.), 1980b. Yogaratnakara. Choukhambha Sanskrit Sansthan, Varanasi, p. 144.

Sharma, P.V., 1982. Dalhana and His Comments on Drugs. Motilal Banarsidas Publishers, Delhi, p. 117.

Trikamji, Yadavji Acharya (Ed.), 1980a. Sushruta Samhita. Chaukhambha Orientalia, Varanasi, p. 507.

Trikamji, Yadavji Acharya (Ed.), 1980b. Sushruta Samhita. Chaukhambha Orientalia, Varanasi, pp. 534–535.

Trikamji, Yadavji Acharya (Ed.), 1980c. Sushruta Samhita. Chaukhambha Orientalia, Varanasi, p. 537.

Trikamji, Yadavji Acharya (Ed.), 1980d. Sushruta Samhita. Chaukhambha Orientalia, Varanasi, p. 565.

Trikamji, Yadavji Acharya (Ed.), 1992a. Caraka Samhita. Munshilal Manoharlal Publishers Pvt. Ltd, Delhi, p. 19.

Trikamji, Yadavji Acharya (Ed.), 1992b. Caraka Samhita. Munshilal Manoharlal Publishers Pvt. Ltd, Delhi, pp. 22–23.

Trikamji, Yadavji Acharya (Ed.), 1992c. Caraka Samhita. Munshilal Manoharlal Publishers Pvt. Ltd, Delhi, p. 23.

Trikamji, Yadavji Acharya (Ed.), 1992d. Caraka Samhita. Munshilal Manoharlal Publishers Pvt. Ltd, Delhi, p. 25.

Trikamji, Yadavji Acharya (Ed.), 1992e. Caraka Samhita. Munshilal Manoharlal Publishers Pvt. Ltd, Delhi, p. 282.

Trikamji, Yadavji Acharya (Ed.), 1992f. Caraka Samhita. Munshilal Manoharlal Publishers Pvt. Ltd, Delhi, p. 322.

Trikamji, Yadavji Acharya (Ed.), 1992g. Caraka Samhita. Munshilal Manoharlal Publishers Pvt. Ltd, Delhi, p. 376.

Trikamji, Yadavji Acharya (Ed.), 1992h. Caraka Samhita. Munshilal Manoharlal Publishers Pvt. Ltd, Delhi, p. 471.

Trikamji, Yadavji Acharya (Ed.), 1992i. Caraka Samhita. Munshilal Manoharlal Publishers Pvt. Ltd, Delhi, p. 483.

Trikamji, Yadavji Acharya (Ed.), 1992j. Caraka Samhita. Munshilal Manoharlal Publishers Pvt. Ltd, Delhi, p. 489.

Trikamji, Yadavji Acharya (Ed.), 1992k. Caraka Samhita. Munshilal Manoharlal Publishers Pvt. Ltd, Delhi, p. 573.

Trikamji, Yadavji Acharya (Ed.), 1992l. Caraka Samhita. Munshilal Manoharlal Publishers Pvt. Ltd, Delhi, p. 677.

Trikamji, Yadavji Acharya (Ed.), 1992m. Caraka Samhita. Munshilal Manoharlal Publishers Pvt. Ltd, Delhi, p. 727.

Trikamji, Yadavji Acharya (Ed.), 1992n. Caraka Samhita. Munshilal Manoharlal Publishers Pvt. Ltd, Delhi, p. 735.

Tripathi, Indradeva (Ed.), 1991. Chakradatta. Chaukhambha Sanskrit Sansthan, Varanasi, p. 225.

Vaidya, Harisastri (Ed.), 2002a. Ashtanga Hridayam. Chaukhambha Orientalia, Varanasi, pp. 124–133.

Vaidya, Harisastri (Ed.), 2002b. Ashtanga Hridayam. Chaukhambha Orientalia, Varanasi, p. 127.

Vaidya, Harisastri (Ed.), 2002c. Ashtanga Hridayam. Chaukhambha Orientalia, Varanasi, p. 166.

Vaidya, Harisastri (Ed.), 2002d. Ashtanga Hridayam. Chaukhambha Orientalia, Varanasi, p. 214.

Vaidya, Harisastri (Ed.), 2002e. Ashtanga Hridayam. Chaukhambha Orientalia, Varanasi, p. 711.

WHO, 2013. WHO Traditional Medicine Strategy 2014–2023. WHO, Geneva, p. 25.

32

Mushroom Intoxication in Mesoamerica

Carl de Borhegyi[1] and Suzanne de Borhegyi-Forrest[2]
[1]Southwestern Michigan College, Dowagiac, Michigan, United States [2]B.S. Biological Sciences, Ohio State University, Columbus, Ohio, United States

Bernal Díaz del Castillo, a foot soldier in the army of Conquistador Hernán Cortés, was one of the first Spanish chroniclers to view and describe Tenochtitlan, the Aztec capital and site of modern-day Mexico City. He tells us that the men marveled at the city's size, magnificence, and cleanliness. But while the Spaniards stood in awe of the Aztec's artistic and cultural achievements, they were appalled at their religious ceremonies. These ceremonies included human sacrifice and the consumption of hallucinogenic mushrooms called teonanácatl or "God's flesh." The Spanish friars perceived these entheogenic (god generating) ceremonies to be a horrific, Satan-inspired, and misinterpretation of Christian communion. In the years that followed, the Catholic Church waged a ruthless campaign to stamp out all traces of native religion, and most specifically mushroom worship. During this cultural holocaust they tore down temples, destroyed idols, and burned thousands of colorful hand-painted manuscripts, called codices, dealing with indigenous history and mythology.

The mushroom encountered and described by the Spanish chroniclers was in all likelihood the *Psilocybe cubensis*, the most common of the 40 or so species of *Psilocybe* mushroom. The hallucinogenic ingredient of this mushroom, as in others of this genus, is psilocybin or psilocin.

Toxicology in Antiquity
DOI: https://doi.org/10.1016/B978-0-12-815339-0.00033-0

459

Somewhat surprisingly, however, this mushroom was not the first and most commonly used hallucinogen. Judging from the many illustrations of sacred mushrooms in ancient Mesoamerican art, some dating back to Olmec times, that distinction may go to the *Amanita muscaria*, or "fly agaric" mushroom of European folklore (Borhegyi de, 2013). The *Psilocybe* and *Amanita* mushrooms, along with the equally well known peyote cactus, are only part of a large pharmacopeia of hallucinogens used in Mesoamerica. These include *Datura inoxia* and other species of the genus *Solandra*; a potent species of tobacco called *piciétl* (*Nicotiana rustica*); and the seeds from two morning glories, the white flowered *Turbina corymbosa*, and the blue or violet flowered *Ipomoea violacea*, whose active principles are closely allied to synthetic LSD-25. Their psychedelic seeds were considered sacred to the point of divinity (Evans and Webster, 2010). There is mixed evidence for the use of highly poisonous toad toxin. In the mid-seventeenth century the English Dominican friar, Thomas Gage, reported that the Pokoman Maya of Guatemala steeped toads and tobacco in the fermentation of their ritual drink (Sharer, 1983). Whether this mixture was hallucinogenic rather than fatal is, however, a matter of some dispute (Fig. 32.1).

FIGURE 32.1   "Hidden in plain sight," the ceramic pre-Columbian mask depicts the transformation of a human into a "were-jaguar," a half-human, half-jaguar deity. The were-jaguar appears in the art of the ancient Olmecs as early as 1200 bc The mask symbolizes the soul's journey into the underworld where it will undergo ritual decapitation, jaguar transformation, and spiritual resurrection. *Photo courtesy of the International Museum of Ceramics in Faenza, Italy.*

FIGURE 32.2 Nine Preclassic mushroom stones found in a cache along with nine miniature metates at the highland Maya archeological site of Kaminaljuyu. The contents of the cache were dated by Stephan de Borhegyi at 1000–500 bc The tall jaguar mushroom stone on the left, also from Kaminaljuyu, was excavated separately. *Photo property of author.*

In South America the Indians, even more than in Mesoamerica, discovered and experimented with the hallucinogenic properties of plants. They successfully combined unrelated species to activate their psychoactive properties or heighten their effects. Lacking the painstaking ethnography of another chronicler of the Spanish Conquest, Franciscan friar Bernardino de Sahagún, and the medical and botanical compilations of the royal physician Francisco Hernández, anthropologists and mycologists have had to rely heavily on the post-Conquest treatises of Jacinto de la Serna and Ruiz de Alarcón for an understanding of the continued functions of plant hallucinogens during the early Colonial period (Fig. 32.2).

The *Amanita muscaria*, described as poisonous in most scientific literature, is a powerful hallucinogen known from Paleolithic times in northern Europe. The substances muscarine and ibotenic acid are responsible for this mushroom's powerful psychoactive effects. Knowledge of the psychoactive properties of both the Amanita and toad toxin, as well as that of other hallucinogens, was presumably brought to the New World from northeastern Asia by early settlers as a shamanic cult. The magical importance of the Amanita to Mesoamerica is clear from its many portrayals in all Mesoamerican art forms.

In the early 1950s the author's father, archeologist Stephan F. de Borhegyi, better known by his contemporaries as Borhegyi, wrote the first of a series of articles on the enigmatic small stone sculptures called "mushroom stones" that had turned up in collections and in a few archeological excavations in the central highlands of Guatemala. Although these artifacts had been known for a number of years, Borhegyi was the first to suspect that they may have been used in an ancient hallucinogenic mushroom cult. After cataloging them by type and provenience, he dated their earliest appearance to approximately 1000 bc According to their archeological context, mushroom stones were associated from their first appearance with ritual human decapitation, a

FIGURE 32.3   A pre-Columbian ceramic Moche portrait vessel from Peru wearing a headdress encoded with two Amanita muscaria mushrooms, together with a mushroom-shaped axe. The Moche culture reigned on the north coast of Peru from 100 to 600 ad.

trophy head cult, warfare and the Mesoamerican ballgame. While mycologists readily accepted this proposition, archeologists, in general, were dubious. Such reluctance, in the face of much evidence to the contrary, may derive from the deep-seated Western cultural fear of mushroom poisoning combined with a distaste for the excesses of "magic" mushroom experimentation (Fig. 32.3).

In an ironic compensation for the terrible loss of indigenous history inflicted by the Catholic Church, Spanish chronicler Bernardino de Sahagún compiled a 12-volume history of New Spain which we know as the Florentine Codex, or *Historia General de las Cosas de Nueva España*. His manuscript is one of the most remarkable ethnographic studies ever assembled, these volumes were the result of decades of research during which Sahagún was aided greatly by his native students. They are located in the Laurentian Library in Florence where they were probably sent by the Inquisition for their content of pagan rituals. When they were discovered in 1883, they became a priceless source of indigenous history, customs, and beliefs.

Sahagún was very likely the first to record the use of hallucinogenic mushrooms in Mesoamerican religion. He writes in Book 9 that merchant groups known as the *pochteca* were devout followers of the god Quetzalcoatl (Sahagun de Bernardino, 1950–1982). A passage from the book reads: "... the Indians gathered mushrooms in grassy fields and pastures and used them in religious ceremonies because they believed them to be the flesh of their gods (teonanácatl)." Mushroom intoxication, according to Spanish reports, gave sorcerers the power to seemingly change themselves into animals. The powerful visions and voices the mushrooms produced were believed to be from God. In the Florentine Codex, Sahagún writes:

It was said that they did not die {the Indians}, but wakened out of a dream they had lived, this is the reason why the ancients said that when men died, they did not perish, but began to live again, waking almost out of a dream, and that they turned into spirits or gods ... and so they said to the dead: "Lord or Lady, wake, for it begins to dawn, now comes the daylight for the yellow-feathered birds begin to sing, and the many colored butterflies go flying," ... and when anyone died, they used to say of him that he was now teotl, meaning to say he had died in order to become a spirit or god.

Among the many gems of native history preserved by Sahagún is a description of the death of Quetzalcoatl Topiltzin. This great ruler of the city of Tula, the capital of the Toltec empire that preceded the Aztec Empire, was named after their legendary culture hero Quetzalcoatl, also known as the Plumed Serpent. It was the god Quetzalcoatl who they believed had brought all learning and wisdom to the Aztecs and their predecessors.

According to legend, Quetzalcoatl Topiltzin disgraced himself and was banished from his beloved Tula. He traveled southward with a band of his followers, eventually reaching the shores of Yucatan, where he immolated himself and ascended into the heavens as Venus, the Morning Star. The account of Quetzalcoatl's death recorded by Sahagún is quite different and may be closer to historical truth. Sahagún records that, being very ill, Quetzalcoatl was visited by a necromancer who brought him a potion to drink. At first he refused, but after much persuasion he agreed to swallow the potion (Sahagun de Bernardino, 1932). Finding the liquid very tasty and pleasant he drank more, only to realize later that he had made a fatal mistake. The contents of the potion have never been identified, but judging from the fact that it caused his death, new evidence would suggest that it was likely the hallucinogenic/toxic *Amanita muscaria* mushroom mixed with honey.

In a guide for missionaries written before 1577, Francisco Hernández de Córdoba, physician to the king of Spain, writes that at the time of the Spanish Conquest, the Aztecs revered three different kinds of narcotic mushrooms (Serna de la, 1900). The Spanish Jesuit scholar Jacinto de la Serna (The Manuscript of Serna) later added these words concerning mushroom use in divination:

These mushrooms were small and yellowish and to collect them the priest and all men appointed as ministers went to the hills and remained almost the whole night in sermonizing and praying (Wasson and Pau, 1962).

Another renowned Spanish chronicler, Fray Diego Durán, writes in his *Histories of New Spain (1537–1588)* that the practice of human sacrifice was the custom that the Spanish considered most shocking. He also writes that mushrooms were used in connection with human sacrifice. Rather than being a punishment, sacrifice was a sacred gift. As he explains, the word for sacrifice, *nextlaoaliztli* in the Nahuatl language of the Aztecs, meant either "payment," or the act of payment (for blessings received). Young children were taught that death by the obsidian knife was a most honorable way to die, as praiseworthy as dying in battle or for a mother and child to die in childbirth. Those who were sacrificed were assured a place in Omeyocan, the paradise of the sun, the afterlife. Quoting Durán:

The Indians made sacrifices in the mountains, and under shaded trees, in the caves and caverns of the dark and gloomy earth. They burned incense, killed their sons and daughters and sacrificed them and offered them as victims to their gods; they sacrificed children, ate human flesh, killed prisoners and captives of war .... It was common to sacrifice men on feast days as it is for us to kill lambs or cattle in the slaughterhouses .... I am not exaggerating; there were days in which two thousand, three thousand or eight thousand men were sacrificed .... Their flesh was eaten and a banquet was prepared with it after the hearts had been offered to the devil ... to make the feasts more solemn all ate wild mushrooms which make a man lose his senses .... One thing in all this history: no mention is made of their drinking wine of any type, or of drunkenness. Only wild mushrooms are spoken of and they were eaten raw ....

Durán called these mushroom ceremonies the "Feast of the Revelations." He also tells us that wild mushrooms were eaten at the ceremony commemorating the accession of the Aztec King Moctezuma in 1502. After Moctezuma took his Divine Seat, captives were brought before him and sacrificed in his honor. He and his attendants then ate a stew made from their flesh.

And all the Lords and grandees of the province ... all ate of some woodland mushrooms, which they say make you lose your senses, and thus they sallied forth all primed for the dance .... With this food they went out of their minds and were in worse state than if they had drunk a great quantity of wine. They became so inebriated and witless that many of them took their lives in their hands. With the strength of these mushrooms they saw visions and had revelations about the future, since the devil spoke to them in their madness (Wasson, 1980).

In 1951, Borhegyi's research on mushroom stones brought him into contact with amateur ethnomycologist Robert Gordon Wasson, and the two began an extensive correspondence. They both wondered at the pervasiveness of toad effigies on mushroom stones, and the toad's association with the "toadstool" and the ubiquitous appearance of the red and white Amanita muscaria mushroom in northern European folk art. Wasson also called Borhegyi's attention to early Guatemalan dictionary sources describing a mushroom called *xibalbaj ocox*, meaning "underworld mushroom," and *k'aizalab ocox*, meaning "lost judgment mushroom" (Sharer, 1983, p. 484).

Soon after their first meeting, Wasson published his account of the modern-day ritual use of hallucinogenic mushrooms among the Mazatec Indians of Oaxaca, Mexico. Wasson's report was followed by similar accounts of hallucinogenic mushroom use among the Zapotec, Chinantec, Mazatec, and Huichol Indians of Mexico. Oddly, although Maya "mushroom stone" sculptures had first suggested the existence of an ancient mushroom cult, evidence for the existence of modern-day mushroom ceremonies in the Maya area appeared to be lacking. Then, in 1972, ethnoarcheologist Peter Furst reported that Lacandon Maya shamans consumed hallucinogenic mushrooms in the course of their religious ceremonies; and in 1978 ethnomycologist Bernard Lowy reported their intimate association with a creation myth among the Tzutuhil Maya (Lowy, 1980).

At about the same time, Michael Coe reported his discovery of numerous toad bones in Olmec burials at San Lorenzo. He suggested that the Olmecs may have used hallucinogenic toad toxin in ritual practices as early as 1200 bc Tatiana Proskouriakoff, graphic artist and illustrator for the Carnegie Institution of Washington archeological research team, demonstrated that in Mayan glyphs the toad is a symbol of rebirth (Sharer, 1983, p. 528). Furst suggested that the Olmecs may have carried elements of their religion, including a shamanistic supernatural jaguar cult, as well as a mushroom cult, throughout Mesoamerica and even into South America.

In a letter written in 1954 to Wasson, Borhegyi mentions two interesting passages from native chronicles relating to indigenous use of mushrooms in Guatemala: A passage from the *Popol Vuh*, the Quiche Maya Book of Creation reads:

And when they found the young of the birds and the deer, they went at once to place the blood of the deer and of the birds in the mouth of the stones that were Tohil, and Avilix. As soon as the blood had been drunk by the gods, the stones spoke, when the priest and the sacrificers came, when they came to bring their offerings. And they did the same before their symbols, burning pericon (?) and holom-ocox (the head of the mushroom, holom = head, and ocox = mushroom).

He also cites a passage from *The Annals of the Cakchiquels* which records:

At that time, too, they began to worship the devil. Each 7 days, each 13 days, they offered him sacrifices, placing before him fresh resin, green branches, and fresh bark of the trees, and burning before him a small cat, image of the night. They took him also the mushrooms, which grow at the foot of the trees, and they drew blood from their ears (Borhegyi and Wasson, 1954).

Both Wasson and Borhegyi proposed that the mushroom stones were modeled after the *Amanita muscaria* mushroom. Unfortunately, after Borhegyi's premature death in an automobile accident in 1969, further inquiry into the subject on the part of archeologists came to a virtual halt. It was left to ethnohistorian Peter Furst and a few mycologists, most notably Bernard Lowy and Gaston Guzmán, to continue to make important contributions to the scientific literature.

Gordon Wasson may have provided an important explanation for this lack of interest among archeologists. He and his Russian wife Valentina had observed that, across the globe, cultures appeared to be divided into those who loved and revered mushrooms, and those who dismissed and feared them. The first group of cultures they labeled "mycophiles," while the latter were "mycophobes." In the New World, it appears that all of the native cultures were, and still are, unquestionably mycophilic. In contrast, the great majority of archeologists and ethnologists who studied and described them, and who traced their cultural origins to Western Europe, were decidedly mycophobic. This major difference in cultural background may be responsible for what I believe should be seen as a lamentable gap in our understanding of indigenous New World magicoreligious origins (Furst, 1976, p. 84−85).

In 1998, I decided to follow up on my father's earlier research on the role of mushrooms in Mesoamerican religion. In this I was aided immensely by Justin Kerr's remarkable compilation and database of roll-out photographs of Mesoamerican art. As a result of this study of Mesoamerican figurines, sculpture, murals, and vase paintings, along with an abundance of evidence from other scholars, I have been able to expand this subject far beyond my father's and Wasson's pioneering efforts.

While I found no mention of images of mushroom stones, pottery mushrooms, or images of actual mushrooms in Kerr's extensive index, I did discover a significant number of images, both realistic and abstract, of both the *Amanita muscaria* and the Psilocybin mushroom. However, it was easy to understand why the imagery had not been noted earlier. On many vases, the images of mushrooms, or images related to mushrooms, were so abstract, and so intricately interwoven with other complex and colorful elements of Mesoamerican mythology and iconography, that they were, quite literally, "hidden in plain sight." This, I believe, was no accident but a deliberate effort to conceal sacred information from the eyes of the uninitiated. As a result, the sacred mushrooms, so cleverly encoded in the religious art of the New World, escaped prior detection. Now after more than a half century of virtual denial by the anthropological community, we have clear visual evidence that two varieties of hallucinogenic mushrooms, the *Amanita muscaria* and the Psilocybin mushroom, as well as the peyote cactus, were worshiped and venerated as gods in ancient Mesoamerica.

Much of the mushroom imagery I discovered was associated with an artistic concept I refer to as underworld jaguar transformation. Under the influence of the hallucinogen, the "bemushroomed" individual is depicted with feline fangs, claws, and spots; attributes of the jaguar as the Underworld Sun God. This esoteric association of mushrooms and jaguar transformation was earlier noted by Furst, together with the fact that a dictionary of the Cakchiquel Maya language compiled *circa* 1699 lists a mushroom called "jaguar ear" (Furst, 1976, p. 80) Many of the mushroom images also involved rituals of self-sacrifice and decapitation in the underworld, alluding to the sun's nightly death and subsequent resurrection by a pair of deities associated with the dualistic nature of the planet Venus as both the Morning star and Evening star. The dualistic aspect of Venus, the brightest "star" in the evening or morning sky, is why this planet was venerated as both a God of Life and God of Death. According to *The Title of the Lords of Totonicapan* (Goetz and Chonay, 1974) "they [the Quiche] gave thanks to the sun and moon and stars, but particularly to the 'star' that proclaims the day, the day-bringer, the Morning star."

Mushrooms were also so closely associated with death, underworld jaguar transformation, and Venus resurrection in Maya vase paintings, that I conclude that they were likely believed to be the vehicle through which both occurred. They are also so closely associated with ritual decapitation, that their ingestion may have been considered essential to the ritual itself, whether in real life or symbolically in the underworld. It is also important to note that in many cases the mushroom images appeared to be associated with period endings in the Maya calendar.

Gordon Wasson theorized that the origin of ritual decapitation lay in the mushroom ritual itself. In a letter to Borhegyi in 1954 he writes of the Mixe use of the Psilocybin mushroom:

The cap of the mushroom in Mije (or Mixe) is called kobahk, the same word for head. In Kiche and Kakchiquel it is doubtless the same, and kolom ocox is not "mushroom heads," but mushroom caps, or in scientific terminology, the pileus of the mushroom. The Mije in their mushroom cult always sever the stem or stipe (in Mije tek is "leg") from the cap, and the cap alone is eaten. Great insistence is laid on this separation of cap from stem. This is in accordance with the offering of "mushroom head" in the Annals [of the Cakchiquels] and the Popol Vuh. The writers had in mind the removal of the stems. The top of the cap is yellow and the rest is the color of coffee, with the gills of a color between yellow and coffee. They call this mushroom, pitpa "thread-like," the smallest, perhaps 2 horizontal fingers high, with a cap small for the height, growing everywhere in clean earth, often along the mountain trails with many in a single place. In Mije the cap of the mushroom is called the "head" "kobahk" in the dialect of Mazatlan. When the "heads" are consumed, they are not chewed, but swallowed fast one after the other, in pairs (Borhegyi and Wasson, 1954, 7 June).

In another letter to Borhegyi, Wasson quotes his Mije informant as follows:

The mushrooms may be gathered by anyone at any hour. Often on kneeling to gather one up, the gatherer utters a prayer of thanks for the divine gift. The mushrooms are placed in a jicara, or gourd bowl, and taken to the church. The mushrooms are placed on the high altar, prayers are said, and copal incense is burned. The mushrooms are taken to the house where the session is to be held, perhaps the home of a sick person. The sick person eats the mushrooms, not a curandero: here is the basic difference from the Mazatec practice. If a lost or stolen object is sought, then the suppliant eats the mushroom in the presence of a close member of his family, all others keeping away. The witness is present to give ear to the words of the eater, as he begins to talk under the influence of the inebriating mushrooms. Furthermore, the mush-rooms will not render service if he who eats them has said or thought disrespectful things about them, and if he is guilty of this sin, then the mushrooms cause him to see horrible visions of snakes and such like" (Borhegyi and Wasson, 1954, 19 June).

In 1960, Christian missionary and anthropologist Eunice V. Pike wrote to Borhegyi that Christian missionaries had difficulty in convert-ing the Mazatec Indians because they equated hallucinogenic mush-rooms with Jesus Christ. In an earlier letter written in 1953, she elaborated on the subject of Jesus Christ and the mushroom. Note the association of the mushroom with Christ's blood. In Mesoamerican mythology mushrooms were believed to grow where lightning had struck the ground. Water was believed to be god's great gift to man-kind; while the sacrificial blood of a human was seen as man's great gift to the gods.

The Mazatecs seldom talk about the mushroom to outsiders, but belief in it is widespread. A 20-year-old boy told me, "I know that outsi-ders don't use the mushroom, but Jesus gave it to us because we are poor people and can't afford a doctor and expensive medicine .... Sometimes they refer to it as 'the blood of Christ', because supposedly it grows only where a drop of Christ's blood has fallen. They say that the land in this region is 'living' because it will produce the mushroom whereas; the hot dry country where the mushroom will not grow is called 'dead.' They say that it helps 'good people' but if someone who is bad eats it 'it kills him or makes him crazy'. When they speak of 'badness' they mean 'ceremonially unclean'." (A murderer if he is ceremonially clean can eat the mushroom with no ill effects.) A person is considered safe if he refrains from intercourse five days before and after eating the mushroom. A shoemaker in our part of town went crazy about five years ago. The neighbors say it was because he ate the mushroom and then had intercourse with his wife .... When a family decides to make use of the mushroom they tell their friends to bring them any they see, but they ask only those who they can trust to refrain from intercourse at that time, for if the person who gathers the mushroom has had intercourse, it will make the person who eats it crazy (Pike, 1953)."

In conclusion, while indigenous traditions are being lost at a rapid rate everywhere in the world, in Mesoamerica they have thus far proved to be highly resistant to change. Though weakened and almost seamlessly incorporated into Christian beliefs, the use of hallucinogenic substances in native religion has apparently persisted in the more remote areas where they are used in native curing and divination ceremonies. It is precisely the Maya's strong sense of cultural identity that has enabled their communities in Guatemala and Mexico to survive years of discrimination and brutal persecution. In this sense, it can be said that their old Mesoamerican gods are still helping to protect them from evil.

References

Borhegyi, S.F., Wasson, R.G., 1954. Wasson Archives, Harvard University Botanical Museum.

Borhegyi de, C.R., De Borhegyi-Forrest, S., Rush, J.A., 2013. The genesis of a mushroom/ Venus religion in Mesoamerica. In: Rush, J.A. (Ed.), Entheogens and the Development of Culture: The Anthropology and Neurobiology of Ecstatic Experience. North Atlantic Books, Berkeley, CA, pp. 451–483.

Evans, S.R., Webster, D.L., 2010. Archaeology of ancient Mexico and Central America: an encyclopedia. In: Evans, S.R., Webster, D.L. (Eds.), Archaeology of Ancient Mexico and Central America: An Encyclopedia. Routledge Publishing, New York, NY.

Furst, P., 1976. Hallucingens and Culture. Chandler and Sharp Publishers, San Francisco, CA.

Goetz, D., Chonay, D.J., 1974. The Title of the Lords of Totonicapan. University of Oklahoma Press, Norman, OK.

Lowy, B., 1980. Ethnomycological inferences from mushroom stones, Maya codices, and Tzutuhil legend. Rev. Interam. 10, 94–103.

Pike, E., 1953. Letter to Wasson. Wasson Archives. Harvard University Botanical Museum.

Sahagun de Bernardino, 1932. The History of Ancient Mexico [Bandelier F, Trans.]. Bardwell Printing Co., Nashville, TN, pp. 180–181.

Sahagun de Bernardino, 1950–1982. Florentine Codex (1540–1585), 12 vols. [Anderson, A., Dibble, C., Trans. and Ed.]. School of American Research, Santa Fe, NM.

Serna de la, J., 1900. Manual. Para los ministros de indios para el conocimiento de sus idolatrias y extirpacion de ellas. Anals 6, 261–280.

Sharer, R., 1983. [expanded and revised edition of Morley SG, The ancient Maya, 1946] The Ancient Maya, 4th ed. Stanford University Press, Stanford, CA.

Wasson, R.G., 1980. The Wondrous Mushroom: Mycolatry in Mesoamerica. McGraw-Hill, New York, NY.

Wasson, R.G., Pau, S., 1962. The hallucinogenic mushrooms of Mexico and psilocybin, a bibliography. Harvard Univ. Bot. Leafl. 20 (2), 39.

Index

Printed in the United States
By Bookmasters